Cognitive Robotics: Intelligent Robotic Systems

Cognitive Robotics: Intelligent Robotic Systems

Edited by **Jared Kroff**

Willford Press

New York

Published by Willford Press,
118-35 Queens Blvd., Suite 400,
Forest Hills, NY 11375, USA
www.willfordpress.com

Cognitive Robotics: Intelligent Robotic Systems
Edited by Jared Kroff

International Standard Book Number: 978-1-68285-212-5 (Hardback)

Printed in China.

Contents

Preface

Every book is initially just a concept; it takes months of research and hard work to give it the final shape in which the readers receive it. In its early stages, this book also went through rigorous reviewing. The notable contributions made by experts from across the globe were first molded into patterned chapters and then arranged in a sensibly sequential manner to bring out the best results.

Robotic systems have successfully aided humans across a variety of disciplines. This innovative and comprehensive book integrates the well-developed theory and practical applications of cognitive robotics through lucid discussions of topics like artificial intelligence and fuzzy logic in robotics, mechatronics, biomimetics, wireless sensor network for navigation, modeling of robotic systems, etc. The various studies that are constantly contributing towards advancing technologies and evolution of this field are examined in detail within this book. It will especially serve as a valuable source of reference for graduate and post graduate engineering students.

It has been my immense pleasure to be a part of this project and to contribute my years of learning in such a meaningful form. I would like to take this opportunity to thank all the people who have been associated with the completion of this book at any step.

Editor

1

Ant Robotic Swarm for Visualizing Invisible Hazardous Substances

John Oyekan [1,*] and Huosheng Hu [2]

[1] Centre for Research in Distributed Technologies, University of Bedfordshire, Park Square, Luton, LU1 3JU, UK

[2] School of Computer Science and Electronic Engineering, University of Essex, Colchester CO4 3SQ, UK; E-Mail: hhu@essex.ac.uk

* Author to whom correspondence should be addressed; E-Mails: john.oyekan@beds.ac.uk or oyekanjohn@gmail.com

---

Abstract: Inspired by the simplicity of how nature solves its problems, this paper presents a novel approach that would enable a swarm of ant robotic agents (robots with limited sensing, communication, computational and memory resources) form a visual representation of distributed hazardous substances within an environment dominated by diffusion processes using a decentralized approach. Such a visual representation could be very useful in enabling a quicker evacuation of a city's population affected by such hazardous substances. This is especially true if the ratio of emergency workers to the population number is very small.

Keywords: bio-inspiration; ant agents; optimal coverage; pollution monitoring; bacterial foraging

---

1. Introduction

Monitoring of hazardous substances in our environments is fast becoming a high priority for governments around the world due to global warming [1–3]. One of the projects that have been setup to tackle pollution is the European Union Shoal project, in which a shoal of robotic fishes would be deployed to find the source of pollution in sea ports and the culprit responsible charged accordingly. This work focuses on the spread of pollution by diffusion in the hydrodynamics section of the

European Union Shoal project. Monitoring of hazardous substances in environments other than the sea is also a necessity. This is especially true if the hazardous substance is not visible to humans. Such substances could include nerve gas, Carbon dioxide, which can be fatal if in large quantities, Sarin gas or even nuclear radiation. Taking a scenario in which nerve gas as been released due to a terrorist attack in a city; the population of that city needs to be evacuated. However, as they cannot see the pollution, they do not know whether they are running towards it or away from it. The emergency services have been called in but are overwhelmed by the ratio of number of citizens to crew men. If however a swarm of flying micro robots were able to visualize the pollutant, it is possible for the population to keep away from affected areas and in the process evacuate themselves from dangerous areas.

This can be done by ensuring that the motion of each individual agent in the swarm of robots is controlled so that eventually, the distribution of the swarm converges in a way that areas containing a high amount of pollutants are more densely covered by swarm individuals than areas that have little or no pollutant. This approach leads to efficient use of mobile sensors especially when their numbers are limited.

Furthermore, the robots could be used as measurement stations to collect information about the pollutant and then a map of the pollution could be developed and built at a ground station. This information could then be used to predict the path and behaviour of the invisible pollutant. By observing the behaviour of the pollutant from the map, it is possible to say what type of pollutant it is. In order for the agents to provide such visual information, the agents need to be equipped with a coverage control algorithm that would enable them to perform optimal coverage.

There has been a lot of work done in the area of coverage using multiple agents, such as the use of Voronoi partitions [4–8]. The Voronoi partition method requires a polygon derived environment and a large computational time due to computation of the Voronoi partition in addition to the computation of the global cost function required to drive the mobile sensors to their required cost minimal positions. Nevertheless, Schwager *et al.* was able to use the Voronoi partition method and a Radial basis function machine learning technique on a group of robots to achieve a visual representation in [6,9]. Shucker *et al.* [10,11] used virtual springs to form the distribution of a pollutant by tracking points in it. Their work requires long distance communication and this might not be possible when considering hardware limitations on simple agents.

The use of deterministic annealing as in [12] also requires a large communication burden. Artificial Physics and Computation Fluid Dynamics (CFD) were used by Zarzhitsky *et al.* in [13] to control a group of robots to find the source of a pollutant in the environment. Artificial Physics was used to make the robots form a lattice while CFD was used to navigate up pressure fluxes in the environment towards the source. This approach works on the assumption that the pollutant would most likely to be in the area where there was a negative flux or sink.

Furthermore, work has also been carried out regarding coverage using ant robots. Ant robotics has been inspired by simple minded creatures like ants whereby individuals following simple rules are able to build complex structures and perform complex feats of engineering. Robots inspired by these creatures often have limited memory, computing, sensing and communication resources [14,15]. These properties make them very cheap and hence disposable in hazardous situations whereby they could be thrown away after being used for monitoring harmful substances. Additionally, most of the work in ant robotics tend to focus on simple algorithms to perform tasks such as lawn mowing, mine sweeping, vacuum cleaning, and so on, and make use of the advantage of multiple agents to introduce graceful degradation of the system [15].

In [14], terrain coverage by ant robots was investigated. In their work, they took inspiration from natural ants by using markings left by individual robots to communicate indirectly with other individual robots. Additionally, the markings were also used to determine the next action of the ant due to its ability to sense what is around it. Their ant robots did not need a map of the terrain, localization, path planning nor did they require memory. They were able to achieve terrain coverage by comparing a variety of algorithms such as Learning Real-time A Star (LRTA*), Node counting, Thrun's Value-Update Rule and Wagner's Value-Update Rule.

In the famous "cooperative cleaners" task in [16], a cleaning protocol algorithm called CLEAN was developed to clean up contaminated tiles in the environment. In [17], a dynamic cleaning algorithm called SWEEP was developed to clean up a dynamically spreading target such as fire. However, this work and that in [18] relies on the fact that critical points are not cleaned so that the task can be completed. Furthermore, in [18], a distributed search for evasive targets using ant robots was presented while in [19] analyses of how many pursuers and how they should move was conducted. Similarly, [18] analysed the amount of time it would take for a swarm of robots to clean both a static and expanding region of pollutants while in [15], the authors showed that even if robots having limited capabilities were moved, it is guaranteed that a swarm of them would still cover an area of interest. In most of the work above, algorithms were either developed or applied after understanding the nature of the problem.

This work uses mathematical models of the bacterium, flocking behaviour as seen in starlings and a velocity function to solve the problem of providing a visual representation of an invisible pollutant.

The bacterium behaviour was used to explore the environment, find pollutants and subsequently navigate up the gradient of the pollutant; the flocking behaviour was used to keep agents from colliding from each other and for collective foraging; while the velocity function was used to simulate a dwelling behaviour in which agents tend to dwell in an area that has a rich source of food compared to other areas that lack food.

This work is divided into two parts. The first part uses a swarm of robots to provide a visual representation of a static pollutant and follows in the path of ant robots. Using the criteria in [15], the robots have a finite bounded memory $k = 4$ and communication radius of $r = 50$ in a grid of 2000 by 2000. Additionally, the agents can only sense the pollutant in their immediate locality (x, y) but cannot sense what is beyond this at for example ($x \pm \Delta x$, $y \pm \Delta y$). Using this information, each individual agent moves according to a set of equations that model the bacterium's behaviour when searching for food. The bacterium only needs current local information related to the area in which it presently is; in addition to a history of past local information $k = 4$ in order to determine its next move.

This is unlike previous work in ant robotics where most of the work relied on assuming that the ant knows about the conditions in tiles surrounding it. The conditions in those tiles then determine its next action.

Also, in this present work, cleaning is not our major objective. The effect of cleaning on the proposed algorithm in this work would be studied in future work. Nevertheless, this is the first time that ant robots have been used in the task of visualizing pollutant.

The second part investigates the possibility of enhancing the first part by using a Genetic Algorithm machine learning technique. In this case, the ant robots are equipped with knowledge of the pollutant profile to be mapped. We do this in accordance to [20] in which Arkin discussed how simple reactive models of robot control could be made more effective through the introduction of knowledge about the environment. This was done in this work by updating the parameters of the proposed ant coverage controller through the use of a modified computationally efficient Gaussian model based Genetic Algorithm. The Genetic Algorithm was used to estimate the distribution of the pollutant in the environment and the information used as a cost function to distribute the agents optimally in the environment. However, it could be argued that introducing this additional capability makes them less of ant robots due to the computational expensive nature of Genetic Algorithms.

Furthermore, it should be noted that the method presented in this work does not require the geometry of the environment to be known *a priori* as in the Voronoi partition method [21]. Also, the approach uses computationally efficient ant algorithms to achieve the same level of coverage as was investigated in [22].

The rest of this paper is arranged as follows. In Section 2, the technical approach is described, including the problem description and coverage controller. Section 3 discusses, briefly, about how a Genetic Algorithm was used to estimate the profile of a pollutant in the environment while Section 4 discusses about simulation results. Finally, a brief conclusion and future work are presented in Section 5.

2. Technical Approach

The problem of distributing sensors into an environment under investigation can be seen as a coverage problem in which a pollution profile $F(S(x))$ presented in an area $S(x)$ is to be covered by n number of robotic agents so that the distribution of the robotic agents $A_T(x)$ is close to that of the pollution profile $F(S(x))$ where x is the set of coordinates (x, y) in $S(x)$. The robotic agents used in this work are assumed to have omnidirectional kinematics.

However, the proposed coverage controller can also be used on other types of robotic platforms with different dynamics by adding a low level controller to convert the control outputs into ones that suitable for the platform being used.

As mentioned previously, the coverage controller is made up of a bacteria controller that gives the agents the ability to navigate up a pollution gradient or a pressure flux gradient if required. This controller gives a source directional force to the flock of agents. A flocking controller is used to make the agents move as a group towards the source of the pollutant and to pull each other out of local minimum situations if they occur. Additionally, the flocking controller enables the flock of agents to localise at the source faster than if they were moving individually. The velocity controller is used to tune the distribution of the agents in the environment as will be discussed in Section 2.3. Each controller will now be discussed separately.

2.1. Bacteria Controller

There have been many investigations into the development of pollution source seeking controllers. For example Baronov and Baillieul [23], used a radial function as a potential function to develop an ascending controller. Their algorithm is not fully dependent on gradient but the function to be mapped or transverse must be known *a priori* for tuning the controller. Mayhew *et al.* [24] used a hybrid controller that combines a line minimization-based algorithm and a vehicle path planning algorithm to find the extremum of a function. Their approach does not need GPS point measurements and does not need any knowledge of the function to be mapped *a priori*. Dhariwal [25] used the Keller and Segel bacteria model as a basis for developing their controller whilst Marques *et al.* [26] investigated the use of rule based approach to using bacteria chemotaxis behaviour on a robot. In their experiments, they also investigated the use of the silkworm moth algorithm and a direct gradient following method. Lilienthal *et al.* [27] used a Braitenberg vehicle approach to find an ethanol source in their own experiments.

However this work uses a controller based on the Berg and Brown model in [28]. Models such as the one in [29], and other modern complex models, were not considered because a random biased walk model is sufficient for environmental exploration and source declaration in an environment. A bacterium motion is composed of a combination of tumble and run phases. The frequency of these phases depends on the measured concentration gradient in the surrounding environment. The run phase is generally a straight line while the tumble phase is a random change in direction with a mean of about 68 degrees in the *E. Coli* bacterium. If the bacterium is moving up a favourable gradient, it tumbles less thereby increasing the length of the run phase and vice versa if going down an unfavourable gradient. This behaviour was modelled by Berg and Brown by fitting the results of their experimental observations in [30] with a best fit equation in [28]. This model is shown below:

$$\tau = \tau_0 e^{\alpha \overline{\frac{dP_b}{dt}}} \tag{1}$$

$$\overline{\frac{dP_b}{dt}} = \tau_m^{-1} \int_{-\infty}^{t} \frac{dP_b}{dt'} e^{\frac{(t'-t)}{\tau_m}} dt' \tag{2}$$

$$\frac{dP_b}{dt} = \frac{k_D}{(k_D+C)^2} \frac{dC}{dt} \tag{3}$$

where τ is the mean run time and τ_0 is the mean run time in the absence of concentration gradients, α is a constant of the system based on the chemotaxis sensitivity factor of the bacteria, P_b is the fraction of the receptor bound at concentration C. In our work, C was the present reading taken by our Robotic agent. k_D is the dissociation constant of the bacterial chemoreceptor. $\frac{dP_b}{dt}$ is the rate of change of P_b. $\overline{\frac{dPb}{dt}}$ is the weighted rate of change of P_b, while τ_m is the time constant of the bacterial system.

The above equations determine the time between tumbles and hence the length of runs between tumbles. During the tumble phase, the robot agent can randomly choose a range of angles in the set $\sigma \in \{0..... 360\}$ by randomly choosing angles. This made it possible for the robot agents to backtrack if there is a favourable gradient behind it. The ability of the implemented bacteria chemotaxis controller to work over long distances has been presented in [31,32] and works fine even in the presence of small disturbances.

The above discussion about the bacterium behaviour can be depicted graphically, as shown in Figure 1.

A counter that increments every time tick is implemented. Once the counter is greater than τ a tumble phase is initiated otherwise the bacterium stays in the run phase. Once a tumble phase is completed through the assigning of a new σ, the run phase is resumed. A velocity controller which will be discussed in Section 2.3 was embedded into the bacterium controller.

It could be argued that the bacterium behaviour is a gradient based algorithm and, as such, an ordinary gradient based algorithm could be used in place of it. However, the Equations (1) to (3) have more to them than meets the eye. These equations take noise in the bacterium sensors, dynamics or environment into consideration. Equation (2) is an exponentially weighted low pass filter that is capable of filtering out noise. As a result of the filter, the Berg and Brown derived bacterium controller is not easily affected by noise and hence would not be readily caught in local maximums when they are encountered [33]. This is further aided by the random nature of the algorithm as a result of the tumble phase of the controller.

Figure 1. Showing the transitions between the two bacterium phases.

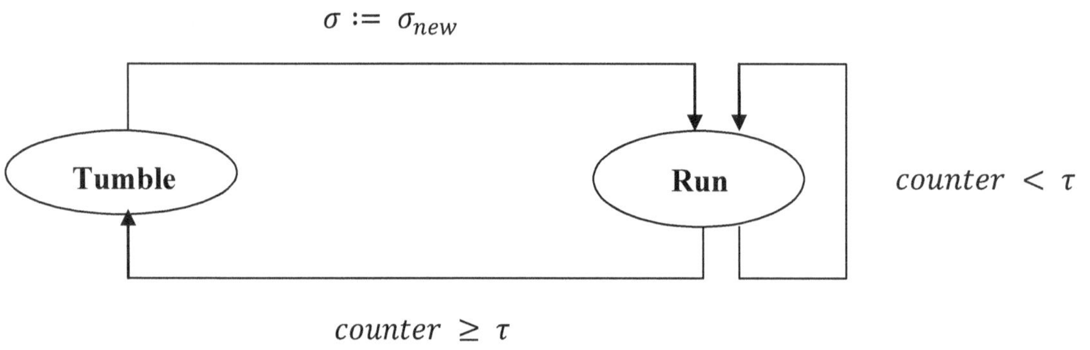

It could be argued as well that the random nature of the algorithm is not beneficial to bacteria. However, as was proven in [34], it is this very random nature that enables them to form a distribution corresponding to the profile of the pollutant in the environment. This is further supported by the works of Mach and Schweitzer, Hamann and Heinz [35,36] in which they discussed how a swarm of stochastic agents having a random component (bacterium tumble phase in this case) and a direct component (bacterium run phase in this case) can be represented by a Fokker Planck equation. The Fokker Planck equation would always converge to its stationary distribution that corresponds to the spatial function of interest.

It is the above properties that make the Berg and Brown derived bacterium controller different from an ordinary gradient ascent algorithm, and make it suited for use on noisy functions as used in the simulations in this work. In this work, the filter as represented by Equation (2) was implemented in discrete form using an ant limited memory resource length $k = 4$ to remember a history of past readings.

2.2. Flocking Controller

There has been a lot of work done on flocking controllers. The first flocking controller was developed by C.W. Reynolds in [37]. In order to achieve flocking, he defined that three rules need to followed. These rules were: Keep as close as possible to your neighbour (Aggregation), do not collide with your neighbour (Collision Avoidance), and move in a similar general heading as your neighbours.

Following this discovery, many researchers such as [38–40] have worked on these three basic rules in investigating various approaches of achieving flocking. Most of the approaches agree that Equation (4) should be satisfied in order to achieve flocking [38].

$$F_{output} = r\,[G_R(||r||) - G_A(||r||)] \tag{4}$$

where r is the distance between agents, $G_R(||r||)$ represents the repulsion function and $G_A(||r||)$ is the attractive function. Following this, we use exponential functions, as in Equation (5), to achieve flocking. This results in the Morse potential as in Equation (5) [41].

$$F^i_{output} = G_G * [(G^i_R - G^i_A) * e^{\frac{||r||}{20}}] \quad (5)$$

$$Output = F^i_{output} * G_F + B^i_{output} * G_B \quad (6)$$

where i is the index of the agents. Gains of 1 for the repulsion term G_R and 0.99 for the attractant term G_A were used. It was discovered that it is possible to control how closely the agents get to each other whilst not colliding by reducing the G_G gain. r in Equation (5) is the distance between the two closest agents. This value was obtained as follows: Each agent had a communication radius $l = 50$ and can only buffer up to $N = 5$ readings obtained from agents within this range. From these readings, the closest agent r is obtained and this value is used in Equation (5) to avoid collisions with other agents. Despite using a limited number N of neighbours, the agents were still able to avoid collisions whilst maintaining cohesion as a flock. Using a buffer of five readings was used only for flocking. This buffer is not necessary if proximity sensors are used instead, making sure that the ant like feature is maintained.

The final output from the flocking controller and bacteria controller is calculated as shown in Equation (6) where G_F and G_B are gains applied to both the flocking controller and bacteria controller output respectively.

2.3. Velocity Controller

The agent velocity updates are such that by controlling them, it is possible for the agents form a distribution of the pollutant in the environment and in works of [32,42] it was obtained by an adaptable velocity that was determined by Equation (7):

$$\beta = \frac{\beta_o * v_k}{C} \quad (7)$$

where β is a dynamic velocity that depends on the present reading of the environmental quantity, β_o is the standard velocity without any reading, v_k is a constant for tuning the dynamic velocity and C is the pollutant reading obtained by the robot's sensor. The separate controllers are combined by using the architecture shown in Figure 2.

The relationship between the bacterium controller and the velocity controller is shown in Equation (8), where $dist_\tau$ is the distance covered by the agent during a run phase and t_τ is the time duration from the last tumble phase.

$$dist_\tau = \tau_o \beta t_\tau \quad (8)$$

By using the velocity Equation (7), it was possible to enable agents achieve various visual distributions of pollution profiles as can be seen in Figures 2 to 5. However, it was also discovered that the constant v_k, can be used to control the spread of the agents in the pollution profile with high v_k values resulting in agents behaving more like gas molecules, covering a smaller area and forming less detailed features of the pollution profile whilst lower v_k values result in agents behaving more like liquid-like molecules, covering a higher area and showing more detailed features of the pollutant profile as can be seen in Figures 6 and 7.

Figure 2. Architecture of coverage controller.

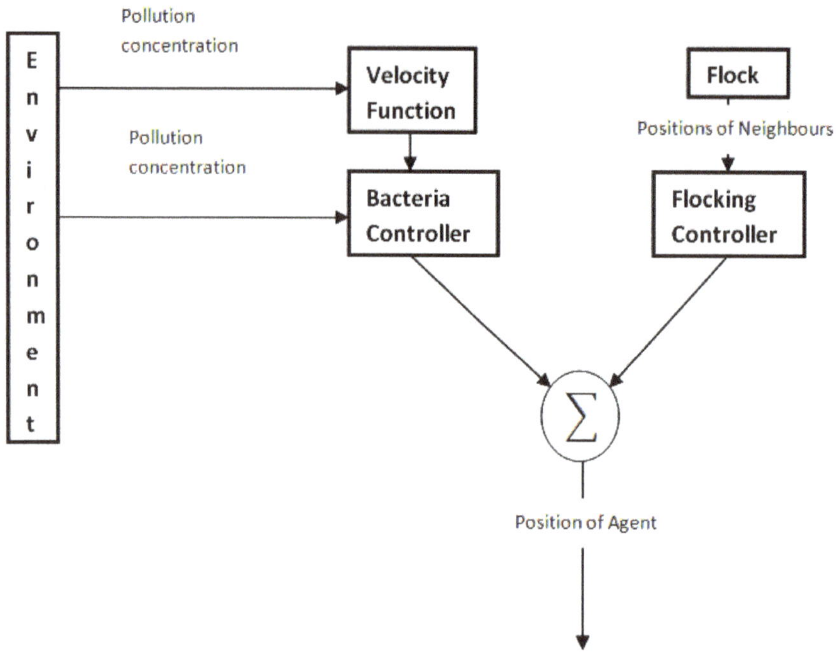

Figure 3. Simulated Arena with Gaussian distributed agents (**a**); Gaussian shaped distribution of pollutant (**b**).

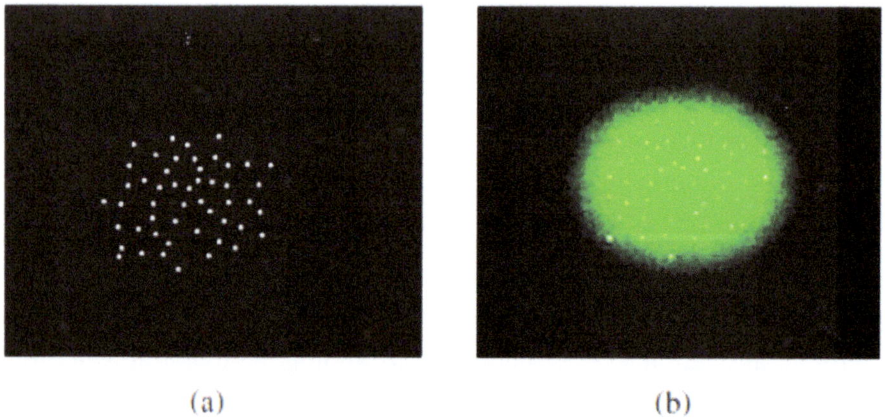

Figure 4. Simulated Arena with double peak Gaussian distributed agents (**a**); Double peak shaped distribution of pollutant (**b**).

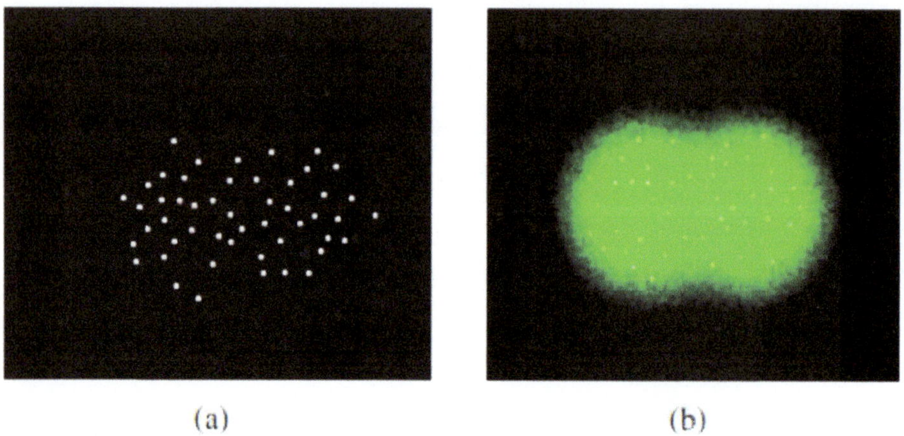

Figure 5. Simulated Arena with doughnut shaped distribution of agents (**a**); Doughnut shaped distribution of pollutant (**b**).

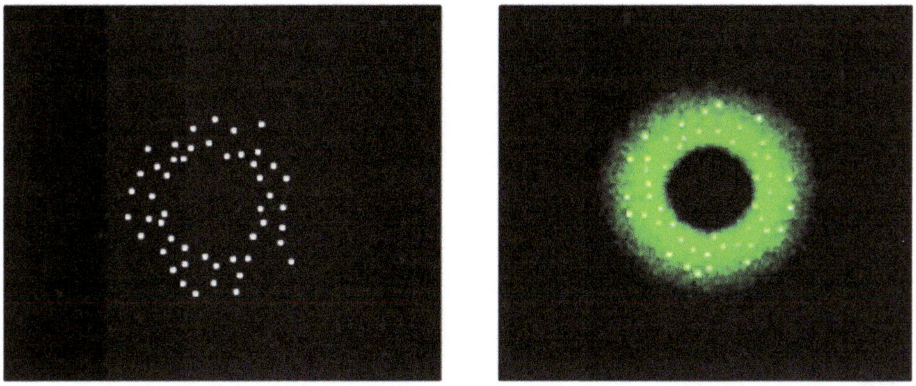

(a) (b)

Figure 6. Simulated Arena with squared shaped distribution of agents (**a**); Square shaped distribution of pollutant (**b**).

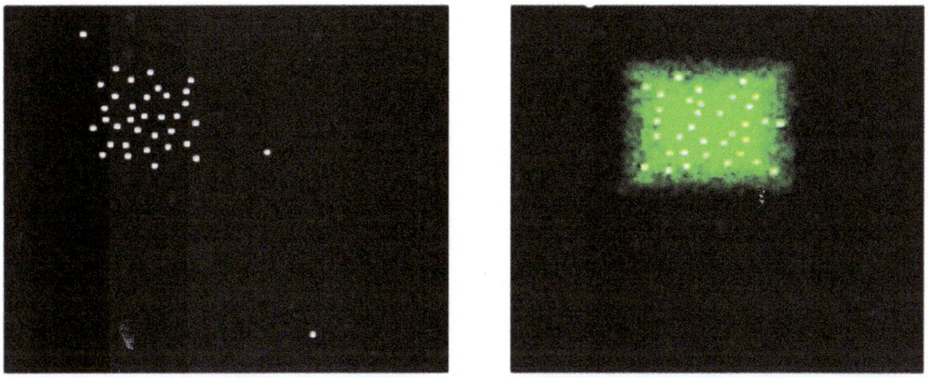

(a) (b)

Figure 7. Showing how the robots are distributed in a simulated skewed Gaussian profile using $\boldsymbol{v_k} = 32$.

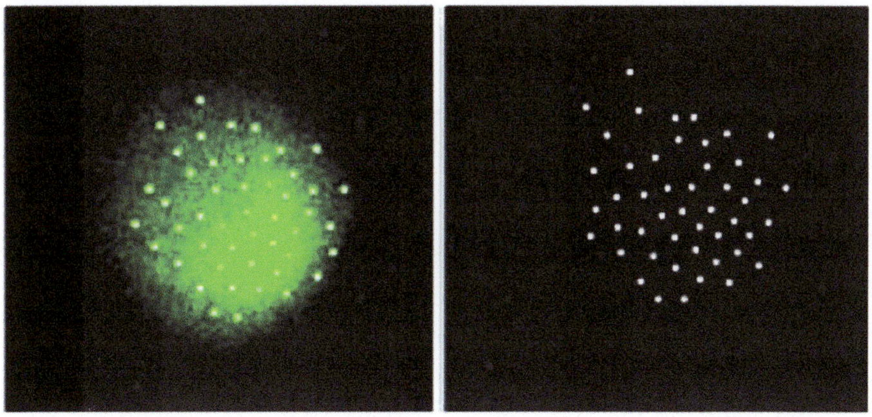

Figure 8. Showing how the robots are distributed in a simulated skewed Gaussian profile using $\boldsymbol{v}_k = 64$.

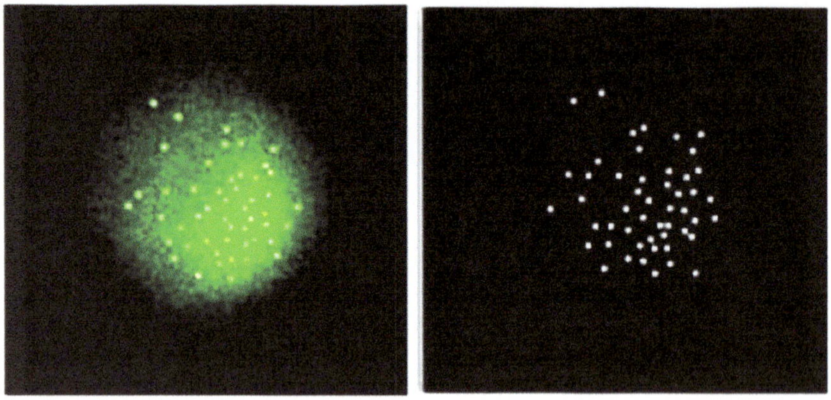

However, as choosing the right v_k value is presently guess work, a way of using machine learning to do this can be investigated. The approach used in this paper relies on using a Genetic Algorithm based on Gaussian models. In order to do this, some modifications are made to Equation (7) as shown in Equation (9):

$$\beta^i(t) = \frac{\beta_o^i * T}{e^{c^i}(t)} \tag{9}$$

where T can be viewed as a system temperature constant and can be used for tuning the system using an adaptable scheme.

By adjusting this constant, the spread of the agents in the environment could be controlled to achieve different effects. $\beta^i(t)$ is a dynamic velocity of agent i that depends on the present reading $C^i(t)$ of the environmental quantity, β_o^i is the standard velocity without any reading. This velocity function was embedded in the bacteria controller. The present reading $C^i(t)$ of the agent adapts the velocity of the agent so that in an area of higher concentration, it moves slowly covering a smaller area and vice versa. How this achieves coverage can be explained using

$$A^i = \beta^i(t)\pi\tau^2 \tag{10}$$

where A^i is the area covered by agent i, τ is the mean run time or mean run length of the bacteria controller and $A = \pi\tau^2$ is the area covered by the agent when the bacteria controller is tuned so that the agents motion is circular. As the agents get closer to the source, the area covered by them decreases due to a lower velocity and higher concentration reading while they cover a larger area if they have a higher velocity due to a lower concentration. The area covered by the pollutant is covered if the individual areas are covered by the agents as follows:

$$A_T = T \sum_{i=1}^{n} A^i \tag{11}$$

This could be used to develop a control law as in Equations (12) and (13) if an estimate of the pollutant profile $F(S(x))$ in the environment $S(x)$ were known and used to tune the system temperature T so that the spread of the agents in the environment is controlled. As a result, Equation (9) becomes Equation (14).

$$E_t(x) = F(S(x)) - A_T(x) \tag{12}$$

$$T_t = (\gamma * \int E_t(x)dt) + (\beth \; * \; E_t\,(x)) \tag{13}$$

$$\beta^i(t) = \frac{\beta_O*[(K_I* \int E_t(x))+ (K_P* E_t(x))]}{e^{C(t)}} \tag{14}$$

where γ is an integral gain and $\beth$ is a proportional gain. A Gaussian model based Genetic algorithm explained in Section 3 was used to estimate parameters of the pollutant profile *F*(*S*(*x*)) locally. The standard deviation parameters of the locally estimated Gaussian are then used in Equation (12) so that it becomes

$$E_t(x)^{std} = F(S(x))^{std} - A_T(x)^{std} \tag{15}$$

where $A_T\,(x)^{std}$ is the standard deviation of the flock positions. From machine learning, combination of Gaussian curves could be used to form various functions [43] so that the agents using the present approach are able to form various complex pollutant shapes. By using machine learning, it is possible to correctly get the shape of the pollutant and increase the rate of convergence. The disadvantage however is that the kernel function used to estimate the pollutant profile must be capable of forming the distribution of the pollutant profile effectively. Investigating the use of machine learning in improving the proposed control system opens up the possibility of studying how various real world conditions could affect the agents' distribution. This knowledge can then be incorporated into the control system so that the flock of agents are better prepared to deal with the environmental conditions as they arise.

3. Pollution Distribution Estimation Using Genetic Algorithm

Genetic Algorithm was used to approximate the distribution *F*(*S*(*x*)) of the pollution in the environment. As the proposed approach is aimed at use in an environment that is dominated by diffusion, we assume that a pollutant can be described using a Gaussian function. This can be explained by observing the behaviour of a drop of ink put into water. It starts out concentrated at the source point and as time progresses, the concentration at the source point will reduce while the surrounding areas become contaminated. This is as a result of diffusion.

This behaviour can be viewed as a Gaussian function of which amplitude at the source reduces from an initial value *A* to a stationary value $A_{stationary}$ as time *t* goes to infinity ($A \rightarrow A_{stationary}$ as $t \rightarrow \infty$) while the standard deviation σ becomes wider and goes to a stationary value $\sigma_{stationary}$ as time *t* progresses to infinity ($\sigma \rightarrow \sigma_{stationary}$ as $t \rightarrow \infty$). This can be modelled as in Equations (16) and (17). Where k_1 is the rate of decay of the amplitude A_t at time *t* and k_2 is a constant that can be used to tune the conversation of the change in amplitude to the spread of the pollutant.

$$\frac{\delta A}{\delta t} = -k_1 A_t \tag{16}$$

$$\frac{\delta \sigma}{\delta t} = \frac{k_1}{k_2}\sigma_t \tag{17}$$

3.1. Estimating Simple Gaussian Functions

In this work, a very slow (near negligible) diffusion was assumed. The Genetic Algorithm was used to estimate the parameters (amplitude, source location, and standard deviation) of a Gaussian function that was used to simulate the pollutant. The estimated standard deviation $F(S(x))^{std}$ was fed back to the robotic flock according to Equations (15), (13) and (14) in order to adjust the velocity of the agents to achieve optimal coverage of the environment.

The Genetic Algorithm was designed with 35 genes on each chromosome, with 7 genes representing each of the Gaussian's amplitude, x and y standard deviations, and x and y means. Each gene could take a value of either 0 or 1. As a result, the estimation range for each of the parameter using the Genetic Algorithm was between 0 to 128. A gene pool size of 400 chromosomes, tournament size of 10, crossover, mutation, reproduction ratios of 10%, 75% and 15% were used respectively with 10 generations per run.

During simulation runs, agents explored the environment using both the bacteria and flocking algorithm whilst collecting spatial pollutant data. The spatial pollutant data was stored in a running memory having n = 20 elements. This data was then passed to the Genetic Algorithm in order to estimate the pollutant parameters.

3.2. Estimating Multi-Modal Gaussian Functions

A pollutant is known to break up into pieces due to the turbulent nature of the environment. Hence, the pollutant's profile becomes one made up of various local maximums and one global maximum. This could also be viewed as noise. Nevertheless, the aim of the method proposed in this paper is to represent a pollutant's profile as closely as possible. This would be a problem for the method discussed in Section 3.1 as it would attempt to estimate a single global estimate from the data collected by the agents.

In order to solve this problem, it is assumed that each agent i transmits its individual estimated Gaussian parameter values to its local neighbours $j \in \{1 \ldots N\}$ present within an l = 5 units radius together with its position x_i. A reduced communication radius value ensures that only local agents to a local maximum contribute to each other's estimation.

Based upon this information, each agent uses an averaging scheme depicted by Equation (18) to reach a consensus as to the parameter values of the local maximum.

The consensus scheme works by using a weighted averaging mechanism so that values closer to what an individual agent i is estimating are given more weight than the values farther from its estimation:

$$k_i^{ave} = \frac{x_i + \sum_{j=1}^{N} x_j * G(\|x_i - x_j\|)}{\sum_{j=1}^{N} G(\|x_i - x_j\|) + G(\|x_i - x_i\|)} \tag{18}$$

where $k^{ave}{}_i$ is the k-average of the parameters estimate by agent i, $G(\|.\|)$ is a Gaussian curve with agent i's estimate as the centre value, N is the number of agents in the neighbourhood of agent i. $N \geq 5$ as this is used to simulate the communication limitation of the agent i in that it can only buffer up to five readings.

By using the parallelism offered by swarm robotics and the consensus scheme as discussed above, it is possible to obtain an estimation of the pollutant's profile in addition to recovering the function representing it.

The overall structure of the algorithm can be presented as in Algorithm 1. Lines four to six are necessary for a moving or changing pollutant profile. This has been tested in simulations with satisfactory results. The use of the model based genetic algorithm with the previous architecture is shown in Figure 9.

Algorithm 1. Using model based GA for agent distribution.

run bacteria and flocking controller
collect data from environment and neighbours
update counter
if counter > noOfRuns **then**
  Use GA to estimate model
end if
use GA estimates to update velocity function.

Figure 9. Architecture of the coverage controller using model based genetic algorithm.

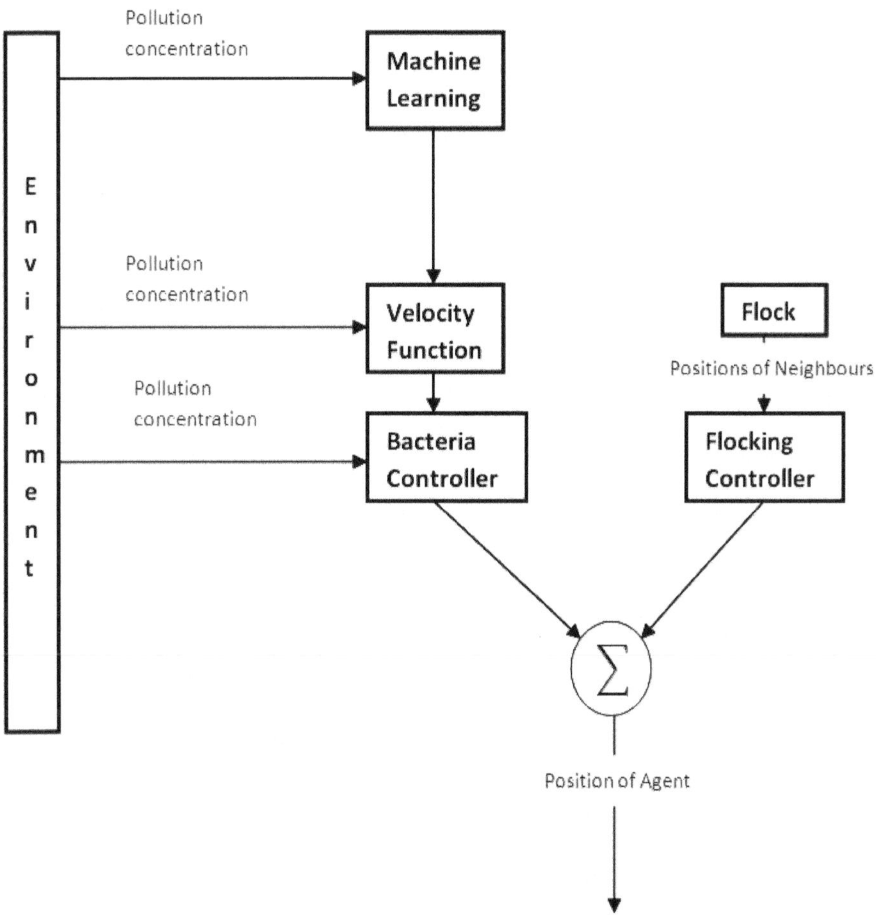

4. Simulation and Results

4.1. Simple Gaussian Functions

In order to test the approach, different pollutant profiles were generated with 50 agents deployed. A Gaussian having the following parameters of mean (x, y) = (400, 200), spread (x, y) = (38, 38) and amplitude of 100 was used. As can be seen in Figure 10, the agents were able to form the distribution even though their motion from Figure 10(b) seems erratic. This behaviour is as a result of the bacteria controller searching for a favourable path towards the optimal of the Gaussian distribution. Figure 10(c) shows how the error between the agents spread and the pollution spread reduces with time. The error was obtained by using Equation (14).

Figure 10. Robots forming a Gaussian distribution. The blue squares are the deployment points while the red dots are the end points of the agents.

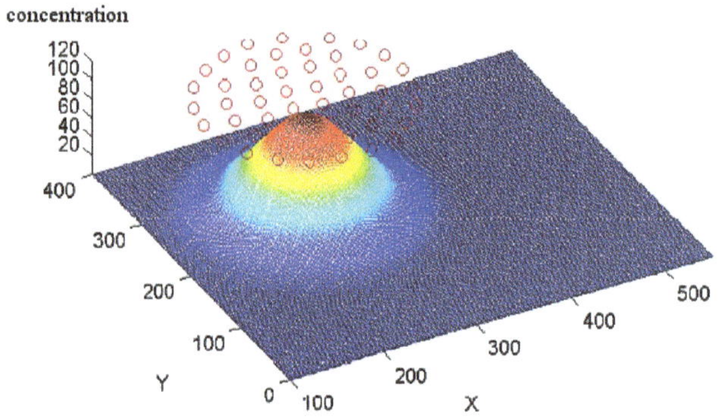

(a)

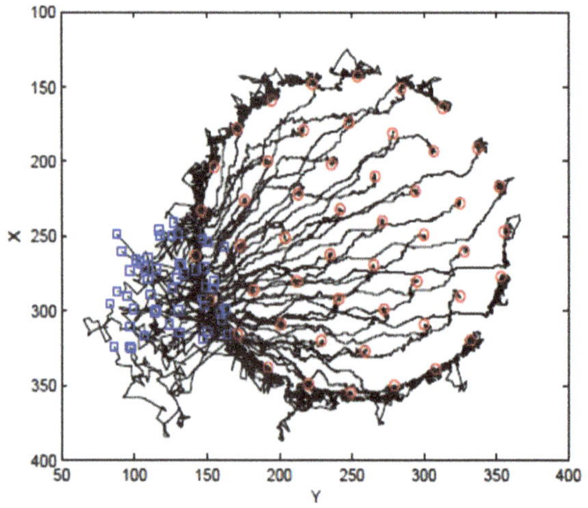

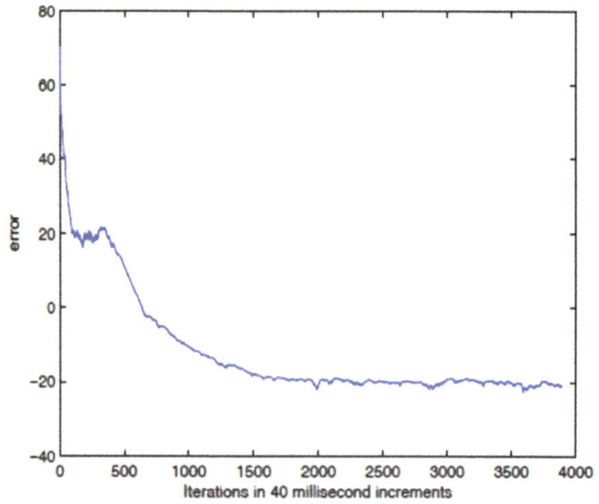

(b) **(c)**

Figure 11. Robots forming a distribution of three Gaussian distributions having means (x, y) = (80, 70), (170, 150), (200, 200), spreads (x, y) = (60, 40), (30, 80), (40, 80) and Amplitudes 100, 45, and 80 respectively.

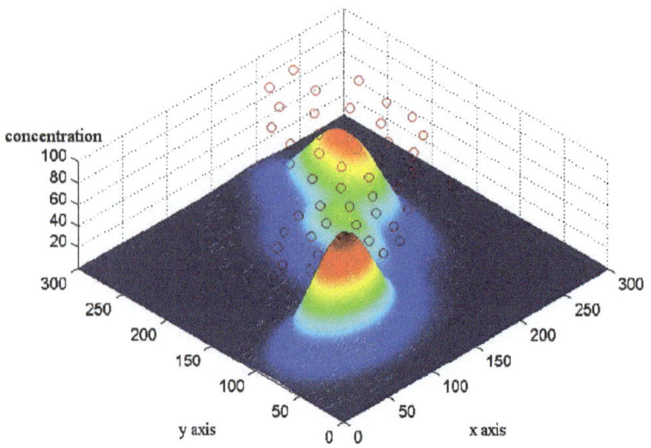

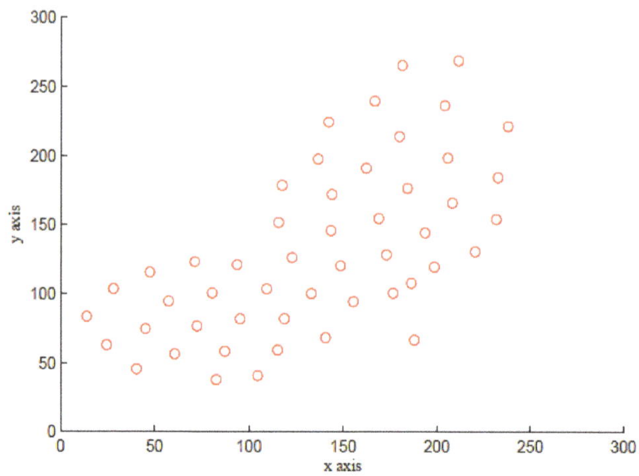

(a) **(b)**

Another simulated pollutant profile generated by adding three Gaussian functions of randomly chosen parameters: means (x, y) = (80, 70), (170, 150), (200, 200), spreads (x, y) = (60, 40), (30, 80),

(40, 80) and amplitudes = 100, 45 and 80 respectively were used as in Figure 11. It was observed that the approach enabled the agents to distribute according to the simulated pollutant's profile.

4.2. Multi-Modal Gaussian Function

Using the approach discussed in Section 3.2, tests on multimodal Gaussian functions generated by $C = 160 + (x^2 - 150\cos(2\pi x P_1) + y^2 - 150\cos(2\pi y P_2))$ were carried out with $P_1 = P_2 = 0.065$. During these experiments, 100 agents were deployed with sensor readings for the agents normalized between the ranges of 0 to 100 as that is the range that the Genetic Algorithm was designed to handle. A higher number of agents were deployed because of the nature of the pollutant profile.

As can be observed in Figure 12, 60 of the agents were able of form the shape of the pollutant to a limited extent with the dips in the multimodal function seen in the agents distribution at (x, y) = (2.5, 16) and (x, y) = (15, 16). The rest of the agents were exploring the environment.

Figure 12. Robots forming a distribution of a complex multimodal function having global maximum (x, y) = (10, 10).

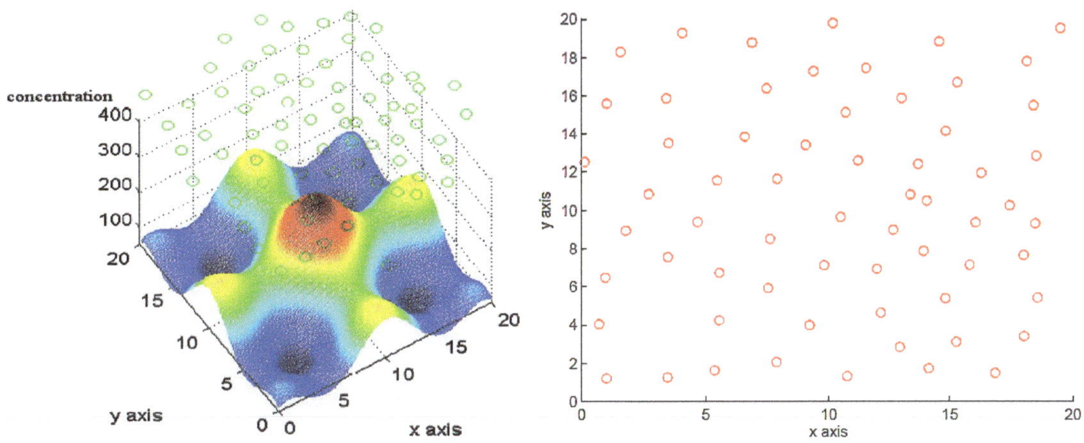

(a) (b)

5. Conclusion

As discussed in the introduction of this paper, this work consisted of two parts: the first part uses a swarm of robots to provide a visual representation of a noisy static pollutant. For this, the method in this work relied on using the bacterium behaviour, flocking behaviour, and a velocity function. However, this part relied on choosing an appropriate value for v_k in the velocity function. It was shown that the proposed method could be classified into the ant robotics category due to its limited communication capability, memory, and computational requirement. As a result, according to present knowledge, this is one of the first pieces of work that use ant robotics to create a visual representation of an invisible pollutant.

The second part of this work is the development of a feedback system that used the error between the standard deviation of the swarm member positions and the pollutant profile to choose an appropriate value for v_k. The standard deviation of the pollutant profile was estimated by using a computationally efficient model based Genetic Algorithm that relied on the parallelism offered by the swarm to reach a consensus among swarm members when the pollutant profile became more challenging.

However, the implementation of Genetic Algorithm takes the agents out of the scope of ant robotics because of the amount of memory elements that would be required to store the gene pool alone. In addition, noise would adversely affect the estimation of the Genetic Algorithm which raises the question of whether an alternative scheme that is less affected by noise can be developed. A cooling scheme borrowed from the theory of Simulated Annealing might offer promising results and this would be the task of future investigations.

Other areas to be investigated include: Estimating swarm convergence time to a pollutant profile and studying the effect of dynamic cleaning of the pollutant on the proposed approach upon encounter.

Acknowledgments

This research work was financially supported by European Union FP7 program, ICT-231646, SHOAL, and the UK EPSRC Global Engagements grant EP/K004638/1.

References

1. Ramana, M.V.; Ramanathan, V.; Kim, D.; Roberts, G.C.; Corrigan, C.E. Albedo, atmospheric solar absorption and heating rate measurements with stacked UAVs. *Quart. J. Roy. Meteorol. Soc.* **2007**, *133*, 1913–1931.
2. Giudice, A.G.; Melita, L.C.D.; Orlando, M.A. An overview of the "Volcan Project": An UAS for exploration of volcanic environments. In *Unmanned Aircraft Systems*; Springer: Berlin, Germany, 2009; pp. 471–494.
3. Puzis, R.; Altshuler, Y.; Elovici, Y.; Bekhor, S.; Shiftan, Y.; Pentland, A. Augmented Betweenness Centrality for Environmentally-Aware Traffic Monitoring in Transportation Networks. Available online: http://web.media.mit.edu/~yanival/JITS-environmental.pdf (accessed on 8 November 2012).
4. Cortes, J.; Martinez, S.; Karatas, T.; Bullo, F. Coverage control for mobile sensing networks. *IEEE Trans. Robotics Automat.* **2004**, *20*, 243–255.
5. Schwager, M.; Slotine, J.; Rus, D. Consensus Learning for Distributed Coverage Control. In *Proceedings of International Conference on Robotics and Automation*, Pasadena, CA, USA, 19–23 May 2008; pp. 1042–1048.
6. Schwager, M.; Mclurkin, J.; Slotine, J.E.; Rus, D. From theory to practice: Distributed coverage control experiments with groups of robots. *Springer Tracts Adv. Robotics* **2009**, *54*, 127–136.
7. Cortez, R.A.; Tanner, H.G.; Lumia, R. Distributed robotic radiation mapping. *Springer Tracts Adv. Robotics* **2009**, *54*, 147–156.
8. Pimenta, L.C.A.; Schwager, M.; Lindsey, Q.; Kumar, V.; Rus, D.; Mesquita, R.C.; Pereira, G.A.S. Simultaneous coverage and tracking (SCAT) of moving targets with robot networks. *Springer Tracts Adv. Robotics* **2010**, *8*, 1–16.
9. Schwager, M.; Slotine, J.; Rus, D. Decentralized, Adaptive Control for Coverage with Networked Robots. In *Proceedings of IEEE International Conference on Robotics and Automation*, Roma, Italy, 10–14 April 2007; pp. 3289–3294.
10. Shucker, B.; Murphey, T.; Bennett, J.K.; Member, S. Convergence preserving switching for topology dependent decentralized systems. *IEEE Trans. Robot.* **2008**, *24*, 1–11.
11. Shucker, B.; Murphey, T.; Bennett, J.K. An Approach to Switching Control beyond Nearest Neighbor Rules. In *Proceedings of American Control Conference*, Minneapolis, MN, USA. 14–16 June 2006.

12. Kwok, A.; Martinez, S. A Distributed Deterministic Annealing Algorithm for Limited-Range Sensor Coverage. In *Proceedings of American Control Conference*, St. Louis, MI, USA, 10–12 June 2009; pp. 1448–1453.
13. Pang, S.; Farrell, J.A. Chemical plume source localization. *IEEE Trans. Syst. Man Cybern. B Cybern.* **2006**, *36*, 1068–1080.
14. Koenig, S. Terrain Coverage with Ant Robots: A Simulation Study. In *Proceedings of the International Conference on Autonomous Agents*, Montreal, Canada, 28 May–1 June 2001; pp. 600–607.
15. Koenig, S.; Szymanski, B.; Liu, Y. Efficient and inefficient ant coverage methods. *Ann. Math. Artif. Intell.* **2001**, *31*, 41–76.
16. Wagner, I.A.; Altshuler, Y.; Yanovski, V.; Bruckstein, A.M. Cooperative cleaners: A study in ant robotics. *Int. J. Robot. Res.* **2008**, *27*, 127–151.
17. Altshuler, Y.; Yanovsky, V.; Wagner, I.A.; Bruckstein, A.M. Swarm Robotics for a Dynamic Cleaning Problem. In *Proceedings of the IEEE Swarm Intelligence Symposium*, Pasadena, CA, USA, June 2005; pp. 1–14.
18. Altshuler, Y.; Bruckstein, A.M. Static and expanding grid coverage with ant robots: Complexity results. *Theor. Comput. Sci.* **2011**, *412*, 4661–4674.
19. Borie, R.; Tovey, C. Algorithms and Complexity Results for Pursuit-Evasion Problems. In *Proceedings of the Twenty-First International Joint Conference on Artificial Intelligence*, Pasadena, CA, USA, 11–17 July 2009; pp. 59–66.
20. Arkin, R.C. *Behaviour-Based Robotics*; The MIT Press: Cambridge, MA, USA, 1998.
21. Schwager, M.; Slotine, J.; Rus, D. Unifying Geometric, Probabilistic, and Potential Field Approaches to Multi-Robot Coverage Control. In *Proceedings of the International Symposium on Robotics Research*, Lucerne, Switzerland, 31 August–3 September 2009.
22. Oyekan, J.; Hu, H.; Gu, D. A Novel Bio-Inspired Distributed Coverage Controller for Pollution Monitoring. In *Proceedings of the 2011 IEEE International Conference on Mechatronics and Automation*, Beijing, China, 7–10 August 2011; pp. 1651–1656.
23. Baronov, D.; Baillieul, J. Autonomous Vehicle Control for Ascending/Descending along a Potential Field with Two Applications. In *Proceedings of the American Control Conference*, Seattle, WC, USA, 11–13 June 2008; pp. 678–683.
24. Mayhew, C.G.; Sanfelice, R.G.; Teel, A.R. Robust Source-Seeking Hybrid Controllers for Nonholonomic Vehicles. In *Proceedings of the American Control Conference*, Seattle, WC, USA, 11–13 June 2008; pp. 2722–2727.
25. Dhariwal, A.; Sukhatme, G.S.; Requicha, A.A.G. Bacterium-Inspired Robots for Environmental Monitoring. In *Proceedings of IEEE International Conference on Robotics and Automation*, New Orleans, LA, USA, 26 April–1 May 2004; pp. 1436–1443.
26. Marques, L.; Nunes, U.; Almeida, T.D. Olfaction-based mobile robot navigation. *Thin Solid Films* **2002**, *418*, 51–58.
27. Lilienthal, A.; Duckett, T. Experimental Analysis of Smelling Braitenberg Vehicles. In *Proceedings of IEEE International Conference on Advanced Robotics*, Coimbra, Portugal, 30 June–3 July 2003; pp. 375–380.

28. Brown, D.A.; Berg, H.C. Temporal stimulation of chemotaxis in *Escherichia coli*. *Proc. Natl. Acad. Sci. USA* **1974**, *71*, 1388–1392.
29. Yi, T.; Huang, Y.; Simon, M.; Doyle, J. Robust perfect adaptation in bacterial chemotaxis through integral feedback control. *Proc. Natl. Acad. Sci.* **2000**, *97*, 4649–4653.
30. Berg, H.C.; Purcell, E.M. Physics of chemoreception. *Biophys. J.* **1977**, *20*, 193–219.
31. Oyekan, J.; Hu, H. Bacteria Controller Implementation on a Physical Platform for Pollution Monitoring. In *Proceedings of IEEE International Conference on Robotics and Automation*, Anchorage, AK, USA, 3–8 May 2010; pp. 3781–3786.
32. Oyekan, J.O.; Hu, H.; Gu, D. Bio-inspired coverage of invisible hazardous substances in the environment. *Int. J. Inform. Acquis.* **2010**, *7*, 193–204.
33. Oyekan, J.; Gu, D.; Hu, H. Hazardous Substance Source Seeking in a Diffusion Based Noisy Environment. In *Proceedings of the 2012 IEEE International Conference on Mechatronics and Automation*, Sichuan, China, 5–8 August 2012; pp. 708–713.
34. Oyekan, J.; Gu, D.; Hu, H. Visual imaging of invisible hazardous substances using bacterial inspiration. *IEEE Trans. Syst. Man Cybern. Syst. Hum.* **2013**, in press.
35. Hamann, H.; Heinz, W. A framework of space-time continuous models for algorithm design in swarm robotics. *Swarm Intelligence* **2008**, *2*, 209–239.
36. Schweitzer, F. Brownian Agent Models for Swarm and Chemotactic Interaction Brownian Agents. In *Proceedings of the Fifth German Workshop on Artificial Life*, Lübeck, Germany, 12–18 March 2002; pp. 181–190.
37. Reynolds, C.W. Flocks, Herds and Schools: A Distributed Behavioral Model. In *Proceedings of the 14th Annual Conference on Computer Graphics and Interactive Techniques*, New York, NY, USA, July 1987; Volume 21, pp. 25–34.
38. Hanay, Y.S.; Ilter, M. Aggregation, Foraging, and Formation Control of Swarms with Non-Holonomic Agents Using Potential Functions and Sliding Mode Techniques. In *Proceedings of European Control Conference (ECC 2007)*, Kos, Greece, 2–5 July 2007; Volume 15, pp. 149–168.
39. Gazi, V. Coordination and Control of Multi-Agent Dynamic Systems: Models and Approaches. In *Proceedings of the 2nd International Conference on Swarm Robotics (SAB '06)*, Rome, Italy, 30 September 30–1 October 2006; pp. 71–102.
40. Olfati-saber, R. Flocking for multi-agent dynamic systems: Algorithms and theory. *IEEE Trans. Automat. Contr.* **2006**, *54*, 401–420.
41. Smith, J.A. Comparison of Hard-Core and Soft-Core Potentials for Modelling Flocking in Free Space. Available online: http://arxiv.org/abs/0905.2260 (accessed on 8 November 2012).
42. Oyekan, J.; Hu, H.; Gu, D. Exploiting Bacterial Swarms for Optimal Coverage of Dynamic Pollutant Profiles. In *Proceedings of the IEEE International Conference on Robotics and Biomimetics*, Tianjin, China, 14–18 December 2010; pp. 1692–1697.
43. Sanner, R.; Slotine, J.J. Gaussian networks for direct adaptive control. *IEEE Trans. Neural Networks* **1992**, *3*, 837–863.

2

Hormone-Inspired Behaviour Switching for the Control of Collective Robotic Organisms

Tüze Kuyucu [1,*], Ivan Tanev [2,*] and Katsunori Shimohara [2]

[1] Engineering Department, Griffith Textile Machines, Washington NE38 8QA, UK
[2] Department of Information Systems Design, Doshisha University, Kyotanabe 610-0321, Japan; E-Mail: kshimoha@mail.doshisha.ac.jp

* Authors to whom correspondence should be addressed; E-Mails: tuze@gtm-ltd.co.uk (T.K.); itanev@mail.doshisha.ac.jp (I.T.)

Abstract: Swarming and modular robotic locomotion are two disconnected behaviours that a group of small homogeneous robots can be used to achieve. The use of these two behaviours is a popular subject in robotics research involving search, rescue and exploration. However, they are rarely addressed as two behaviours that can coexist within a single robotic system. Here, we present a bio-inspired decision mechanism, which provides a convenient way for evolution to configure the conditions and timing of behaving as a swarm or a modular robot in an exploration scenario. The decision mechanism switches among two behaviours that are previously developed (a pheromone-based swarm control and a sinusoidal rectilinear modular robot movement). We use Genetic Programming (GP) to evolve the controller for these decisions, which acts without a centralized mechanism and with limited inter-robot communication. The results show that the proposed bio-inspired decision mechanism provides an evolvable medium for the GP to utilize in evolving an effective decision-making mechanism.

Keywords: collective robotics; modular robotics; swarm robotics; hormone-inspired; XGP; Genetic Programming; reconfigurable robotics; pheromone-inspired

---

1. Introduction

Modular robots present potential robustness characteristics beyond the capabilities of wheeled vehicles, such as: the ability to traverse challenging terrain and insignificant performance degradation when partial damage is inflicted. On the other hand, they lack the emergent intelligence, manoeuvrability and flexibility of swarms. Thus, modular robotic movement and swarming can be viewed as two complementing behavioural traits rather than mutually exclusive ones. In this article, we present a hormone- and emotion-inspired control mechanism for a collective reconfigurable robotic system. The objective of the proposed control mechanism is to decide whether a robotic organism that is deployed for the exploration of an unknown environment will benefit from switching between states of swarming into moving as a single modular robot and *vice versa*. Although a single decision at its core, the aforementioned mechanism's result is a determining factor in a switch between different and conflicting behaviours. Furthermore, the behavioural switching is time-consuming and a determining factor in the success and survival of the robotic organism. Such a decision can not be easily made via a predetermined algorithm or predetermined protocol, because of the environment being unknown, unpredictable and only partially observed by the perception subsystems of the robotic modules. Moreover, the collective decision, of the robotic system as a whole, to switch from a certain type of behaviour into another one when each individual robot makes its own decisions is even more difficult to formulate. Rather, it could emerge as a higher-level property of the complex, non-linear robotic system, as a result of the interaction between the lower-level entities (modules) and the environment [1]. In terms of the practical application being considered, the work by Berend *et al.* is the most related: to design a system that is reliant on online evolution in achieving organism formation via a swarm of robots in order to increase survivability of the robots [2]. The model presented here relies on offline evolution and is aimed at addressing predetermined objectives. The adaptability of the system is obtained as an emergent property derived from both the interactions (i) between the robotic modules and (ii) between modules and the environment.

Hormone- and emotion-inspired decision models in collective robotics have been developed by various authors to address a range of tasks in the control of a group of robots. The two terms (hormone and emotion) are often used in highly related systems; thus, we consider models from both sources of inspiration relevant and use the two terms interchangeably. The hormone-inspired decision systems model the behavioural effects of emergent emotional changes, due to the hormonal fluctuations that can be observed in biological organisms.

The models developed differ in the style of implementation, as well as the point of view taken on how these hormone-inspired models function. Often, the difference in the modelling comes from whether a low- or high-level point of view is used. In a system where the low-level interactions of hormones and chemical pathways are modelled, the aim is to create a tightly-coupled control mechanism that can define detailed behaviours in a robot. From a different point of view, the fluctuations in hormones are associated with certain emotions leading to high-level behaviours (often used as a behaviour switching mechanism among pre-defined behaviours rather than as a control mechanism). Shen *et al.* [3] present a control model for multiple robot coordination that is based on detailed hormone interactions. Their mechanism, based on Turing's reaction-diffusion model [4], defines the coordinated movements of robots via hormone-messages passed between neighbouring robots. Another reaction-diffusion-based complex coordination mechanism is used by Hamann *et al.* [5]. These two mechanisms presented by Shen *et al.* and Hamann *et al.* behave as detailed coordination mechanisms inspired by the micro-level interactions of the hormones in biological organisms. These mechanisms are designed to be applied as controllers of low-level actions (such as controlling the actuators of wheeled robots) that lead to an emergent higher-level behaviour. Due to the gap between the low-level properties of the micro-entities and the emergent high-level properties of the system as a whole, these mechanisms would be difficult to configure (or evolve).

Utilizing a simpler hormone-inspired decision mechanism limits the flexibility and creativity of the underlying system. However, it drastically simplifies the control models and allows them to be easy to modify, as well as easy to integrate with other control models. Murphy *et al.* [6] achieved multi-robot coordination in a small team of heterogeneous robots using a simple emotion-based control with a limited amount of coding. Moioli *et al.* [7] showed that a simple hormone-inspired approach to task switching in a swarm of robots works well with tasks featuring conflicting objectives (such as exploration and energy preservation). The ALLIANCEarchitecture [8], as well as a similar approach by Walker and Wilson [9] also provides simple, yet efficient, behaviour switching mechanisms based on motivations, such as impatience and acquiescence, similar to the hormone-inspired model proposed in our work. Both of these models use continuous broadcasts between robots, since they require the robots to be aware of others' tasks. This presents problems with scalability, as well as reliability issues as the number of modules in the swarm grows. Although not comprehensive enough to cover all the related hormone-inspired control mechanisms, an analysis of a large number of emotion-inspired mechanisms can be found in [10]. The proposed hormone-inspired mechanism is similar to the simple behaviour switching mechanisms used by many other authors. The novelty of our approach is in its area of application, integration, as well as its implementation.

The proposed decision mechanism is inspired by the fluctuations in the hormonal signals of biological organisms that cause changes in the emotional states, which, in turn, determine the actions for many crucial decisions. Our work is inspired by the drastic behavioural fluctuations often observed in biological organisms, due to the changes in hormonal regulators [11]. We propose an automated design of the decision mechanism via simulated evolution, because we believe that an *a priori* hand-coded solution would not be adaptable, due to the unknown and unpredictable nature of the environment. Moreover, such a solution would not be an optimal one, due to the inherent complexity of the modular robotic system. Indeed, the analytical models for such complex, modular robotic systems do not exist, and often, the desired high-level properties of the system as a whole cannot be directly inferred from the hand-coded low-level behaviour and morphology of its entities. We use XML-based Genetic Programming (XGP) for the evolution of the control mechanism (Tanev [12]). The evolved model is intended to decide the timing of switching between the swarming and snake-like locomotion behaviours, and *vice versa*. Both behaviours were developed in our earlier works [13,14]. The task is for a group of robots that approach a corridor as a swarm to overcome the various obstacles presented and explore as much of the area as possible in the process. The experiments demonstrate the intuitive behaviour achieved by the robots as a group by switching among the main behaviours (swarming, modular robot reconfiguration and movement) via the use of the presented hormone-inspired decision mechanism.

This article is organized as follows. In Section 2, we describe the adopted robotic modules and their simulated environment. Section 3 details the algorithms executed for all three behaviours that the robotic system can perform. Section 4 explains the decision mechanism designed for switching among various behaviours (described in Section 3). In Section 5, we describe the experimental setup and present the empirical results obtained from the hand-tuned version of the hormone-inspired algorithm. In the same section, we provide the values of the major parameters of XGP and the results obtained by the simulated evolution. Section 6 provides a comparative analysis on the results, presented in the preceding section, and Section 7 draws a conclusion of our work.

2. Robots and Their Environment

We use the commercially available robot simulation platform, Webots, in our experiments, which realistically models the physics of the adopted mobile robots, their interactions with each other and with the environment [15]. Table 1 details the parameter settings used for the Webots simulation platform. The robots are spherically shaped two-wheeled robots with a differential drive and $7\,cm$ in size. Robots feature a set of a simulated radio emitter/receiver and four infra-red (IR) sensors in front, as shown in Figure 1. The radio emitter and receiver set allow the robots to establish direct one-to-one and one-to-many communication with other robots. The radio communication models a simple RFmodule with a custom communication protocol, *i.e.*, without a known communication protocol, such as WiFi or Zigbee. The robots are equipped with two actuators—one in front (vertically oriented) and one at the back (horizontally oriented)—which have magnetic connectors attached that allow them to connect with other robots. The front actuator has an "active" connector (an electromagnet) that can be activated or de-activated by the robot. The rear actuator has a "passive" connector (a permanent magnet), which is always in the same state and can only be connected with an "active" type connector (since two passive connectors have the same polarity). Thus, the robot that is connecting with its "active" (front) actuator has the initiative in establishing a robot-to-robot connection.

Table 1. Webots related parameters of the simulated robots.

World Properties		Robot Physics Properties	
CFM	0.00001	density	−1
ERP	0.2	mass	0.1
basicTimeStep	32	centerOfMass	0 −0.03 −0.031
bounce	0.5	bounce	0.5
bounceVelocity	0.01	bounceVelocity	0.01
coulombFriction	0.5	coulombFriction	1
forceDependentSlip	0.004	forceDependentSlip	0
runRealTime	true		

The robots are designed to reflect a realistic size of a robot with similar capabilities (e.g., epuck, Symbrion robot [16]). They are aimed at being capable of forming a swarm of wheeled robots and achieve common tasks, such as exploration of unknown environments. The robots are also designed to substitute the spherical modules used in the sidewinding Snakebots in our earlier work [17]. The latter is the reason why the robots have a spherical casing. The robots here do not use ball joints as before (due to the limitations of the Webots simulation platform), but utilize two separate joints.

Figure 1. The differential-wheeled robot used in simulation experiments with four infra-red (IR) sensors of 7 cm range pointing forward and two connection points that allow docking with other robots.

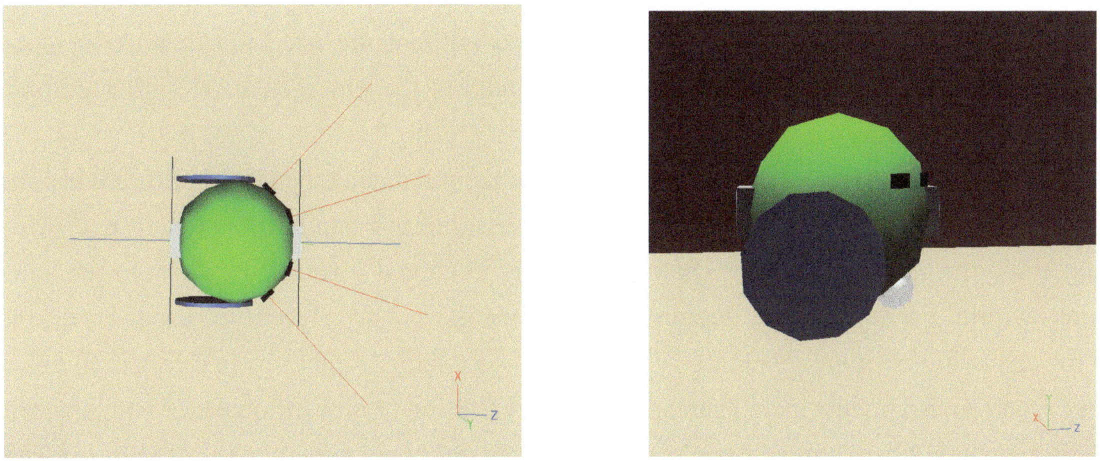

3. Collective Behaviours

To behave as a collective robotic organism, the robots need to be able to achieve two behavioural states: (i) swarming; and (ii) modular robot. These two states require completely different behaviours from the individual robots, and thus, they are coded as separate control mechanisms. In order to achieve a transition between these two states, the robots need to be able to "reconfigure" themselves, which is the third behavioural state: (iii) reconfiguration. All three behavioural states are illustrated in Figure 2 to demonstrate an example case where these three behaviours would be highly beneficial if used correctly. All these behaviours are previously developed and are not the main focus of this work. Therefore, in the following subsections, we will only provide a brief explanation of these two behaviours.

Figure 2. An example scenario of the collective robotic organism utilizing all three behaviours.

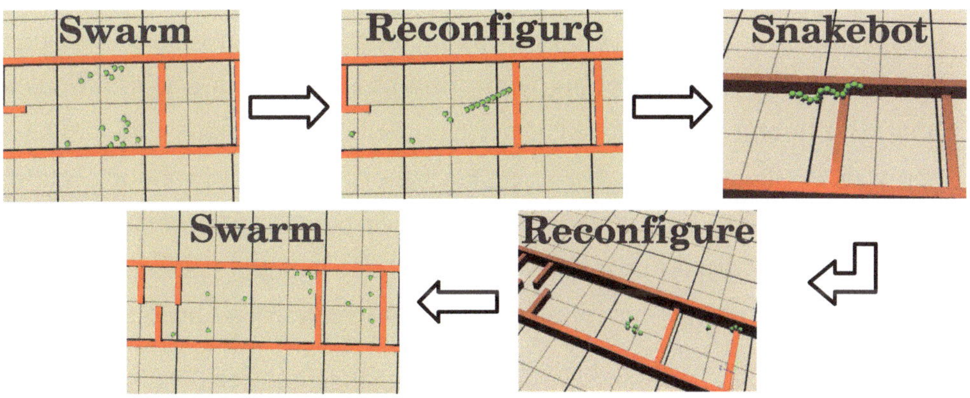

3.1. Swarming

Swarm intelligence is an emergent behaviour from the decentralized interactions between many systems. In robotics, it is the coordinated (explicitly or implicitly) movement and actions of many robots that contribute to achieving tasks via the emergent behaviour. We use the pheromone-inspired swarm control [14], which does not require any direct interaction within the robots to achieve the multi-robot coordination required. Pheromones can be used to provide a stigmergic medium of communication, which influences the future actions of a single individual or a group of individuals via changes made to the environment. A term first introduced by Grasse [18], stigmergy as a communication mechanism allows the history of an individual's actions to be tracked without the need to construct a model of the environment within the individual's own memory; making the emergence of higher complexity behaviours from a group of simple individuals possible.

The swarm control algorithm used here is a pheromone-inspired mechanism that builds environmental information by chemical gradients. The environment is assumed to handle the storing and diffusion of chemicals; thus, the robot controllers do not store any chemical information, except the sampled concentrations within the immediate vicinity of the robot. The algorithm used is developed specifically for achieving optimum exploratory behaviour via a large number of real robots in unknown environments. Such environments include areas of high devastation (e.g., earthquake or tsunami stricken settlements) or distant and dangerous missions. Exploring an unknown area quickly is a mission-critical objective in rescue operations. Such operations can face a list of limitations, such as the lack of a terrain map, the failure of previously established communication networks and the lack of reliable GPS tracking. In such missions, the first task is to search the area in question as quickly as possible and locate targets. The robots would be required to be capable of various functionalities other than area exploration; therefore, it is desired that the integration to a swarm and the ability to explore are seamless and do not consume a large amount of the robot's resources. Utilizing a real stigmergic communication would be an efficient method of achieving such emergent behaviour with low overhead.

Robots that utilize stigmergic trails to communicate with each other have been shown to effectively coordinate and quickly explore a given terrain [19,20]. The use of stigmergic pheromone-based communications in robotics has a range of other potential advantages, such as the possible ability to adjust the range and persistence of a pheromone, not being limited to line-of-sight, the ability of pheromones to propagate through the environment (while forming gradients) and freeing the individual robots from the burden of communication management and processing—usually involved with other forms of commonly used communications (such as radio, IR, visual, audio, *etc.*). On the other hand, the use of physical substances for pheromone-based communication within robots is problematic and poorly understood. However, there is undergoing work in improving their use with promising results, and it is predicted that with improvements in sensing technology, it may be possible that a robot could carry a lifetime supply of chemicals [21]. The use of olfactory sensors in robotics is a developing area with promising demonstrations of distinguishing multiple odours by mobile robots, e.g., [22], and recent development of high resolution olfactory sensors for robotic systems [23]. With these developments, the use of real chemicals for inter-robot communication can become as convenient as any other conventional method, such as infra-red or radio.

The swarming behaviour used here is similar to the "node counting" algorithm described in [20]; however, in our case, the goal is to achieve a quick "survey" of an area by visiting key locations as quickly as possible, instead of "sweeping" by visiting all the available locations in the environment. The developed algorithm assumes that a more probabilistic approach to exploration is highly beneficial in realistic situations and that the common strategy of sweeping of the whole area for the exploration of unknown environments is highly inefficient and unrealistic. The algorithm is implemented as a three-layered subsumption architecture, where each layer implements one basic behaviour: *random walk* ("exploration") is the behaviour corresponding to the lowest priority layer, while *pheromone-based coordination* is the middle layer, and the *wall avoiding* behaviour is realized by the highest priority architectural level. In the pheromone-inspired algorithm, the robots explore a given environment by dropping artificial pheromones in their environment (to mark the visited locations) and sampling the pheromones dropped by other robots. Each pheromone has the ability to diffuse and evaporate, and they are simulated in a grid-like environment, where each grid point is the size of a robot.

The pheromone-inspired swarming (a completely separate mechanism from the hormone-inspired mechanism, which is the main focus of this article) is explained in detail in [14].

3.2. Modular Snakebot

Due to its simple chain-like geometry, the automated formation and configuration of a modular Snakebot is relatively easy. Some useful features of snake-like robots include the smaller size of the cross-sectional areas (in comparison to other modular robots), stability, ability to operate in difficult terrain, good traction and complete sealing of the internal mechanisms [24,25]. Moreover, due to the modularity and homogeneity of their design, the snake-like robots have high redundancy, which, in turn, provides it with inherent fault tolerance and adaptability properties [17].

In our model, once the robots connect to each other to form a long chain, they can use the actuators between the neighbouring robots to achieve locomotion. To achieve coordinated movement amongst the modules, the robots synchronize their internal timers and assign themselves individual IDs depending on their location in the Snakebot. The head of the snake being zero, the tail, n − 1 (in a Snakebot with n modules). It was noted in an earlier work that a sidewinding locomotion is the fastest and most efficient form of locomotion for a snake-like robot [17]. Even though the results obtained in those early studies are used here, we implement a rectilinear locomotion instead. The current robots achieve a stable rectilinear locomotion, due to the presence of wheels. In fact, the addition of wheels also makes it difficult to use the previously evolved sidewinder controls, since the wheels can change the geometry of the module. Equation (1) (adapted from the evolved solutions in [13]) is used for the movement of vertical actuator in each robot, which successfully achieves a moving sine-wave within the Snakebot. The horizontal actuators are locked at their initial perpendicular position. Although the motion gait, considered in this work, is the rectilinear one (which is also similar to the locomotion gait of caterpillars) instead of sidewinding, as exhibited in our previous implementations of the Snakebot, we use the term Snakebot to refer to the specific morphology of the modular robot (rather than particular locomotion gait) in order to provide a clear connection to the source and inspiration of this work [17]. A rectilinear locomotion is preferred in these preliminary experiments, because, compared to sidewinding, it offers the advantages of (i) simplicity of implementation, (ii) simplicity of steering and (iii) a much reduced cross-sectional area of the moving Snakebot.

$$V_p = -0.1 \times (id - 16) \times \pi sin(id + 3t - 1) \qquad (1)$$

Equation (1) is used with a maximum of 15 modules in the experiments. V_p is the position of the vertical actuator in radians. This equation is only used for achieving locomotion within a modular robot and does not have an impact on the hormone-inspired mechanism described here.

3.3. Reconfiguration

In order to switch from a swarm to a single, yet modular, robot (and *vice versa*), the robotic modules need to coordinate their motion and assemble together (or disassemble, respectively). This process is referred to as reconfiguration, and it can be initiated by one or more robots. The remaining robots then have the freedom to join or not join the initiating robot. When there are no robots willing to join the modular robot, the connected robots (if more than one) may then start to move as a single, snake-like modular robot. In a reverse situation, the robots simply decide to disconnect, while moving as a Snakebot, and, once separated, to start moving as entities of a robotic swarm. For the experiments presented in Section 5, the initiating robot first turns towards east, which is the direction that the experimental corridor stretches. This is done to make sure that the Snakebot is facing the correct direction once it is formed. Reconfiguration is the only stage the presented robots communicate directly via wireless messages. These wireless messages involve broadcasts of a robot's request for potential partners to dock and the positive responses that the other robots may give.

Self-assembly and automatic reconfiguration can be quite a challenging task on its own depending on the number of the degrees of freedom of the modular robot (e.g., [16]), and sophisticated solutions that rely on evolutionary computational techniques have been developed, e.g., [26]. The focus of our work, however, is not in the self-reconfiguration of modular robots; hence, the complexity of the reconfiguration scenario is kept to a minimum. Since the considered modular robots have only two points of connection (front and back), the only shape that can be achieved is a snake-like shape, which simplifies the possible control mechanisms required for the locomotion of the formed modular robot. The tasks of robotic modules in achieving a reconfiguration of a swarm of robotic modules into a single modular Snakebot involve:(i) finding robots in the swarm that want to dock; (ii) locating and docking to each other without collisions; (iii) generating the correct IDs (used for Snakebot locomotion) for each robot; and (iv) deciding when the reconfiguration is complete. The initial decision to initiate a reconfiguration or to become available for reconfiguration participation is decided by a separate mechanism. This mechanism is the primary focus of our work and will be elaborated upon later in Section 4.

Once a robot decides to initiate a reconfiguration process, it broadcasts a message (via radio communication) to all other robots to signal its decision. The robots that are available for docking respond with messages addressed to the initiating robot. If no responses are received (implying that no robots are available for docking), the initiating robot terminates the reconfiguration process and returns back to the swarm. Otherwise, the initiating robot picks the closest one of the responding robots by estimating the distances from the signal strength of the received responses. Once chosen, the initiating robot waits for the other to dock, occasionally exchanging messages in order to: (i) maintain connection;

(ii) provide the docking robot with the direction the waiting robot is facing; and (iii) allow the docking robot to have a rough estimation of the distance between them. If the connection between the two robots is lost or the docking robot takes a long time to accomplish the procedure, the waiting robot terminates the procedure and looks for a different robot to connect with. If the docking process completes successfully, the docked robot takes the role of the initiating robot and looks for other robots to connect with. However, if in this case, the waiting robot is unable to get a response from other robots for docking, it then signals all the other robots that are part of the same modular robot to switch into modular robot mode (by synchronizing timers) and start moving as a Snake.

Disassembly is a much simpler process, which does not require any coordination among the robots. Again, the decision is made by the mechanism explained in detail in Section 4. The disassembly results in the deactivation of the front ("active") connector of the corresponding robot.

4. Hormone-Inspired Behavioural Switching Based on Patience

The decision to switch states (disconnect from a modular robot and join a swarm or start forming a modular robot while swarming) is made on an "impatience" value that the robots increase or decrease depending on the environmental factors. This is modelled after the biological organisms, namely animals, which maintain emotions, such as anxiety, tolerance, restlessness and eagerness, that contribute to large changes in behaviour via hormonal feedback in their bodies. Hormones in biological organisms are known to achieve coordination amongst different members and are a determining factor for social behaviours [11]. We are inspired by this duty of hormones in biology and believe that a similar decision mechanism could provide intuitive changes in robotic behaviour. Here, we provide a simplistic model of a hormone, where we do not try to provide a biologically-plausible implementation. The main source of inspiration is not how the hormonal chemical networks work in biology, but the situational uses (*i.e.*, behavioural switching) of slow, but smooth, chemical gradients instead of fast and sharp logical decisions. The model presented provides some of the basic functions of a hormonal network in biological organisms, namely: (i) storage and secretion of hormones; (ii) recognition and processing of the hormone; and (iii) degradation of the hormone. The hormones in our model are not transported to other "cells", since they are used in the regulation of a single-cellular entity (*i.e.*, the individual robot) that does not share its hormonal state with other robots. However, the resulting behavioural shifts in a robot are shared with other members of the group, which affects the status of the other individuals. Thus, the hormones produced indirectly control the behavioural state of the whole group.

The impatience value used to model the behaviour of hormones has a certain range, and it constantly degrades over time (multiplied by 0.95 every time step in the presented experiments), even when not in use. The production (secretion) of the impatience value is determined by a separate logic (hand-coded or evolved), which only controls the incremental changes in the value of the impatience value, as elaborated in Section 5. We define a threshold level for the impatience value, and if this value is reached, an action takes place depending on the robot's state. If the impatience value is high enough to initiate an action, it is reset to zero (*i.e.*, it is "consumed"). The three major behaviours described in Section 3 define the states that a robot can be in. We spread these three behaviours to a total of five states: **State 0** is swarming only, **State 1** is Snakebot locomotion only, **State 2** is swarming, but ready to join a Snakebot,

State 3 is initiate reconfiguration for a Snakebot and **State 4** is docked and looking for other robots to form a Snakebot. The three behaviours are spread over five states in order to simplify the reconfiguration process and the integration of the three distinct behaviours and encourage gradual change in behaviours rather than abrupt switching. For the experiments presented in Section 5, the robots are initialized in State 2. The decision mechanism determines the changes between these states that place the robot in a different behavioural zone. When a change in state is triggered, the state transitions are as follows: State 0 –> State 2; State 2 –> State 3; states 1, 3 or 4 –> State 0. Figure 3 illustrates the state machine built to accomplish the desired behaviours, with the aforementioned five states.

Figure 3. The states and the possible transitions in the main controller of the robots. The solid lines illustrate the state transitions, due to the decisions made by the hormone-inspired behaviour switching, whereas the dashed lines represent the state transitions, due to changes in the environmental conditions.

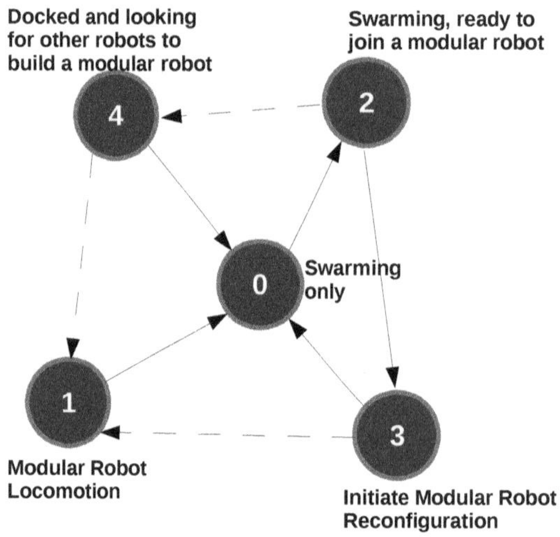

These states and the state transitions define the behaviours and the order of change between these behaviours, which are easy to define. The most difficult task, however, is to decide when to initiate these state transitions. Table 2 shows a list of variables that each robot has access to and we believe are sufficient for the robots to analyse their environments and adjust their internal states. All these variables are either acquired without any communication or via the existing communication taking place among the robots during the reconfiguration procedure.

Table 2. Perception information available to each robot relevant for determining the change in the impatience value.

Perception	Description
lowestPH	Value of lowest pheromone conc.requests
No. largestModuleSize	No. of modules in the largest active Snakebot.
moduleImpact	$largestModuleSize - currentSnakebotSize$

The pheromone levels, which are stored by the environment, provide information to individual robots about the environment and whether the goal of exploration is being satisfied [14]. The $requests$, $largestModuleSize$ and $moduleImpact$ are affected by the behavioural state of the other robots in the environment. These variables inform the individual robots about the state of the other robots and create the means for social monitoring and pressure. Thus, the hormonal changes within the other robots have the possibility to cause hormonal changes within.

5. Experiments

An experimental environment is designed to develop and test the decision mechanism for its ability to provide efficient and intuitive switching between behaviours in order to accomplish quick exploration of unknown areas. The environment is designed to test the ability of the robotic group to demonstrate all three behavioural states to successfully explore a given environment without any information about the area. The environment is a long corridor ($24\,m$) with two types of obstacles: low continuous walls and high walls with gaps. The low walls present a challenge to individual robots, which are too small to overcome these obstacles, but these low walls can be climbed over by a Snakebot. The high obstacles cannot be overcome by either the individual robots or a Snakebot. However, the high obstacles are arranged to have small offset gaps that give individual robots the opportunity to circumnavigate. Figure 4 provides different views of the environment.

Fifteen robots are initialized at the left end of the corridor as a swarm. The goal for the robots is to clear all the obstacles in the way and explore as much of the environment as they can. The robots are expected to explore this section as well as they can before joining together to overcome the low obstacles blocking their way. Once formed, the modular robots that overcome the low obstacles are expected to partially disassemble after the first low obstacle, while the rest disassemble when they reach the high obstacles, where they can no longer move forward. The disconnected individual robots that reach the high obstacles are then expected to find their way to the other side of these high obstacles. This requires the robots to form a Snakebot and then go back to being a swarm at least once.

We use a simple hand-coded algorithm shown in Algorithm 1 to successfully clear all the obstacles shown in Figure 4 with appropriate task switching in approximately 7.5 min. Different behaviour control patterns can be obtained by adjusting the constants used for the impatience increment, as well as the constant values, $PHEROMONECEILING$, $REQUESTCEILING$ and $NHEADMODULES$. The values used for the latter three in our experiments are 600, 100 and 1, respectively. The constants are picked and optimized via numerous trial and error runs in the experimental environment. For example, $PHEROMONECEILING$ is set to 600, as this is the commonly encountered minimum pheromone concentration surrounding a robot in an already explored area. In cases where a smaller value is detected, the area is likely to be only partially explored, and for higher values, there is a chance of long delays in behaviour switching, even after the area is fully explored (thus, 600 is picked as a sub-optimal compromise). The value of the impatience increment in the case of high pheromone levels is set to 2, which is not a large enough increase to bring the impatience levels high enough (50) to trigger a change in behaviour, since the impatience value is multiplied by 0.95 every time step (thus, the maximum impatience value attainable is $\frac{2}{0.05} = 40$). Most of the other sub-conditions under the initial condition (of high pheromone values or requests) increase the impatience value with larger increments of 5.0. In these cases, a quick change in behaviour is required, and we aim to bring them about with large increments. The latter is not true for the first condition, which checks if the robot is part of a moving modular robot. In this case, the increments are made in significantly smaller values (0.5001, which means that the maximum value impatience can reach is $\frac{2.0+0.5001}{0.05} = 50.002$) in order to allow the Snakebot the opportunity to move when it is first formed. By the time the newly formed modular robot can start moving, the impatience value is already high (*i.e.*, 40.0); thus, using a large increment value would cause the Snakebot to disassemble before it can start to move.

Figure 4. The experimental environment viewed from three different positions.

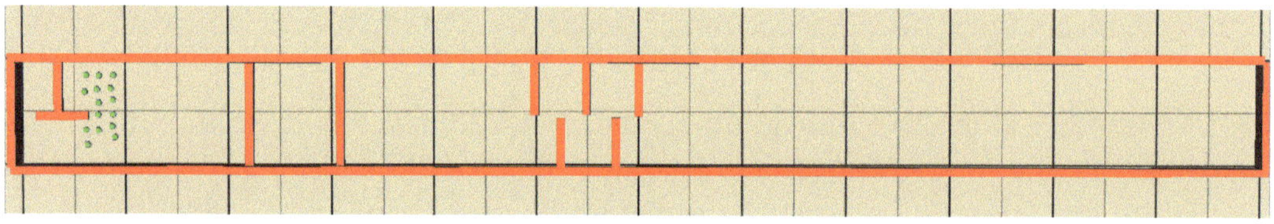

(a)

(b)

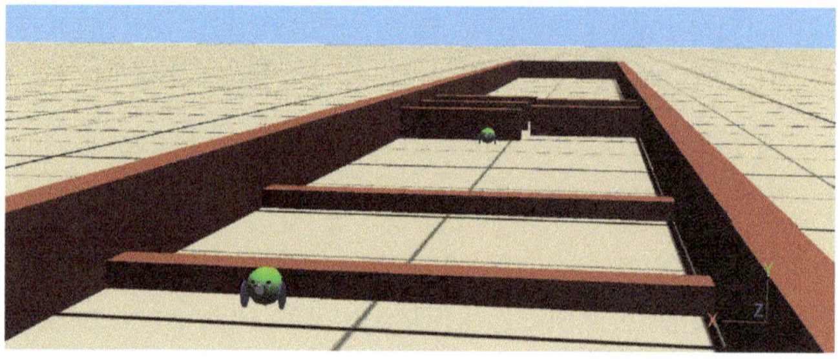

(c)

Algorithm 1 The hand-coded algorithm for controlling behaviour changes by adjusting the impatience value.

if $lowestPH > PHEROMONECEILING$ **or** $requests > REQUESTCEILING$ **then**
    $impatience \leftarrow impatience + 2.0$
    if $state = 1$ **and** $impatience >= 39.9$ **then**
        $impatience \leftarrow impatience + 0.5001$
    else if $(state = 1$ **or** $state = 3$ **or** $state = 4)$ **and** $moduleImpact > 0$ **then**
        $impatience \leftarrow impatience + 5.0$
    else if $state = 0$ **and** $largestModuleSize < 6$ **then**
        $impatience \leftarrow impatience + 5.0$
    else if $state = 2$ **and** $largestModuleSize < NHEADMODULES$ **then**
        $impatience \leftarrow impatience + 5.0$
    end if
end if

An example run in the environment shown in Figure 4 with 15 robots is illustrated in Figure 5.

As mentioned earlier, it takes a total of 7.5 min for the first robot to clear all the obstacles. The algorithm is written to ensure that there is only one robot initiating the configuration of a Snakebot. Although the latter ensures that the Snakebot formed is as large as it can be to have the best chances of overcoming the obstacles, it is time-consuming, and the configuration process takes more than three minutes. The amount of time it takes for the first robot to clear all the obstacles can be reduced by allowing two robots to initiate the Snakebot configuration process (by setting $NHEADMODULES$ to 2). In the latter case, the first robot clears all the obstacles within 5 min. However, neither of the Snakebots disassemble between the two low obstacles; thus, that area remains unexplored. If, however, the number of robots that can initiate the Snakebot configuration process is set to 3 or more, no Snakebots that can cross the low barriers form (they are too small); thus, none of the robots can clear the obstacles.

Figure 5. Snapshots of a run using the hand-crafted Algorithm 1. (**a**) 10 s; (**b**) 1 min 18 s; (**c**) 4 min 20 s; (**d**) 4 min 35 s; (**e**) 5 min 3 s; (**f**) 7 min 33 s.

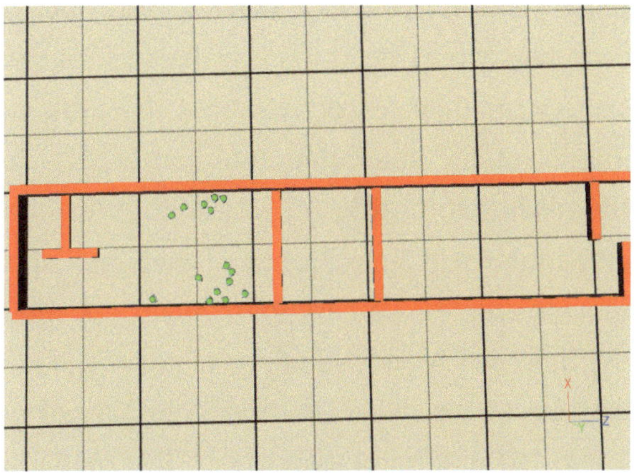

(**a**)

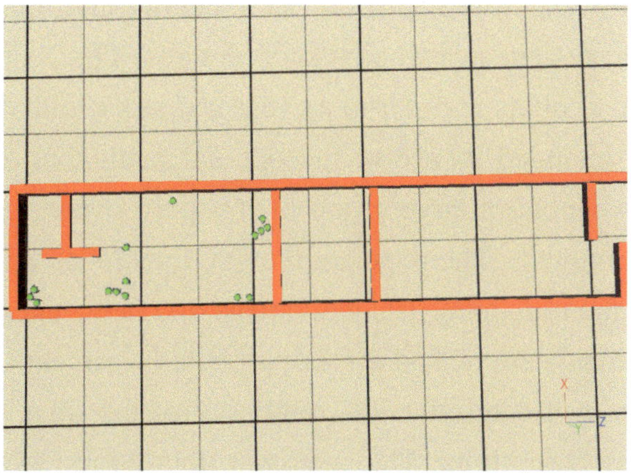

(**b**)

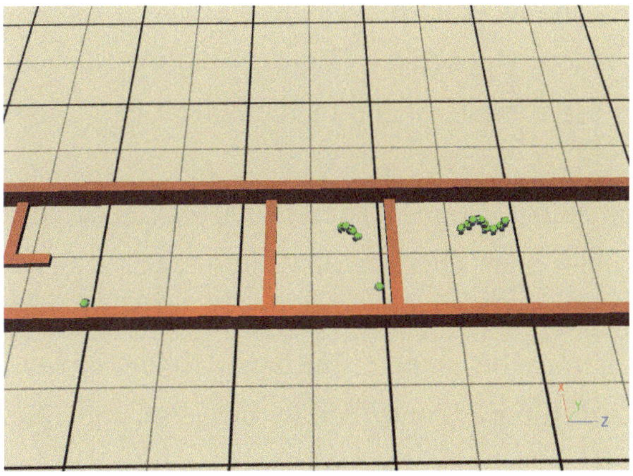

(**c**)

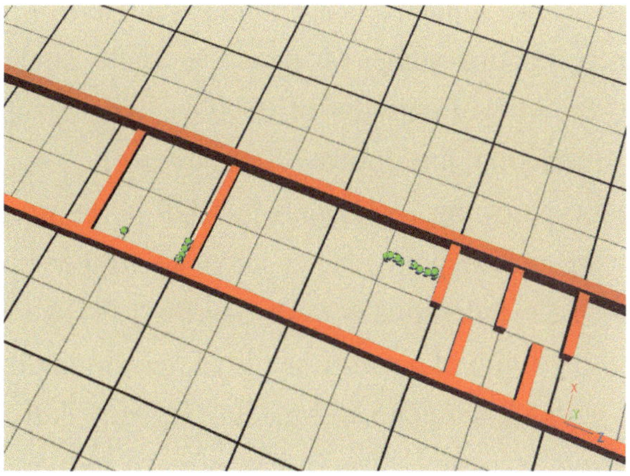

(**d**)

Figure 5. *Cont.*

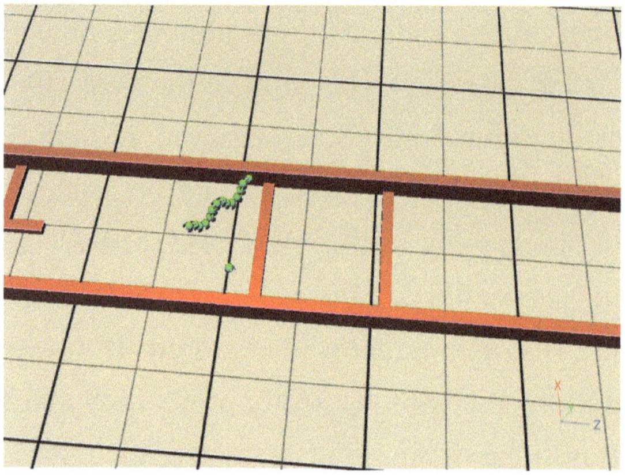

(**c**)

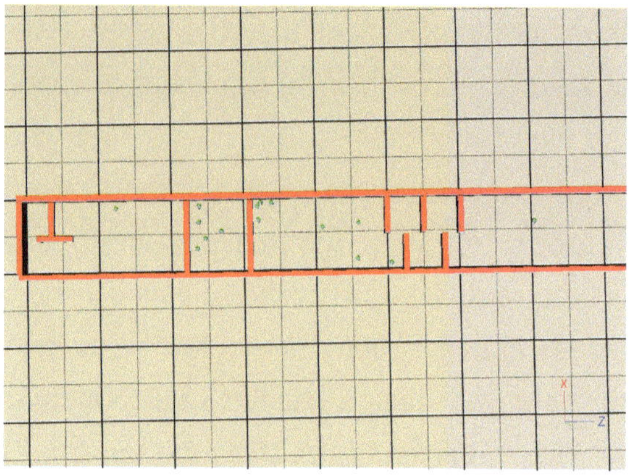

(**d**)

We believe that this is a well-fitting problem for GP, as it only involves 4 perception inputs and a single output; yet, it is a difficult problem to solve using hand-coded logic without any map-specific information. The mechanism of incrementing the impatience value that yields a desired behaviour of the robotic system as a whole is not obvious. The changes in behaviours need to be well synchronized amongst the robots in order to prevent fruitless oscillations between their respective states. By utilizing XGP, we intended to evolve the optimal mathematical model of the conditions that can trigger the transitions among a large group of robots to overcome the obstacles in order to successfully explore their environment. The population of XGP includes 200 individuals with an elite size of 10 individuals. To create the remaining 190 individuals of a new generation, we employ a binary tournament selection: two individuals are picked at random. Ninety-percent of the time, a new individual is created via single point crossover (reproduction), and 10% of the time, the fittest of the two is chosen to be passed on to the next generation. The crossover point is randomly selected within the genotype. The mutation randomly alters 2% of the newly created individuals (all except the elites). Each run lasts 40 evolutionary generations. Table 3 illustrates the main parameters of XGP. The set of terminal symbols of XGP consists of the four perception values (as shown in Table 2), the randomly generated floating point constants of [0..1], and the integer constant of [0..100]. The function set consists of the mathematical operations, addition, subtraction, multiplication and division.

The genotype of the individuals in XGP is represented as parse trees. These threes are evolved to increment the impatience value in order to trigger beneficial state changes in various environmental conditions. Each individual is evaluated for 1,000 s in the Webots simulation platform, which roughly corresponds to 5 min of a real-time run of the simulation platform on average. Each experiment involves 15 robots, with a homogeneous breeding strategy in that each robot is controlled by the same individual of XGP being evaluated. The fitness of each individual is determined by the number of checkpoints visited and the number of robots that clear all the obstacles. Checkpoints are placed every 25 cm in the environment, and they are meant to encourage exploratory behaviour, as well as prevent an over-fit solution to overcoming the obstacles in the corridor (*i.e.*, filter out the solutions that start reconfiguration to form a modular robot before any exploration is done). The fitness value is evaluated according to Equation (2).

$$fitness = 10 \times NoRobotsClearObstacles + checkPointsVisited \quad (2)$$

Table 3. Main parameters of XML-based Genetic Programming (XGP).

Category	**Value**
Genotype	Impatience increment (parse tree)
Population Size	200 individuals
Selection	Binary tournament selection; ratio: 0.1 Reproduction ratio: 0.9
Elitism	10 individuals
Mutation Rate	0.02
Trial Interval	1,000 simulated seconds
Termination Criteria	No. of generations = 40

After a total of 20 independent evolutionary runs, various successful control mechanisms evolve that can achieve the desired behaviour. Figure 6 shows the average fitness convergence over the 20 evolutionary runs. In 10 of the runs, the robotic system, controlled by the evolved mechanism of reconfiguration, was able to clear all the obstacles in the experimental environment. Out of the 10 successful runs, 6 provided robust solutions, where the re-runs could achieve good results. The reason for the latter is the over-fitting of the environmental conditions created during the evolutionary runs. Although half of the evolved controllers were able to cross both the low and high obstacles, only two got stuck at the tall obstacles, 1 got stuck in between the low obstacles and the rest could not cross the low obstacles. We can conclude that the successful formation of a Snakebot in time to cross the low obstacles was the major crux of the problem for evolution.

We expected to obtain a control mechanism that can form a single Snakebot from all 15 modules to carry them over the two low obstacles and then disassemble when the high obstacles are encountered. Although such control mechanism emerged, the most common evolved behaviour was the formation of multiple Snakebots (mostly two or three Snakebots) to overcome the obstacles. The latter was implicitly favoured by evolution, due to the gains in speed in carrying the robots over the obstacles, which allowed them a longer time to explore the much larger area beyond the high obstacles. By having multiple Snakebots configured at the same time, not only the reconfiguration process was parallelised, but also the problem with narrowing spaces behind the large Snakebots that slow down the docking robots were alleviated. Furthermore, the formation of multiple robots is a more robust approach, since the Snakebots are prone to falling over (which immobilizes the Snakebot) when using the rectilinear locomotion. In the case of a single Snakebot, this meant that the robots are unable to cross the low obstacles if the Snakebot falls over shortly after they are formed, whereas, in the case of multiple Snakebots, there is a smaller chance of all of the modular robots toppling over. Figure 7 shows snapshots from some of the successful runs, all of which demonstrate different behaviour and utilize different environmental information in making the decisions.

Figure 6. The fitness convergence of the runs over the evolutionary generations. The dashed lines at fitness 60 and 80 roughly indicate the fitness values when the first two low obstacles are crossed. The thick dashed line is the average fitness convergence of all the runs. The standard deviation starts at around 10 for the first generation and reaches and stays constant at 90 after the 5th generation.

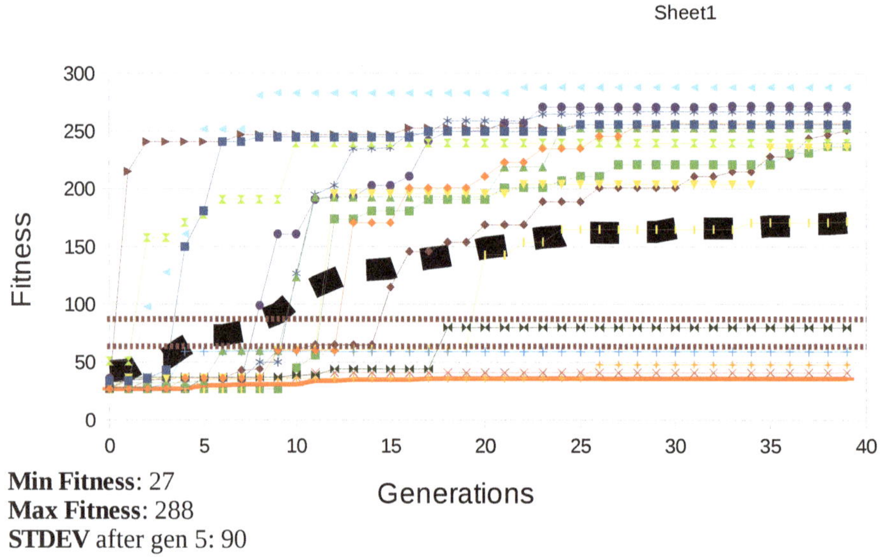

Figure 7. Snapshots of three different control mechanisms evolved using XGP. Runs 1 and 2 illustrate the formation of two separate Snakebots, and Run 3 shows a large Snakebot being formed to cross the low obstacles. The controllers evolved for the first and third runs dismantle the Snakebot when an unexplored area is encountered, whereas the evolved controller used in the second run utilizes a strategy of disassembling the Snakebot after some time passes from the discovery of an unexplored area. In the latter case, the Snakebots cross the low obstacles and get stuck at the high obstacles for a while until there is a high enough pheromone concentration nearby. (**a**) Run 1; (**b**) Run 1; (**c**) Run 1; (**d**) Run 2; (**e**) Run 2; (**f**) Run 2; (**g**) Run 3; (**h**) Run 3; (**i**) Run 3.

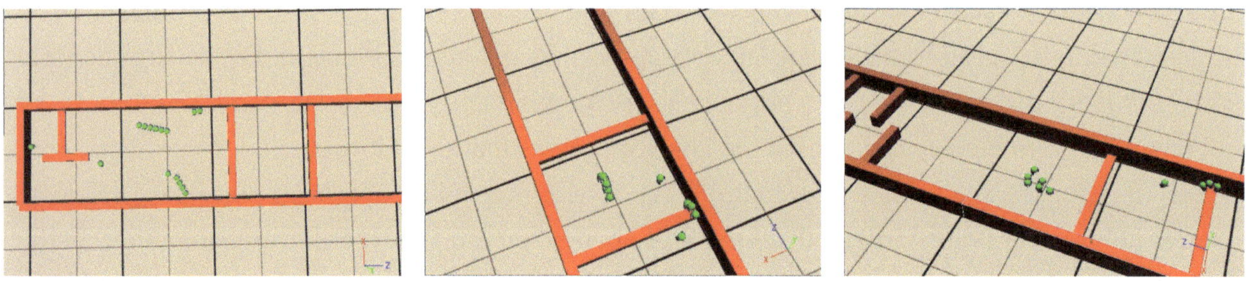

(**a**) (**b**) (**c**)

Figure 7. *Cont.*

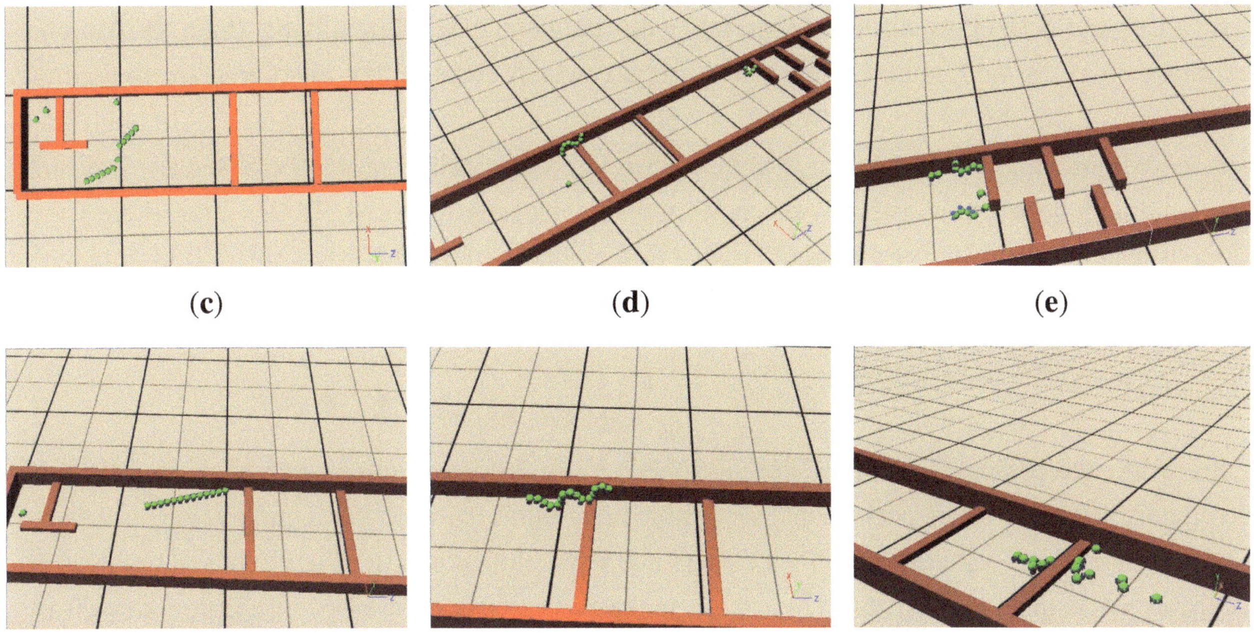

(**c**) (**d**) (**e**)

(**f**) (**g**) (**h**)

6. Discussion

In our observations, the resulting genetic programs mainly utilize the $lowestPheromone$ variable to initiate state changes in the robots. The remaining inputs are used to limit the frequency and number of robots changing into particular states (such as initiating modular robot reconfiguration). In all cases, we observed that when the $lowestPheromone$ level reached above a certain threshold, the robots under the influence would start reconfiguration (if they were swarming). On the other hand, the variables, $largestModuleSize$ and $moduleImpact$, were used for preventing the robots from initiating too many reconfigurations. The use of randomized inputs were rare. The initial idea of using random variables as an input was to encourage specialization among robots and avoid switching to same states at the same time. In our attempts to construct some simple hand crafted controllers prior to the evolution runs, the most common problem was preventing global state changes (for example, when all the robots want to initiate reconfiguration). The inputs listed in Table 2 seem to be sufficient in preventing this; thus, a random input may not be beneficial.

The procedure for disassembling from a modular robot was significantly different among the various controllers evolved. One of the two common solutions involved detecting low pheromone concentrations. This proved to be a good approach to ensuring that all the newly discovered areas can be efficiently explored by a swarm. This, however, caused two problems: (i) early disassembly, where some of the modules might be left behind or on top of an obstacle (e.g., Figure 7(c)); or (ii) the inability of some robots to disconnect from the Snakebots once the pheromone levels increase in the environment. The latter could be solved by not allowing the robots to drop pheromones while they are part of a modular robot. This may, however, lead to other unanticipated problems and needs experimenting. The release of pheromones by robots in modular mode help them keep the other robots away during reconfiguration. This helps to reduce collisions and to speed-up reconfiguration process. The second solution to disassembly of Snakebots involved using pheromone values within a certain range, which made sure that the Snakebots do not disassemble the moment they start moving (high pheromone concentration area), as well as cover some new ground, which ensured that all robots are carried as far as they can be via the Snakebot (no disassembly in the low concentration areas). Although this solution scored quite high, due to the large number of robots that could clear all the obstacles, it meant that the small area between the two low obstacles remains partially explored. In this case, there were also some robots that could never disassemble, due to the pheromone levels becoming too high, which left them defunct.

7. Conclusions

In this work, we presented a simple decision mechanism for behaviour switching in a collective robotic organism. The decision mechanism uses simple rules and is based on the accumulation, as well as decay of an "impatience" value inspired by the hormonal regulation of emotions in biological organisms. The decision mechanism is used to orchestrate a range of previously developed behaviours (such as swarming, modular robotic locomotion and reconfiguration) in accomplishing a challenging task for a group of 15 robots. The control of the decision mechanism is achieved via an evolved Genetic Program (GP), which utilizes various information that is readily available to the robots from their interactions with the environment or other robots. The controllers developed for the behaviour switching relied heavily on the information gathered via the pheromones dropped by swarming robots, which show that the use of simple chemical gradients in the environment can be useful for coordinating behaviours other than swarming.

The evolved solutions present multiple ways of achieving the target goal of exploration of the whole corridor. The robots are able to clear the obstacles quickly ($\approx$4 min when three Snakebots are formed and $\approx$6 min when a single Snakebot is formed). The presented hormone-inspired behaviour switching mechanism provides a gradual, yet constrained, way to control the changes in the emergent behaviours. The use of XGP yielded quick results in achieving the desired behaviours, demonstrating that the underlying mechanisms are evolvable. The evolved solutions outperformed the hand-coded controller by a large margin and provided a greater diversity of strategies. The solutions indicate a few remaining uncertainties, such as the need for a Snakebot to release pheromones, and the use of random variables, which require further investigation.

Acknowledgments

The presented work is part of a fellowship funded by JSPS.

Conflict of Interest

The authors declare no conflict of interest.

References

1. Jin, Y.; Guo, H.; Meng, Y. A hierarchical gene regulatory network for adaptive multi-robot pattern formation. *IEEE Trans. Syst. Man Cybern. B Cybern.* **2012**, *42*, 805–816.
2. Weel, B.; Haasdijk, E.; Eiben, A. The Emergence of Multi-Cellular Robot Organisms through On-Line On-Board Evolution. In *Applications of Evolutionary Computation*; Springer: Berlin/Heidelberg, Germany, 2012; Volume 7248, pp. 124–134.
3. Shen, W.; Will, P.; Galstyan, A.; Chuong, C. Hormone-inspired self-organization and distributed control of robotic swarms. *Auton. Robots* **2004**, *17*, 93–105.
4. Turing, A. The chemical basis of morphogenesis. *Philos. Trans. R. Soc. Lond.* **1952**, *237*, 36–73.
5. Hamann, H.; Stradner, J.; Schmickl, T.; Crailsheim, K. A Hormone-Based Controller for Evolutionary Multimodular Robotics: From Single Modules to Gait Learning. In Proceedings of the IEEE Congress on Evolutionary Computation (CEC'10), Barcelona, Spain, July 2010.
6. Murphy, R.; Lisetti, C.; Tardif, R.; Irish, L.; Gage, A. Emotion-based control of cooperating heterogeneous mobile robots. *Robot. Autom. IEEE Trans.* **2002**, *18*, 744–757.
7. Moioli, R.C.; Vargas, P.A.; Husbands, P. A Multiple Hormone Approach to the Homeostatic Control of Conflicting Behaviours in an Autonomous Mobile Robot. In Proceedings of the Eleventh Conference on Congress on Evolutionary Computation (CEC'09), Trondheim, Norway, 18–21 May 2009; IEEE Press: Piscataway, NJ, USA; pp. 47–54.
8. Parker, L. ALLIANCE: An architecture for fault tolerant multirobot cooperation. *Robot. Autom. IEEE Trans.* **1998**, *14*, 220–240.
9. Walker, J.; Wilson, M. A Performance Sensitive Hormone-Inspired System for Task Distribution Amongst Evolving Robots. In Proceedings of the IROS 2008, IEEE/RSJ International Conference on Intelligent Robots and Systems, Nice, France, September 2008; pp. 1293–1298.
10. Rumbell, T.; Barnden, J.; Denham, S.; Wennekers, T. Emotions in autonomous agents: Comparative analysis of mechanisms and functions. *Auton. Agents Multi-Agent Syst.* **2012**, *25*, 1–45.
11. Adkins-Regan, E. *Hormones and Animal Social Behavior*; Princeton University Press: Princeton, NJ, USA, 2005.
12. Tanev, I. DOM/XML-based portable genetic representation of the morphology, behavior and communication abilities of evolvable agents. *Artif. Life Robot.* **2004**, *8*, 52–56.
13. Kuyucu, T.; Tanev, I.; Shimohara, K. Incremental Genetic Programming via Genetic Transpositions for Efficient Coevolution of Locomotion and Sensing of Simulated Snake-Like Robot. In Proceedings of the European Conference on Artificial Life, Paris, France, August 2011; pp. 439–446.
14. Kuyucu, T.; Tanev, I.; Shimohara, K. Evolutionary Optimization of Pheromone-Based Stigmergic Communication. In *Applications of Evolutionary Computing, EvoApplications 2012*; Springer Verlag: Malaga, Spain, 2012; Volume 7248, pp. 61–70.
15. Michel, O. Webots: Professional mobile robot simulation. *Int. J. Adv. Robot. Syst.* **2004**, *1*, 39–42.

16. Liu, W.; Winfield, A.F.T. Autonomous Morphogenesis in Self-Assembling Robots Using IR-Based Sensing and Local Communications. In Proceedings of the 7th International Conference on Swarm Intelligence (ANTS'10), Brussels, Belgium, September 2010; Springer-Verlag: Berlin/Heidelberg, Germany; pp. 107–118.
17. Tanev, I.; Ray, T.; Buller, A. Automated evolutionary design, robustness and adaptation of sidewinding locomotion of simulated snake-like robot. *IEEE Trans. Robot.* **2005**, *21*, 632–645.
18. Grassé, P.P. La reconstruction du nid et les coordinations inter-individuelles chez Bellicosi-termes natalensis et Cubitermes sp. La theorie de la stigmergie: Essai d'interpretation des termites constructeurs (in French). *Insectes Sociaux* **1959**, *6*, 41–83.
19. Beckers, R.; Holl, O.E.; Deneubourg, J.L.; Bielefeld, Z.U.; Bielefeld, D. From Local Actions to Global Tasks: Stigmergy and Collective Robotics. In *Artificial Life IV (1996)*; MIT Press: Cambridge, MA, USA, 1994; pp. 181–189.
20. Svennebring, J.; Koenig, S. Building terrain-covering ant robots: A feasibility study. *Auton. Robot.* **2004**, *16*, 313–332.
21. Purnamadjaja, A.; Russel, R. Bi-directional pheromone communication between robots. *Robotica* **2010**, *28*, 69–79.
22. Distante, C.; Indiveri, G.; Reina, G. An application of mobile robotics for olfactory monitoring of hazardous industrial sites. *Ind. Robot Int. J.* **2009**, *36*, 51–59.
23. Misawa, N.; Mitsuno, H.; Kanzaki, R.; Takeuchi, S. Highly sensitive and selective odorant sensor using living cells expressing insect olfactory receptors. *Proc. Natl. Acad. Sci. USA* **2010**, *107*, 15340–15344.
24. Dowling, K. Limbless Locomotion: Learning to Crawl. In Proceedings of IEEE International Conference on Robotics and Automation, Detroit, MI, USA, May 1999; Volume 4, pp. 3001–3006.
25. Hirose, S. *Biologically Inspired Robots: Snake-Like Locomotors and Manipulators*; Oxford University Press: Oxford, UK, 1993.
26. Meng, Y.; Zhang, Y.; Jin, Y. Autonomous self-reconfiguration of modular robots by evolving a hierarchical mechanochemical model. *IEEE Comp. Int. Mag.* **2011**, *6*, 43–54.

3

Illumination Tolerance for Visual Navigation with the Holistic Min-Warping Method

Ralf Möller [1,2,*], Michael Horst [1] and David Fleer [1]

[1] Computer Engineering, Faculty of Technology, Bielefeld University, Bielefeld D-33594, Germany; E-Mails: mhorst@ti.uni-bielefeld.de (M.H.); dfleer@ti.uni-bielefeld.de (D.F.)

[2] Center of Excellence "Cognitive Interaction Technology", Bielefeld University, Bielefeld D-33594, Germany

* Author to whom correspondence should be addressed; E-Mail: moeller@ti.uni-bielefeld.de

---

Abstract: Holistic visual navigation methods are an emerging alternative to the ubiquitous feature-based methods. Holistic methods match entire images pixel-wise instead of extracting and comparing local feature descriptors. In this paper we investigate which pixel-wise distance measures are most suitable for the holistic min-warping method with respect to illumination invariance. Two novel approaches are presented: tunable distance measures—weighted combinations of illumination-invariant and illumination-sensitive terms—and two novel forms of "sequential" correlation which are only invariant against intensity shifts but not against multiplicative changes. Navigation experiments on indoor image databases collected at the same locations but under different conditions of illumination demonstrate that tunable distance measures perform optimally by mixing their two portions instead of using the illumination-invariant term alone. Sequential correlation performs best among all tested methods, and as well but much faster in an approximated form. Mixing with an additional illumination-sensitive term is not necessary for sequential correlation. We show that min-warping with approximated sequential correlation can successfully be applied to visual navigation of cleaning robots.

Keywords: visual robot navigation; image distance function; correlation

---

1. Introduction

We first compare feature-based to holistic methods (Section 1.1), recapitulate the holistic min-warping method (Section 1.2), discuss the approaches to illumination invariance taken by feature-based and holistic methods (Section 1.3), and outline the contributions and the structure of the paper (Section 1.4).

1.1. Feature-Based vs. Holistic Methods

In many application domains—e.g., stereo matching, motion detection, pose estimation, construction of 3D models, pattern matching, image registration, or image retrieval—computer and robot vision is presently dominated by "feature-based" approaches. Feature-based methods (i) detect key-points like corners and only extract feature descriptors from their vicinity; (ii) perform complex transformations into feature descriptors and match feature descriptors between images; and (iii) typically require high-resolution images [1–5]. A number of feature detectors and descriptors have been developed, examples being the wide-spread FAST detector [6] and the SIFT [7] and SURF detectors and descriptors [8], and the modern binary descriptors such as BRIEF [9,10], ORB [11], or FREAK [12]. In robot navigation, matches between features are either used for estimating ego-motion between camera postures from two images (such as in visual odometry, see [13,14]), for estimating the metrical position of the corresponding landmark in a geometrical map [15–18], or for place recognition [19]. Together with subsequent outlier processing (like RANSAC) and n-point methods [14], feature-based methods can reliably estimate the relative posture between two images in 5 dimensions (up to scale).

For visual robot navigation, an emerging alternative to feature-based methods are "holistic" methods which (i) use the entire image (without feature extraction); (ii) compare images as they are (with only moderate preprocessing) by pixel-wise distance measures; and (iii) work on low-resolution images (typically on panoramic images). Most holistic methods have their origin in models of insect visual navigation; also the holistic min-warping method [20] used as an example in this paper has a biologically plausible counter-part [21]. Social insects such as bees, ants, and wasps return to important places (nest locations, food sites) by using visual information (reviews: [22,23]). Their tiny brains with mostly local interconnections and the low visual resolution of their compound eyes make it unlikely that insects employ feature-based methods for navigation. There is a general agreement among modelers that insects probably use holistic methods [21,24–30].

We presented a classification of holistic local visual homing methods in [20] which we recapitulate here in modified form for the reader's convenience. Local visual homing methods take a panoramic current view and a snapshot captured at the goal location and compute an azimuthal rotation estimate between the two images and a home vector pointing from the current to the snapshot location. At the present stage of development, all methods assume a movement of the agent in the plane. There are presently five classes of holistic methods: parameter methods, DID methods, visual compass methods (which are part of local visual homing methods), warping methods, and multi-snapshot methods:

- In *parameter methods*, entire images are reduced to parameter vectors, and homing to a goal location is achieved by a gradient descent in a distance function between parameter vectors.

An extremely parsimonious example is the Average Landmark Vector (ALV) model where the parameter vector comprises just two coordinates and a compass angle [31,32]. This model explains some observations in experiments with bees [33], crickets [34], and desert ants [27]. Our experience shows that parameter methods do not achieve the same navigation quality in robot experiments as other holistic methods. However, these methods may be useful for loop-closure detection [35,36].

- *DID methods* (descent in image distances) are based on the observation that, if compass-aligned images are used, the image distance (e.g., pixel-wise Euclidean distance) between two panoramic images increases smoothly with increasing spatial distance between their capture points [25,26,37,38]. Therefore, a gradient descent in the spatial image distance function between current images and snapshot image will lead the agent back to the snapshot location (robot experiments: [39]). Since the gradient has to be estimated by sampling the image distance function at three non-collinear points, test steps need to be inserted into the homing process [25]. This can be avoided by capturing just a single image and predicting two images as if two short test steps (in perpendicular directions) had been executed from the capture position of the real current image [40–42].
- Several *visual compass methods* have been developed—most of them in the DID framework—which extract the azimuthal orientation difference from two panoramic images [25,26,35–38,43–47]. Two panoramic images are rotated relative to each other in azimuthal direction and a pixel-wise distance is computed. The minimum location in this rotational image distance function approximately corresponds to the difference in azimuthal orientation between the two images and can be used to align the two images as if they had been captured in the same azimuthal (compass) direction. These methods can be used for aligning images prior to a descent in image distances (see above), for loop-closure detection [35,36], and for multi-snapshot methods (see below).
- *Warping methods* [20,21,48–51] distort one of the images according to a simulated movement and compare the distorted image and the second image by some pixel-wise distance measure. A search for the minimal distance through all movement parameters yields both the estimated home vector and the estimated orientation difference between the two images. Warping has been applied to navigation in topological maps [50,52,53]. The min-warping method [20] addressed in this paper has successfully been applied to cleaning robot control [54,55].
- *Multi-snapshot methods* [28,56] form a novel class which takes a middle ground between visual compass and warping methods. Instead of characterizing a goal location by a single panoramic snapshot image, multiple snapshots are collected in the vicinity of the goal location. Each snapshot is captured while the agent faces the goal location. A rotational image distance function (visual compass) is computed between a current view and all snapshots. The best minimum among all snapshots it taken to orient the agent towards the goal. It is our impression that these methods are currently the most convincing model of insect navigation, but it still has to be explored how they perform under changes in illumination.

We feel that, despite the dominance and success of feature-based methods in navigation, holistic methods deserve attention for several reasons:

(1) Even at the present, relatively early stage in their development, holistic method are almost competitive with feature-based methods with respect to their *computational effort*. Two properties of holistic methods have a conflicting influence: Low visual resolution promises short computation times whereas the fact that the entire image is used increases the computational effort. The min-warping method—which is among the most complex holistic methods due to its systematic parameter search—requires about 16 ms to compute a home vector and compass estimate from two low-resolution panoramic images (288×30 pixels) on a modern CPU (Intel i7, 2.3 GHz); for a partial search around a coarse estimate ($1/4$ of the search range for each parameter), less then 6 ms are required (96 steps in both parameters, single search). Calonder *et al.* [10] report times between 3 ms and 7 ms for different versions of their BRIEF descriptor (detection, descriptor computation, and matching of 512 key points, but without RANSAC and n-point method; 2.66 GHz Intel x86-64 CPU).

(2) Holistic methods exhibit a very regular algorithmic structure—e.g., distances between *all* pairs of image columns are computed in the first phase of min-warping—and are therefore perfectly suited for an *efficient implementation on SIMD units* of modern CPUs (SSE, AVX; NEON). For these two reasons, holistic methods may be particularly suitable for consumer market robots (e.g., cleaning or lawn-mowing robots) and for flying robots which typically have limited onboard processing power due to constraints in cost, weight, and battery power.

(3) Despite their simplicity, some holistic navigation models have been successfully applied to real-world images from outdoor environments [25,26,56] and for indoor navigation of cleaning robots [54,55]. At least for rotation estimation it has been observed that holistic methods perform better than feature-based approaches [57] which seem to be susceptible to calibration errors. Milford [58] demonstrated that place recognition is possible with extremely low visual resolution if the temporal sequence of images is considered. For these reasons we are optimistic that holistic methods can be competitive with feature-based methods with respect to their *navigation performance*. However, an extensive comparison of feature-based *vs.* holistic methods is still missing.

(4) Holistic methods use all available information from an image pair and may therefore be superior to feature-based methods in *environments with few suitable features* (e.g., corners), such as natural outdoor environments; however, this assumption needs experimental verification.

(5) We know that min-warping *works well in very different visual environments*, e.g., in computer simulations with random wall textures without easily discernible landmark features and in different, unmodified real rooms (labs, apartments) [55].

In a recent paper on visual place recognition, Milford [58] discusses advantages and drawbacks of feature-based methods and of alternative methods using low-resolution images (which we would call "holistic"). According to Milford, feature-based approaches "exhibit desirable properties such as scale and rotation invariance, and a limited degree of illumination invariance". However, "feature-based approaches require a feature detector that is suitable for the type of environment", thus requiring a selection or training phase. He points out that "feature-based techniques may also fail since visual features can change radically over day-night, weather and seasonal cycles". These techniques "generally require relatively high-resolution, crisp and noise-free imagery". In contrast, pixel-by-pixel comparison techniques (as they are also presented in this paper) only need low-resolution images and have the advantage of "not requiring assumptions about the type of visual features in the environment". However, additional stages are required where the images are transformed (e.g., shifted by a range of offsets) before the pixel-wise distance measures can be applied—in min-warping, this is achieved by computing matches between *all* pairs of image columns in the two images in multiple versions with different vertical scaling; the systematic search in the second phase selects a subset of these column distances.

Having to apply image transformations before a pixel-wise matching is possible explains a limitation of existing holistic methods: The number of parameters of this transformation has to be kept small, e.g., by limiting movement of the agent to the plane (two parameters since estimation is only possible up to scale). This is also the case in min-warping which assumes a panoramic camera with vertical optical axis with respect to the plane and a flat environment without additional degrees of freedom (changes in height, tilt of the robot) and is thus currently limited to wheeled robots and non-natural environments. In contrast, featured-based methods in combination with n-point methods can estimate relative movement in five degrees of freedom (up to scale).

Holistic methods have received far less attention by the computer-vision community than feature-based methods. We feel that it may still be too early in their development to venture a performance competition with sophisticated feature-based methods—judging from an earlier study where SIFT was applied to local visual homing [59], we actually consider it likely that, at least on the indoor image databases used in this paper, feature-based methods would perform as well as or better than min-warping. However, we think that the insights on illumination tolerance for different distance measures gained in this paper are an important step to make holistic methods more competitive, at least for the min-warping method investigated here. Moreover, for feature-based methods, a number of studies have been dedicated to the comparative evaluation of different feature detectors and descriptors for different applications [1,2,5,17–19] indicating that there may be no single method suitable for all applications. A systematic study exploring the application of feature-based methods for local visual homing with panoramic images is not yet available, and would also be beyond the scope of this paper.

1.2. Min-Warping, a Holistic Local Visual Homing Method

The holistc "min-warping" method [20] studied in this paper interrelates two panoramic, monochrome images under the assumption that they have been captured by a panoramic camera with an optical axis vertical to the movement plane and that the robot carrying the camera moves in a flat environment. Min-warping is a local visual homing method: From the two panoramic images captured at different points, it computes two angles, a bearing angle describing the direction from one capture point to the other (alternatively expressed as a unit-length "home vector"), and a rotation angle expressing the azimuthal rotation between the two images (also called "visual compass"). Home vector and visual compass can either be used to approach a target location where the snapshot was taken (by following the home vectors in Section 5), or they can be used to obtain a position estimate by triangulating multiple snapshots with known positions (see Section 7). The latter application of min-warping is the basis of the visual lane-control method for domestic cleaning robots presented in [54,55]. Here min-warping performed well both in computer simulations with different environments (walls with random textures, walls with photos of real environments) and in robot experiments in unmodified lab rooms and apartments, at least under constant conditions of illumination.

Figure 1. Min-warping, top view of the plane in which the robot is moving. A panoramic reference image (snapshot) is captured at **S**; the robot's orientation at **S** is indicated by the thick arrow. A simulated movement in direction α and with orientation change ψ and (unknown) distance d is executed. After the simulated movement, the robot is located at the current position **C** and assumes the orientation indicated by the thick arrow. A landmark at **L** (represented by a pixel column of a panoramic image) appears under an azimuthal angle $\Theta = \alpha + x$ in the snapshot, and under an angle $\Theta' = \alpha - \psi + x + y$ in the current view. Since neither the distances to the landmark (r, r') nor the covered distance d are known, the landmark could be located at all positions on the beam originating at **S** (indicated by three example locations). Min-warping searches for best-matching image columns over a range of the angle Θ' (indicated by three example angles). Figure and caption modified from [55].

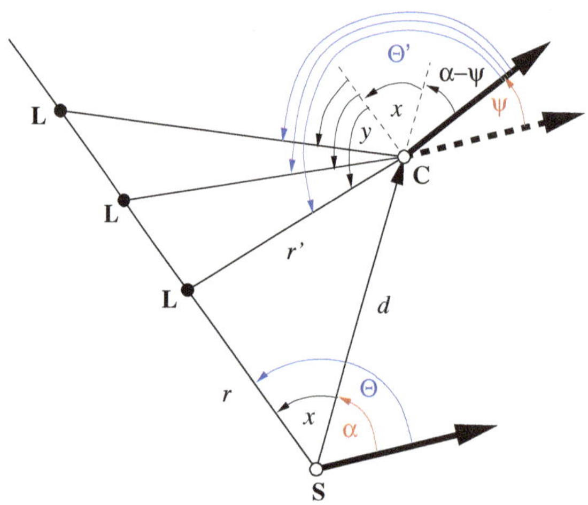

In the following we recapitulate the min-warping algorithm; for details, the reader is kindly referred to [20]. Min-warping rests on the assumption that all features within a column of a panoramic image have the same ground distance from the vantage point. This is the reason why image distances between columns of panoramic images are computed. While this equal-distance assumption is often violated, min-warping nevertheless produces reliable estimates of home vector and compass in indoor environments. Figure 1 presents the geometrical relationships in the plane from which min-warping is derived. At position **S**, a panoramic snapshot has been captured in the orientation indicated by the thick arrow. Min-warping tests how well hypothetical movements of the robot explain the transformation between the snapshot and the current view. Two movement parameters, movement direction α and azimuthal rotation ψ, are varied systematically between 0 and 360 deg. A movement parametrized by (α, ψ) leads to the simulated current position **C**. The traveled distance d is unknown. Assume that a column of pixels belongs to a landmark position **L** in the plane. The column appears under the azimuthal angle $\Theta = x + \alpha$ in the panoramic snapshot image. The distance r to the landmark column is unknown as well. In the current view, the same column could appear at a range of azimuthal angles $\Theta' = \alpha - \psi + x + y$, measured from the assumed orientation indicated by the dashed thick arrow. Also the distance r' is unknown.

Figure 2. Column distances (square images) computed by J_{ASC+} (Section 3.7) between a panoramic snapshot (SS) and a current view (CV or CV cross) from indoor image databases. Snapshot and current view are taken at different (known) locations (grid positions (6,3) and (4,1) in the living1 databases) and differ in their azimuthal orientation. Darker distance pixels correspond to lower distances. *Top left:* Snapshot and current view (CV) from the same database (living1day, taken at day). Good matches are easily discernible. Azimuthal angles Θ and Θ' are indicated. *Top right:* Same snapshot, but current view (CV cross) from different image database in the same environment (living1night, taken at night). Good matches are more difficult to see. *Bottom:* Magnified versions of the snapshot and the two different current views. The current views were taken at the same position but under different conditions of illumination (day, night). Images are shown as they are used by min-warping (low resolution, low-pass filtered).

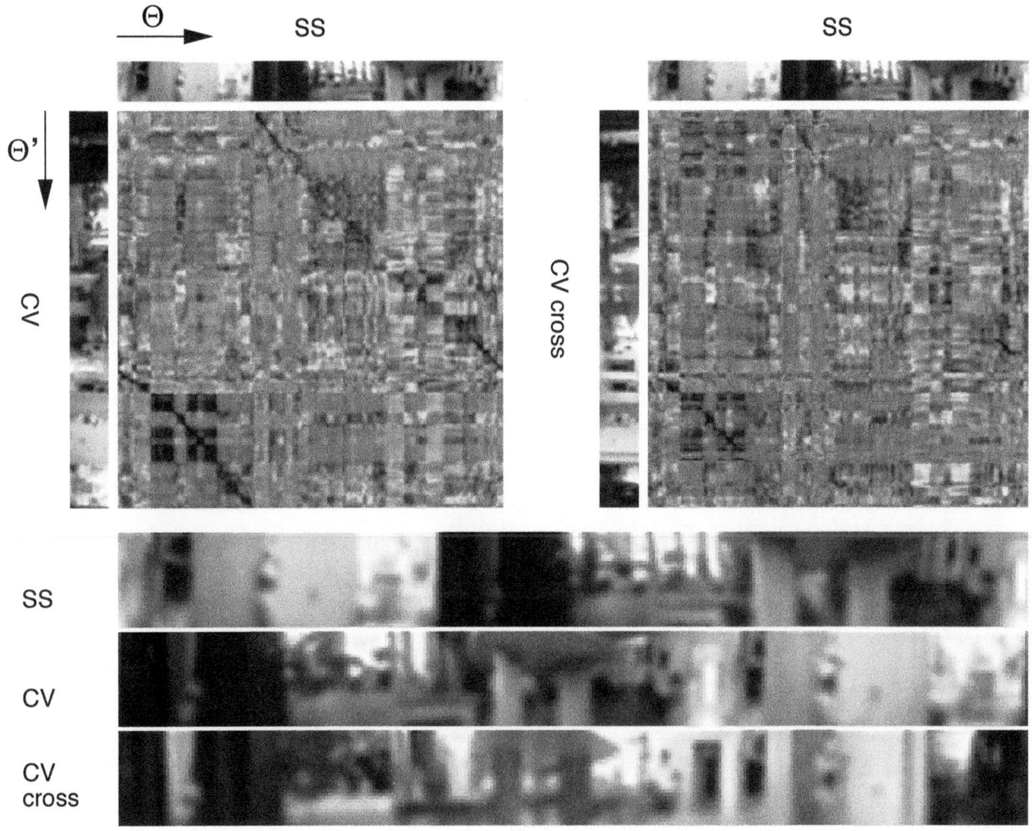

For each simulated movement, min-warping searches through all possible locations (azimuthal angle Θ') where a column in the snapshot (at azimuthal angle Θ) can reappear in the current view and looks for the best match between the snapshot column and the current view column. Even though r' and r are unknown, their ratio $\sigma = r'/r$ can be computed from the law of sines in the triangle **SCL**: $\sigma = (\sin x)/\sin(x+y)$. The ratio σ is referred to as "scale factor" since it determines how the column in the snapshot should have changed in vertical direction when seen from the current location **C**, given the movement parameters (α, ψ) and the azimuthal angles Θ and Θ' under which the column at **L** appears. Min-warping magnifies one of the columns according to the scale factor and compares it with the other column using some distance measure (the latter being the focus of this paper). The minimal distances (therefore *min*-warping), obtained from the best matches between columns at given Θ and varying Θ', are summed over all columns Θ of the snapshot and stored in a distance array. Those movement parameters (α, ψ) which produce the smallest sum are returned as result. They can directly be converted into the estimates of home vector and visual compass.

The min-warping algorithm essentially consists of four nested loops through α, ψ, Θ, and a range of Θ' (where the range depends on x). However, magnifying and matching image columns in the innermost loop of the min-warping algorithm would be too costly. The algorithm is therefore split into two phases. In the *first phase*, min-warping precomputes matches between magnified image columns according to a discrete set of scale factors (later, in the second phase, the computed scale factor σ is matched to the closest scale factor from this set). For each scale factor from the set, distances between all image columns in a pair of panoramic images are computed. Figure 2 shows the results of such a distance computation for two images, called snapshot and current view (here used in non-magnified form). The resulting distance image (square image in Figure 2) is referred to as "scale plane". Figure 3 shows a set of image pairs computed from a set of five scale factors. Computing the distance image between the two images of each pair results in a "scale-plane stack" where multiple scale planes computed over a range of magnification factors are stacked; see Figure 4. In the four nested loops in the *second phase* of min-warping, distances between a snapshot column and a range of current view columns are obtained from the scale-plane stack. Simulated movement (α, ψ) and azimuthal positions (Θ, Θ') uniquely determine the scale change σ of an image column such that the scale plane with the nearest magnification factor from the scale-plane stack can be consulted. The result of the second phase is a distance array over the movement parameters (α, ψ); examples are shown in Figure 5. The white dot in Figure 5 marks the best match (minimum).

Figure 3. Image pairs of snapshot and current view in magnified versions as used in the construction of the scale-plane stack in Figure 4, obtained from the situation shown in Figure 2 (snapshot and current view from same database). The scale factor σ is shown for each image pair. For $\sigma < 1$, the snapshot is magnified (indicated by the arrows), for $\sigma > 1$, the current view is magnified.

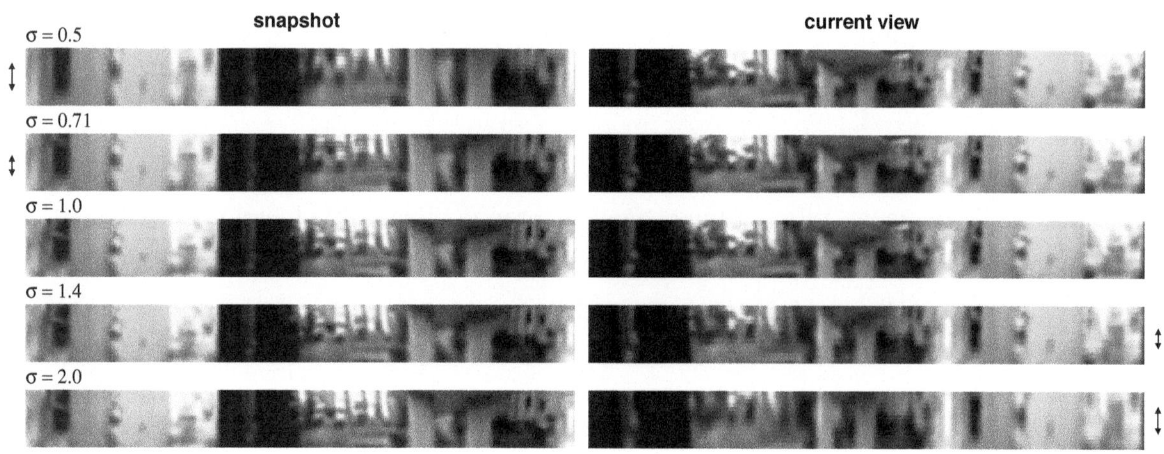

Figure 4. Scale-plane stack obtained from the image pairs in Figure 3. The scale factor σ is shown for each scale plane. Dark values indicate good matches between columns.

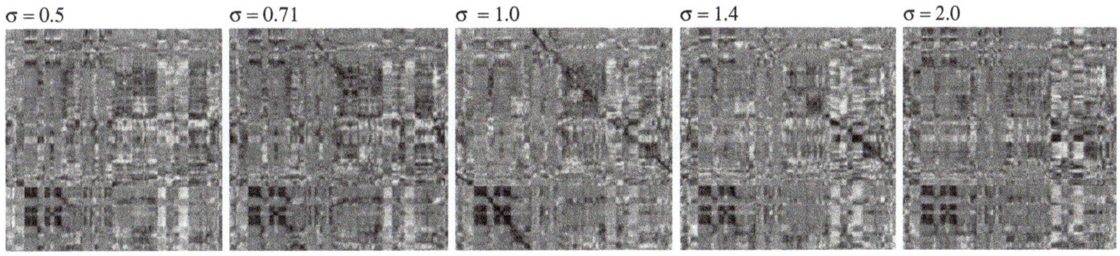

Figure 5. Match-quality arrays for the situation shown in Figure 2. The match quality for each discretized value of the parameters α and ψ (interval $[0, 2\pi)$, 72 steps) is shown color-coded (blue: good matches). The best match is indicated by a white dot. *Left:* Same-database test: snapshot and current view from the same database (day). *Right:* Cross-database test: snapshot and current view from different databases (day/night). Illumination changes in the cross-database test lead to larger uncertainty of the parameter estimates.

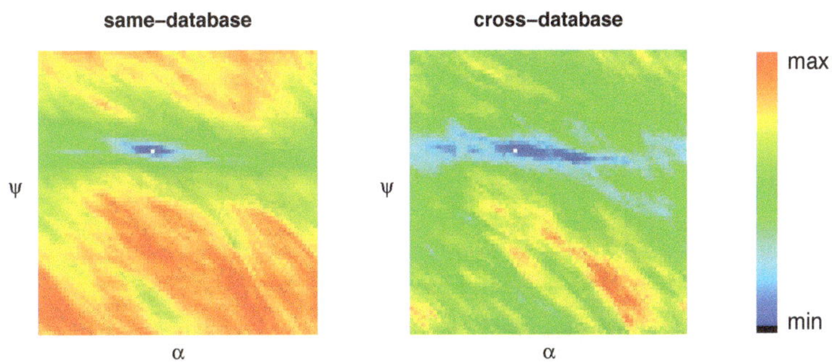

As it is typical for holistic methods, min-warping uses low-resolution input images, and it performs somewhat better if these are slightly low-pass filtered [20,51]. Experiments show that increased image size or inclusion of higher image frequencies have almost no effect on the homing performance (unpublished data). However, using larger images substantially increases the computation time. The computational effort depends on image width w, image height h, number of scale planes n, and number of search parameters n_α and n_ψ. The effort of the first phase is dominated by a term proportional to nw^2h, the effort for the second phase is proportional to $n_\alpha n_\psi w^2/4$. Doubling image width and height will therefore increase the computation time of the first phase by a factor of 8 and of the second phase by a factor of 4.

1.3. Illumination Invariance and Distance Measures

Illumination changes affect feature-based and holistic methods alike. It is not clear, however, whether methods for achieving illumination tolerance have the same effect on feature-based and holistic methods. The major difference is that feature-based methods only consider high-variance image regions (such as corners) whereas holistic methods consider all parts of an image.

We first look at counter-measures to illumination changes in *feature-based methods*. In these methods, an image patch is first transformed into an invariant feature descriptor which is then matched by some simpler distance measure (reviews: [1–5,17–19]); here the burden of invariance (not only against illumination changes, but also against occlusions, spatial distortions, or noise) is imposed on the transformation. *SIFT descriptors* [7] comprise summed gradient magnitudes for different orientations in a histogram-like representation. Tolerance against multiplicative illumination changes is achieved by normalizing the feature vector to unit length. Since its entries are derived from images gradients, the feature vector is already tolerant against additive changes. The authors observed that non-linear illumination changes mostly affect the magnitude, not the direction of the gradient vectors, and therefore apply a threshold to all elements of the feature vector and re-normalize. Feature vectors are matched by the Euclidean distance. *SURF descriptors* [8] are based on Haar wavelets which are tolerant against additive intensity changes. Invariance to multiplicative changes is achieved by normalizing the feature vector to unit length. For matching the descriptors, the Mahalanobis or Euclidean distance is suggested. The original publications on *binary feature descriptors* [9–12] remain tacit on how illumination tolerance is achieved. Binary descriptors are bit vectors where each bit encodes whether the intensity difference between two pixels (or regions) of the pair is positive. As long as illumination changes do not affect the sign of the difference, the corresponding bit is unaffected. The selection of the spatial locations of pixels in each pair affect the performance of the predictor, but the publications do not specifically investigate how this selection is related to illumination tolerance. Binary feature vectors are compared by the Hamming distance. The comparative studies in [1,2,5,17] explicitly investigate illumination tolerance (albeit not for the modern binary descriptors).

Holistic methods belong to a wide range of approaches where *distance measures* (a.k.a. image distance functions or matching costs) are directly computed on intensities or their derivatives (reviews: [60–63]). Different types of invariance (or at least tolerance), among them invariance against illumination changes, have to be achieved by the distance measure itself. A plethora of such distance measures has been suggested, often with specific applications and specific types of invariance in mind. An extensive review and classification (including a performance comparison) is presented in [61,62]; partial reviews and performance comparisons are provided in [60,63–67]. Chambon and Crouzil [62] provide a helpful classification of distance measures: cross correlation-based methods ("cross"), classical statistics-based measures ("classical", including distance metrics), derivative-based measures ("derivative", often based on intensity gradients), non-parametric measures ("non-parametric", including rank-based measures), and robust measures ("robust", including median-based measures). An additional class could be introduced which includes histogram-based measures ("histogram", overview: [68]) including mutual information [66,69].

This paper addresses the question of how to achieve illumination tolerance in the first phase of min-warping where the distances between image columns are computed. This application guides the selection of the distance measures studied here. On the one hand, min-warping works on low-resolution panoramic images where each column typically has only 20–50 elements. On the other hand, the number of comparisons between image columns is large since each column in one image is compared with each column in the other. For images with a width of 360 pixels which are presented in 9 different magnifications [20], $360^2 \times 9 \approx 1.2$ million comparisons between image columns are required. We did not investigate whether it is possible to transform short columns of panoramic images into even shorter feature descriptors to accelerate the large number of matches; instead, we test and develop extremely simple, pixel-wise distance measures. With respect to the classification above, we focus on distance measures directly computed on intensities or their derivatives. All methods studied here belong to the classes "cross" and "classical". Since min-warping compares vertical image columns and not 2D patches, most measures from the "derivative" class which rely on the direction of intensity gradients cannot be applied. We only explore the effect of a vertical edge filter on measures from "cross". Measures from the classes "non-parametric" and "robust" have not been studied since we concluded from the results presented below that the additional tolerance introduced by these methods is likely to have a negative effect on the performance of min-warping. The small number of pixels in our application—distance measures are computed between short image columns—also excludes measures from the "histogram" class.

There is currently only one publication which specifically addresses illumination tolerance in holistic local visual homing methods [26]. Stürzl and Zeil investigate how spatial and rotational image distance functions in the DID model are affected by two different ways of preprocessing: DoG filtering and contrast normalization. These are possible alternatives to the edge filtering used in some of our methods, particularly in the novel sequential correlation methods, but so far we have not explored this direction.

Illumination tolerance and tolerance against long-term scene changes is an actively investigated problem in robotics. Even though they are not directly relevant for the work presented in this paper, we point to some recent publications. Our method uses monochrome input, but including color information offers interesting opportunities such as methods for the removal of shadows [70]. Another possible approach to deal with illumination changes is to learn a dictionary of image regions under different conditions of illumination and then transform image regions according to this dictionary before they are matched between images [71]; this may be a possible approach if min-warping is embedded in a more complex localization framework. Some methods are possibly limited to feature-based approaches, such as learning mechanisms where the stability of a feature is determined by integrating multiple observations under different conditions of illumination; in the application phase, only stable (static) features are considered [72]. Also in the context of feature detection, an illumination-invariant difference-of-Gaussians operator for the construction of scalespaces has been introduced [73]; whether a transfer of insights to holistic approaches is possible has to be investigated. Note that some distance measures described in the following appear as components in many computer vision approaches; subtraction of the mean and normalization by the variance of an image patch, for example, has been used for an appearance-based approach to place recognition [74].

1.4. Contributions and Outline of the Paper

This work focuses on tolerance against changes in illumination in min-warping and makes a minor and a major contribution. The minor contribution lies in the development of *tunable distance measures*. In these distance measures, a portion invariant against multiplicative and/or additive changes in illumination and a portion expressing an overall difference in intensity are superimposed by a weighted average. In experiments on image databases we show that there typically is an optimum of min-warping's navigation performance for some intermediate weight. One of the tunable measures contains the Euclidean distance (which we have used as distance measure for min-warping before [20]) as a special case. At the optimal weight, the illumination-invariant portion is emphasized compared to the Euclidean distance.

The major contribution of this work are two *novel correlation-like measures*: "sequential correlation" and a faster form "approximated sequential correlation". These measures interpret intensities in two vectors as a 2D curve and analyze the direction of curve segments. They are invariant with respect to intensity shifts—in contrast, the well-known normalized cross-correlation is invariant against scaling of intensities, its zero-mean version against scaling and shift. Our experiments show that the performance of sequential correlation is better than that of normalized cross-correlation applied to edge-filtered columns without requiring a second, illumination-sensitive term. A straightforward extension to two-dimensional image patches is described.

Section 2 discusses different types of invariance against shift and scaling of intensities. Section 3 introduces all measures used in this work. Experiments on indoor image databases and their results are presented in Sections 4 and 5. We discuss the results in Section 6. Section 7 demonstrates that min-warping with the novel approximated sequential correlation measure can be applied to the navigation of cleaning robots. Conclusions from the database experiments are drawn in Section 8.

2. Types of Illumination Invariance

In this section, we introduce different types of invariance against shift and scaling of intensities within an image region (in the case of min-warping: a column of a panoramic image). This allows us to classify the distance measures introduced in Section 3 accordingly. For simplicity, the following discussion assumes a one-dimensional, continuous representation of intensities $a(x)$ and $b(x)$ in two image regions. A distance measure $f\{a(x), b(x)\}$ computes some (non-negative) integral over these intensity functions. We assume that a perfect match is indicated by a distance measure of zero, *i.e.*, $f\{a(x), a(x)\} = 0$.

The methods investigated here establish tolerance against intensity shifts (additive changes) or against intensity scaling (multiplicative changes) over the entire region. In addition, as will be discussed below (Section 6), those distance measures which work on edge-filtered images appear to achieve some kind of local invariance within each image region.

This work deals with two types of *invariance against intensity scaling*, *i.e.*, against multiplicative changes in intensity. The first type we call *weak scale invariance* which is achieved if the distance measure exhibits the property

$$f\{ca(x), a(x)\} = 0$$

This means that for an arbitrary intensity scaling of an intensity function by a factor c, the distance measure indicates a perfect match with the original intensity function. The second type, *strong scale invariance*, includes weak scale invariance but in addition ensures that

$$f\{ca(x), b(x)\} = f\{a(x), b(x)\}$$

i.e., scaling of one intensity function has no effect on the distance to the other intensity function. This is the stronger type of scale invariance since it not only guarantees that the match of an image region *with itself* is invariant to scaling, but also that the distance to a *different* image region is unaffected by scaling. We expect that measures with strong scale invariance will perform better than those with weak scale invariance.

Every distance measure can be made *invariant against intensity shifts*, *i.e.*, against offsets in intensity. In this work, shift invariance is accomplished by subtracting the mean value or by performing a first-order differentiation before applying the distance measure, or both. If one of the vectors is shifted by an offset c, the distance measure is not affected if the mean is subtracted beforehand,

$$f\{(a(x)+c)-(\overline{a}+c), b(x)-\overline{b}\} = f\{a(x)-\overline{a}, b(x)-\overline{b}\}$$

or if the distance measure is applied to edge-filtered images,

$$f\left\{\frac{d[a(x)+c]}{dx}, \frac{db(x)}{dx}\right\} = f\left\{\frac{da(x)}{dx}, \frac{db(x)}{dx}\right\}$$

It is clear that if two intensity functions are scaled versions of each other, *i.e.*, $a(x) = cb(x)$, then their derivatives will also be scaled versions of each other

$$\frac{da(x)}{dx} = c\frac{db(x)}{dx}$$

as will be their zero-mean versions,

$$a(x) - \overline{a} = c(b(x) - \overline{b})$$

and the zero mean versions of the derivatives,

$$\frac{da(x)}{dx} - \overline{a'} = c\left(\frac{db(x)}{dx} - \overline{b'}\right)$$

where $\overline{a'}$ and $\overline{b'}$ are the means of the derivatives. Obviously the methods used to introduce additional shift invariance have no effect on scale relationships.

While subtracting the mean removes intensity shifts affecting the entire image region, a first-order differentiation also removes local intensity shifts in sub-regions of constant intensity; only the edge slope between regions of constant intensity is affected. We suppose that this introduces local illumination tolerance (see discussion in Section 6).

3. Distance Measures

In the following subsections we introduce the different distance measures explored in this work. We start with the tunable sum of squared differences since it includes the Euclidean distance—which was originally used by min-warping—as a special case (Section 3.1). A zero-mean version is introduced in Section 3.2. Normalization of the illumination-invariant portion leads to the well-known normalized cross-correlation (Section 3.3). In its tunable form, this measure is combined with the absolute difference of the sum of the intensities; the same illumination-sensitive term is used in all following methods. A zero-mean version of normalized cross-correlation is introduced in Section 3.4. In Section 3.5 we present distance measures where normalized cross-correlation and its zero-mean version are applied to edge-filtered image columns, since also the novel correlation-like measures introduced in this work include an edge filter. Again, tunable versions are defined. We then describe the two novel correlation-like measures introduced in this work, "sequential correlation" (Section 3.6) and its approximated version (Section 3.7), both also available as tunable versions.

In the following we make predictions on the effects of the different distance measures on the performance of min-warping. As performance criterion, we always assume the home vector error as described in Section 4.3.

3.1. Tunable Sum of Squared Differences (TSSD)

In its original form, our visual navigation method "min-warping" computed the Euclidean distance (the square root of the "sum of squared differences", SSD) between two image columns. The indoor image databases used in the evaluation of the method [20] were collected under constant or moderately varying illumination (mostly caused by changing cloud cover over the day), but not under extreme variations in illumination (like changes between day and night or between natural and artificial illumination). Camera-gain control and histogram equalization were applied to reduce the impact of illumination changes (note that, even under constant illumination, these steps introduce moderate intensity changes as well). Under these conditions, SSD produced robust navigation performance.

Unsurprisingly, as the experiments below will demonstrate, the performance of SSD breaks down under stronger changes in illumination. We therefore first attempt to develop a distance measure which is related to SSD—keeping the good performance under the above-mentioned mild conditions—but which is at the same time tolerant against changes in illumination. For this we assume that changes have a *multiplicative* effect on the intensities (with intensities being encoded by non-negative values). A distance measure invariant against multiplicative changes should rate the distance between two intensity vectors as minimal (zero) if one is a scaled version of the other.

As shown in Section 2, additional tolerance against *additive* changes can be achieved by subtracting mean intensities from the intensity vectors. However, it is unclear how this will affect the performance of min-warping: First, invariance against scaling and shift are partly exchangeable—in noisy images it may be difficult to distinguish between moderate scaling and shift. Second, introducing too much tolerance may lead to mismatches. We will explore the influence of double invariance in the experiments.

Applying an illumination-invariant distance measure *alone* may, however, falsely indicate matches for intensity vectors which are representing *different* features in the environment and markedly differ in their overall intensity but which nevertheless are scaled (or shifted) versions of each other. While this may occur only rarely in structured image regions, indoor environments often contain uniformly colored walls, curtains, or furniture which can only be distinguished by their overall intensity. Since min-warping does not preselect features (e.g., high-variance regions around corners), ignoring these overall intensity differences may discard information valuable for visual homing. The illumination-invariant measure should therefore be combined with a second, illumination-sensitive measure expressing the difference in the overall intensities of the two image columns. Depending on a weight factor, the influence of both measures can be tuned. We expect that there will be an optimal, intermediate weight factor with respect to the navigation performance.

In the following derivations, the vectors encoding the intensities of the two image columns are denoted by $\mathbf{a}$ and $\mathbf{b}$, both of dimension n, with $a_i \geq 0$ and $b_i \geq 0$. We start with the derivation of the illumination-invariant part of the measure. We suggest a novel form of the "parametric sum of squared differences" (PSSD, see e.g., [64,65]) where the scale parameter γ is split and distributed to both vectors such that the resulting measure is inversion-symmetric (*i.e.*, if $\mathbf{a}$ and $\mathbf{b}$ are exchanged, the scale factor is inverted but the distance stays the same):

$$\tilde{J}_{\text{PSSD}}(\mathbf{a}, \mathbf{b}, \gamma) = \tfrac{1}{2} \left\| \frac{1}{\sqrt{\gamma}} \mathbf{a} - \sqrt{\gamma}\, \mathbf{b} \right\|^2 \tag{1}$$

We determine the optimal value $\check{\gamma}$ where the distance $\tilde{J}_{\text{PSSD}}$ becomes minimal. It is straightforward to show that the optimal scale factor is

$$\check{\gamma} = \frac{||\mathbf{a}||}{||\mathbf{b}||} \tag{2}$$

If we insert $\check{\gamma}$ into Equation (1) we obtain

$$\begin{aligned} J_{\text{PSSD}}(\mathbf{a}, \mathbf{b}) &= \tilde{J}_{\text{PSSD}}(\mathbf{a}, \mathbf{b}, \check{\gamma}) = ||\mathbf{a}||\, ||\mathbf{b}|| - \mathbf{a}^T \mathbf{b} && (3) \\ &= ||\mathbf{a}||\, ||\mathbf{b}|| \Bigg(1 - \underbrace{\frac{\mathbf{a}^T \mathbf{b}}{||\mathbf{a}||\, ||\mathbf{b}||}}_{J_{\text{NCC}}}\Bigg) && (4) \end{aligned}$$

We see from the second Equation (4) that the measure is closely related to the widely used normalized cross-correlation NCC (see e.g., [60,63–65]), but leaves out the normalization. J_{PSSD} becomes zero if and only if one intensity vector is a scaled version of the other,

$$||c\mathbf{a}||\, ||\mathbf{a}|| - (c\mathbf{a})^T \mathbf{a} = 0,$$

thus the measure exhibits *weak scale invariance*; see Section 2. Since

$$||c\mathbf{a}||\, ||\mathbf{b}|| - (c\mathbf{a})^T \mathbf{b} = c(||\mathbf{a}||\, ||\mathbf{b}|| - \mathbf{a}^T \mathbf{b})$$

the measure does not exhibit strong scale invariance, though. Since the measure was derived from a quadratic term, it is non-negative. This property is also directly visible from Equation (3) by applying the Cauchy-Schwarz inequality.

For the comparison of the overall intensities we use the squared difference of the length (SDL):

$$J_{\text{SDL}}(\mathbf{a}, \mathbf{b}) = (||\mathbf{a}|| - ||\mathbf{b}||)^2$$

The combined distance measure, which we call "tunable sum of squared differences" (TSSD), is

$$J_{\text{TSSD}}(\mathbf{a}, \mathbf{b}, w) = w J_{\text{SDL}}(\mathbf{a}, \mathbf{b}) + (1 - w) J_{\text{PSSD}}(\mathbf{a}, \mathbf{b}) \tag{5}$$

where w is a weight factor from the interval $[0, 1]$. For $w = 0$, TSSD only contains the illumination-invariant portion; for $w = 1$, it only considers overall intensity differences.

Interestingly, the SSD

$$J_{\text{SSD}}(\mathbf{a}, \mathbf{b}) = ||\mathbf{a} - \mathbf{b}||^2$$

is obtained as the special case $w = \frac{1}{3}$ of the TSSD in Equation (5) (up to a constant factor),

$$J_{\text{TSSD}}\left(\mathbf{a}, \mathbf{b}, \tfrac{1}{3}\right) = \tfrac{1}{3} J_{\text{SSD}}(\mathbf{a}, \mathbf{b})$$

The SSD measure can therefore be interpreted as a 2:1 mixture of the fully illumination-invariant PSSD measure and the SDL measure expressing the difference in overall intensities. Reducing the weight w below $\frac{1}{3}$ emphasizes the illumination-invariant portion, and we expect an optimal performance of our navigation method in this range. Note that this quantitative interpretation of the weight is only possible for TSSD, not for the other tunable distance measures presented below.

In our implementation of min-warping, we use $\sqrt{J_{\text{TSSD}}}$ as distance measure to stay close to the Euclidean distance used before (the square root is applied before column distances are summed over the entire image, with the sum indicating the overall match; see [20]).

3.2. Tunable Zero-Mean Sum of Squared Differences (TZSSD)

We also implemented a zero-mean version of TSSD, called TZSSD in the following. Here the illumination-invariant portion J_{PZSSD} is computed for two zero-mean intensity vectors,

$$J_{\text{PZSSD}}(\mathbf{a}, \mathbf{b}) = J_{\text{PSSD}}(\mathbf{a} - \bar{a}\mathbf{1}, \mathbf{b} - \bar{b}\mathbf{1})$$

where $\bar{a}$ and $\bar{b}$ are the means of the vectors $\mathbf{a}$ and $\mathbf{b}$, respectively, and $\mathbf{1} = (1, 1, \ldots, 1)^T$. Since most parts of our implementation of PSSD use *unsigned* integers, we do not subtract the mean in the beginning, but correct for the mean at later stages in the computation (which is possible by expanding the expression for the zero-mean vectors).

The illumination-sensitive portion is determined from the original intensity vectors, not from the zero-mean versions (since it would otherwise be zero). For this and all following tunable measures we use the absolute difference of sums (ADS) as the illumination-sensitive portion

$$J_{\text{ADS}}(\mathbf{a}, \mathbf{b}) = \kappa \left| \sum_{i=1}^{n} a_i - \sum_{i=1}^{n} b_i \right|$$

This measure is somewhat easier to compute in an SSE implementation (see Section 4.1) since it doesn't include multiplications and since we don't have to take the square root. Moreover, the experiments show that this measure alone ($w = 0$) works as well as SDL (see Section 5.2). The factor κ results from factors in our integer-based implementation; for input images with pixel range $[0, 1]$ it is $1/16$ for all correlation-type measures and 0.186 for TZSSD.

The TZSSD measure can be written as

$$J_{\text{TZSSD}}(\mathbf{a}, \mathbf{b}) = wJ_{\text{ADS}}(\mathbf{a}, \mathbf{b}) + (1 - w)\sqrt{J_{\text{PZSSD}}(\mathbf{a}, \mathbf{b})}$$

A subtraction of the mean makes the PSSD measure invariant to shifts in illumination (see Section 2). Since PSSD already exhibits (weak) invariance against multiplicative changes in illumination, the resulting measure is invariant against both scaling and shift. We expect a loss of performance compared to TSSD due to an increased portion of false-positive matches as discussed above.

3.3. Tunable Normalized Cross-Correlation (TNCC)

We already saw that PSSD is related to normalized cross-correlation NCC, except for the missing normalization; see Equation (4). In fact, we could derive a form of NCC — here called NCC+ since it only returns positive values from the interval $[0, 2]$ — from the distance measure

$$\tilde{J}_{\text{NCC+}}(\mathbf{a}, \mathbf{b}, \gamma) = \frac{1}{2} \frac{\left\| \frac{1}{\sqrt{\gamma}} \mathbf{a} - \sqrt{\gamma}\, \mathbf{b} \right\|^2}{\|\mathbf{a}\| \, \|\mathbf{b}\|} \tag{6}$$

We obtain the same optimal value for gamma as in Equation (2), and by inserting the optimal gamma into Equation (6) we get

$$J_{\text{NCC+}}(\mathbf{a}, \mathbf{b}) = 1 - \underbrace{\frac{\mathbf{a}^T \mathbf{b}}{\|\mathbf{a}\| \, \|\mathbf{b}\|}}_{J_{\text{NCC}}}$$

where the original NCC measure (see e.g., [63]) is underbraced. Alternatively — and this makes the principle underlying this measure more clear —, NCC+ can be written as

$$J_{\text{NCC+}}(\mathbf{a}, \mathbf{b}) = \frac{1}{2} J_{\text{SSD}} \left(\frac{\mathbf{a}}{\|\mathbf{a}\|}, \frac{\mathbf{b}}{\|\mathbf{b}\|} \right) = \frac{1}{2} \left\| \frac{\mathbf{a}}{\|\mathbf{a}\|} - \frac{\mathbf{b}}{\|\mathbf{b}\|} \right\|^2$$

where obviously the scaling between the two vectors is eliminated by normalizing them before computing their SSD distance. While normalized cross-correlation can be interpreted as a distance measure, it can also be seen as a transformation (normalization) accomplishing scale invariance followed by a matching using a simpler distance measure, here SSD (for further interpretations of the correlation coefficient, see [75]). NCC+ exhibits *strong scale invariance* since

$$1 - \frac{(c\mathbf{a})^T \mathbf{b}}{\|c\mathbf{a}\| \, \|\mathbf{b}\|} = 1 - \frac{\mathbf{a}^T \mathbf{b}}{\|\mathbf{a}\| \, \|\mathbf{b}\|}$$

thus we expect a better performance than the PSSD measures.

However, as pointed out by [76], normalization in the presence of noise leads to erroneous estimates for small vectors which may negatively affect the performance. In most feature-matching methods (review: [4]), features to be matched are only extracted at selected key points where the intensities in the neighborhood vary strongly (often corners or scale-space extrema are detected). Normalization is unproblematic in this context as the denominator is far from zero. Min-warping, in contrast, matches *all* image columns without any selection, thus also image columns with low intensity will be processed by the distance measure. Since the denominator of NCC approaches zero in regions of low intensity, numerical problems may occur and image noise is amplified, adding to the difficulty to make predictions of the performance of NCC+. In our integer-based implementation (using 16 bit unsigned words), we add 1 to the integer denominator to avoid division by zero (if one of the vectors has zero length, a value of $J_{\text{NCC+}}(\mathbf{a}, \mathbf{0}) = 1$ is obtained).

For the tunable version, we combine NCC+ with the absolute difference of sums:

$$J_{\text{TNCC}}(\mathbf{a}, \mathbf{b}, w) = w J_{\text{ADS}}(\mathbf{a}, \mathbf{b}) + (1 - w) J_{\text{NCC+}}(\mathbf{a}, \mathbf{b})$$

3.4. Tunable Zero-Mean NCC (TZNCC)

We also implemented a zero-mean version of NCC+,

$$J_{\text{ZNCC+}}(\mathbf{a},\mathbf{b}) = J_{\text{NCC+}}(\mathbf{a}-\bar{a}\mathbf{1},\mathbf{b}-\bar{b}\mathbf{1})$$

The tunable form is called TZNCC,

$$J_{\text{TZNCC}}(\mathbf{a},\mathbf{b}) = wJ_{\text{ASC}}(\mathbf{a},\mathbf{b}) + (1-w)J_{\text{ZNCC+}}(\mathbf{a},\mathbf{b})$$

Here the problem of small denominators is even more severe since it occurs not only for vectors with low *intensities*, but for all vectors with low *variance*. This may lead to a performance loss in addition to the problems we expect as a results of the double invariance (see above).

3.5. Tunable NCC on Edge Images (TENCC, TEZNCC)

Edge filtering is integral part of our novel "sequential correlation" methods (Sections 3.6 and 3.7). For comparison, edge filtering should also be considered as preprocessing step for the standard correlation method (NCC). Since min-warping uses image columns as features, a first-order vertical edge detector is applied (subtraction of the intensities of subsequent, neighboring pixels):

$$\begin{aligned} a'_i &= a_{i+1} - a_i, \quad i = 1,\ldots,n-1 \\ b'_i &= b_{i+1} - b_i, \quad i = 1,\ldots,n-1 \end{aligned}$$

We introduce abbreviations for the illumination-invariant terms,

$$\begin{aligned} J_{\text{ENCC+}}(\mathbf{a},\mathbf{b}) &= J_{\text{NCC+}}(\mathbf{a}',\mathbf{b}') \\ J_{\text{EZNCC+}}(\mathbf{a},\mathbf{b}) &= J_{\text{ZNCC+}}(\mathbf{a}',\mathbf{b}') \end{aligned}$$

the bottom one being the zero-mean version, and define the tunable measures TENCC and TZENCC:

$$\begin{aligned} J_{\text{TENCC}}(\mathbf{a},\mathbf{b}) &= wJ_{\text{ADS}}(\mathbf{a},\mathbf{b}) + (1-w)J_{\text{ENCC+}}(\mathbf{a},\mathbf{b}) \\ J_{\text{TEZNCC}}(\mathbf{a},\mathbf{b}) &= wJ_{\text{ADS}}(\mathbf{a},\mathbf{b}) + (1-w)J_{\text{EZNCC+}}(\mathbf{a},\mathbf{b}) \end{aligned}$$

Figure 6. Two identical edge-filtered, zero-mean columns can correspond to widely differing intensity columns.

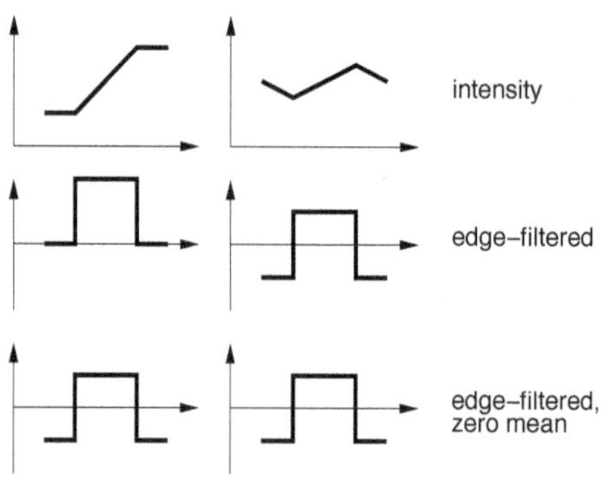

The effect of subtracting the mean is difficult to predict. On the one hand, a column of n intensity differences will already have a small mean with a maximal absolute value of a/n if the intensities come from the interval $[0, a]$, so any effect should be small. On the other hand, very different intensity vectors can have the same zero-mean edge-filtered version, so there is the danger of mismatches; see Figure 6. We therefore expect a small performance loss if the mean is subtracted.

Figure 7. Magnification around the horizon (dashed line) using nearest neighbor interpolation leads to duplicated pixels (gray bars). Intensities in the original (**top**) and magnified 1D image column (**bottom**) are shown as bars in order to visualize the nearest-neighbor selection. Arrows indicate the pixel assignment.

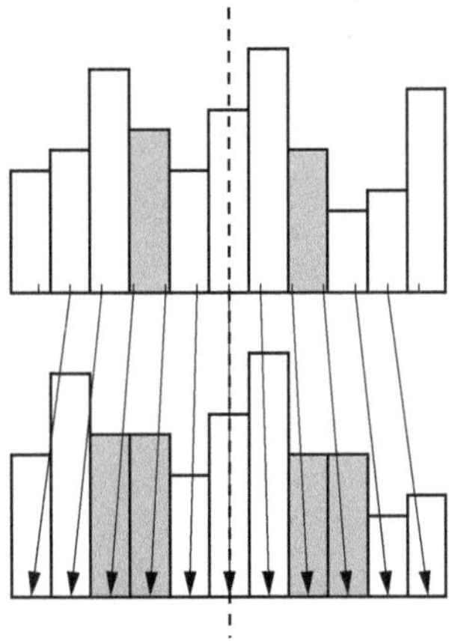

Min-warping does not only compare a single pair of images, but also magnified versions of each image with the original version of the other (see [20]). Magnification only affects the vertical direction and leaves the horizon unchanged. For efficiency reasons, magnification is performed using nearest-neighbor interpolation between image rows which may lead to duplicated image rows, see Figure 7. If a first-order vertical edge filter is applied to a magnified column, erroneous zero entries may be produced. We therefore decided to apply the vertical edge filter to the *original* two images, and then magnify the *edge-filtered* image. However, one has to be aware that exchanging the order of the operations affects the result: If an image (for simplicity in continuous representation) is first magnified (M) and then edge-filtered (ME), the derivatives are effectively multiplied by the magnification factor and therefore scaled down (magnification is achieved for $0 < \sigma < 1$). Assuming a horizon at $x = 0$, this can be written as

$$f(x) \quad \rightarrow \quad f_M(x) = f(\sigma x) \quad \rightarrow \quad f_{ME}(x) = f'_M(x) = \sigma f'(\sigma x) \tag{7}$$

This has to be considered if the image is first edge-filtered (E) and then magnified (EM), since there the derivatives would remain unchanged:

$$f(x) \quad \rightarrow \quad f_E(x) = f'(x) \quad \rightarrow \quad f_{EM}(x) = f_E(\sigma x) = f'(\sigma x)$$

Here we have to scale the derivatives by multiplying $f_{EM}(x)$ by σ to arrive at the same result as in Equation (7). We will test both versions (with and without scaling of the derivatives).

3.6. Sequential Correlation (SC, TSC)

Sequential correlation and its approximated form (next section) are the major novel contributions of this work. In most of the well-known distance measures like NCC, SSD, PSSD, as well as in the TSSD, TNCC, TZSSD, and TZNCC measures introduced above, the order of the elements in the intensity vectors is of no concern. The fact that the corresponding pixels are adjacent in the image is ignored, and all pixel pairs could as well have been sampled in arbitrary order from the two image columns. For our novel "sequential correlation" measure (SC, or ASC for the approximated version), however, we request that the vectors are ordered such that adjacent elements correspond to adjacent pixels in the image. In natural images, neighboring pixels are known to be strongly correlated, so there is usually a smooth transition of intensities between them, even more so as min-warping prefers slightly low-pass filtered images [20,51]. Please note that the "image Euclidean distance" (IMED) [77,78] was also motivated by the fact that the traditional Euclidean distance ignores the spatial relationship between pixels (leading to large Euclidean distances for small spatial deformations). Interestingly, IMED is equivalent to computing the Euclidean distance on low-pass filtered images [77]. The spatial relationship between pixels is ignored by most other distances measures where a two-dimensional *point cloud* $\{(a_i, b_i)\}$ is analyzed, whereas our measures interpret the sequence (a_i, b_i), $i = 1 \ldots n$ as a sampled two-dimensional *curve*. Essentially, it is a vertical edge filter which introduces the neighborhood relationship between pixels in our novel measures, so the same effect could be achieved by applying other distance measures to edge-filtered images (which we test for normalized cross-correlation; see Section 3.5). Figure 8 (left) visualizes d_i.

Figure 8. Visualization of d_i used in sequential correlation (**left**) and of $\hat{d}_i$ used in approximated sequential correlation (**right**). The zero plane is shown in both plots.

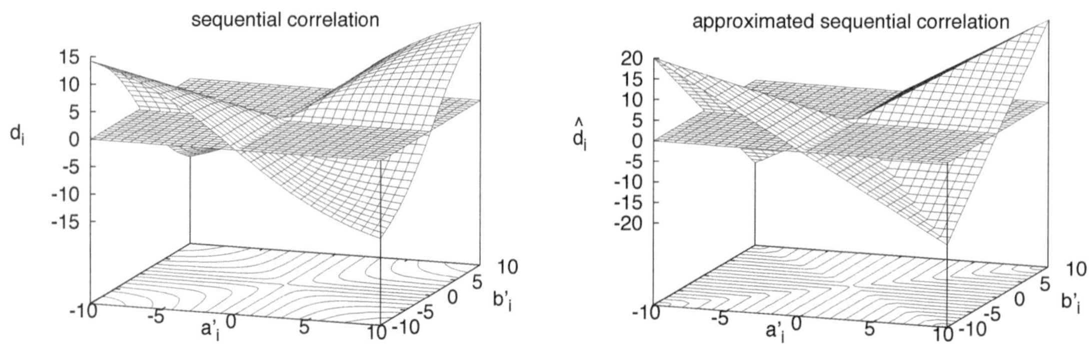

For the definition of the sequential correlation measure, we start by defining the segment vector between adjacent sequence points,

$$\mathbf{s}_i = \begin{pmatrix} a'_i \\ b'_i \end{pmatrix} = \begin{pmatrix} a_{i+1} - a_i \\ b_{i+1} - b_i \end{pmatrix}, \quad i = 1, \ldots n-1$$

This discrete differentiation step introduces shift invariance; see Section 2. For each segment i of the sequence, we compute the direction term

$$d_i = \frac{2a'_i b'_i}{||\mathbf{s}_i||} = ||\mathbf{s}_i|| \sin(2\alpha_i) \tag{8}$$

where α_i is the orientation of segment i with respect to the a_i axis (see Figure 9). The second form shows that (for constant $||\mathbf{s}_i||$) d_i will be maximal for $\alpha_i = 45°$ and $\alpha_i = -135°$ (intensity changes in both columns have the same sign and absolute value), and minimal for $\alpha_i = -45°$ and $\alpha_i = 135°$ (intensity changes have the same absolute value but different sign). The first form in Equation (8) is used in the implementation (since the second contains a slow trigonometric function). Division by $||\mathbf{s}_i||$ may lead to numerical problems (in our integer implementation, where numerator and denominator are integers, we add 1 to the denominator to avoid this), but the second form shows that at least image noise will not be amplified by this normalization.

Using the sums

$$D = \sum_{i=1}^{n-1} d_i, \quad S = \sum_{i=1}^{n-1} ||\mathbf{s}_i|| \tag{9}$$

We define sequential correlation as

$$J_{\text{SC}} = \frac{D}{S} \in [-1, 1], \quad \text{or} \quad J_{\text{SC+}} = 1 - J_{\text{SC}} \in [0, 2] \tag{10}$$

where the second form is used in our implementation. With the standard correlation coefficient (NCC), J_{SC} shares the range $[-1, 1]$ and the symmetry property. SC is invariant against shifts, here achieved by the transition from intensities to intensity differences, but not invariant against scaling. In the integer-based implementation, we add 1 to the denominator to avoid division by zero in Equation (10). We test versions with and without scaling of the derivatives; see Section 3.5. Again, we introduce a tunable form called TSC:

$$J_{\text{TSC}}(\mathbf{a}, \mathbf{b}) = wJ_{\text{ADS}}(\mathbf{a}, \mathbf{b}) + (1 - w)J_{\text{SC+}}(\mathbf{a}', \mathbf{b}')$$

Figure 9. Visualization of sequential correlation. Top row: intensity sequence (a_i, b_i). Segment vectors are color-coded according to $\sin(2\alpha_i)$ (see circle on the bottom right). The value of J_{SC} is given in the upper left corner. Bottom row: corresponding individual intensity curves a_i and b_i over i. Left column: positive sequential correlation, center column: approximately zero sequential correlation, right column: negative sequential correlation.

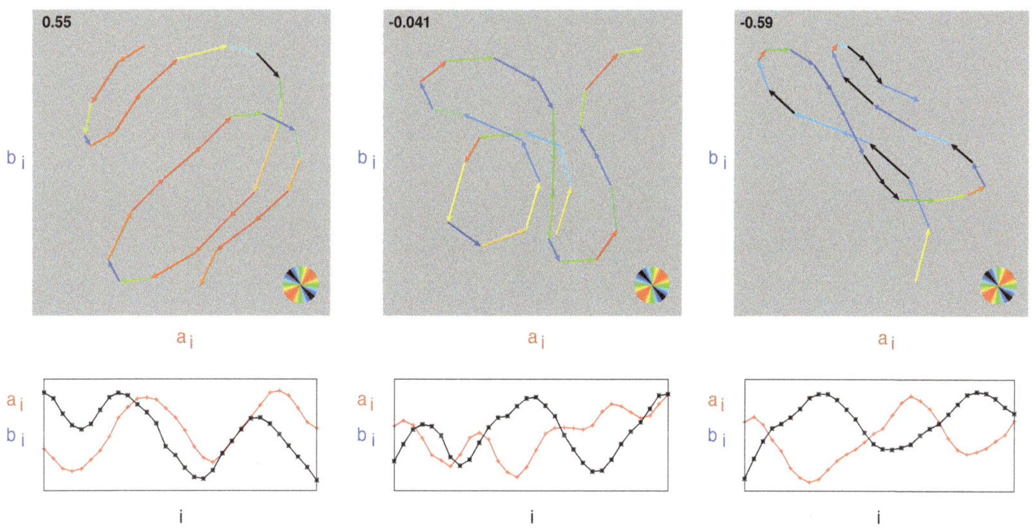

Sequential correlation provides a clear graphical interpretation of correlation on edge-filtered data. In Figure 9 (top row), the intensity sequence (a_i, b_i) is shown for three different examples: positive, approximately zero, and negative sequential correlation. The segment vectors are color-coded according to $\sin(2\alpha_i)$ (red $+1$, blue -1). The total length of all segments corresponds to S, the value obtained by summing the product of segment length $\|\mathbf{s}_i\|$ times $\sin(2\alpha_i)$ gives D. Their ratio $J_{\mathrm{SC}} = D/S$ is specified in the upper left corner.

3.7. Approximated Sequential Correlation (ASC, TASC)

In the implementation of SC, the first form in Equation (8) is used. This computation is, however, awkward due to the square root required to compute the vector length and due to the division. Our implementation uses 16-bit integer arithmetic with eightfold SIMD (single instruction multiple data) parallelity as offered by Intel's SSE2 instruction set, but parallel versions of square root and division are not available for this format. Since these operations have to be performed for each segment vector $\mathbf{s}_i$, the detour via packed floats (4 floats that can be processed in parallel by SSE instructions) results in a marked increase of computation time compared to normalized cross-correlation.

We therefore developed an approximated version of SC, called "approximated sequential correlation" (ASC) which can be implemented almost fully in parallel using just integer instructions. The term d_i in Equation (8) was designed such that, for constant $\|\mathbf{s}_i\|$, it becomes maximal at $\alpha_i = 45°$ and $\alpha_i = -135°$, and minimal at $\alpha_i = -45°$ and $\alpha_i = 135°$; zero values are achieved for the axis directions ($a'_i = 0$ or $b'_i = 0$). Moreover, in each direction from the origin, the direction term d_i is proportional to $\|\mathbf{s}_i\|$. A function with the same properties, but with piecewise linear transition is

$$\hat{d}_i = |a'_i + b'_i| - |a'_i - b'_i| \tag{11}$$

see Figure 8 (right). This approximation of d_i (up to a constant factor) eliminates square root and division and allows us to implement this computation with eightfold integer parallelity. An equivalent expression is

$$\hat{d}_i = 2\max\{\min(a'_i, b'_i), -\max(a'_i, b'_i)\} \tag{12}$$

which saves one operation on CPUs which have instructions for maximum and minimum (as is the case for SSE2 instructions on Intel CPUs); note that the factor 2 can be factored out in Equation (13) such that it only has to be applied once per column pair. We define

$$\hat{D} = \sum_{i=0}^{n-1} \hat{d}_i, \quad \hat{S} = \sum_{i=0}^{n-1} |a'_i| + \sum_{i=0}^{n-1} |b'_i| \tag{13}$$

and it can easily be shown by applying the triangle inequality that the measure

$$J_{\mathrm{ASC}} = \frac{\hat{D}}{\hat{S}} \tag{14}$$

lies within the range $[-1, 1]$. Proof: Triangle inequality: $|x + y| \leq |x| + |y|$. Therefore $\sum_i(|x_i + y_i| - |x_i - y_i|) \leq \sum_i(|x_i| + |y_i|) - \sum_i |x_i - y_i| \leq \sum_i(|x_i| + |y_i|)$ which ensures $J_{\mathrm{ASC}} \leq 1$. Define $y'_i = -y_i$. Then $-\sum_i(|x_i + y_i| - |x_i - y_i|) = \sum_i(|x_i + y'_i| - |x_i - y'_i|) \leq \sum_i(|x_i| + |y'_i|) - \sum_i |x_i - y'_i| \leq \sum_i(|x_i| + |y'_i|) = \sum_i(|x_i| + |y_i|)$ which ensures $J_{\mathrm{ASC}} \geq -1$.

For our implementation, we use

$$J_{\text{ASC+}} = 1 - J_{\text{ASC}} \in [0, 2]$$

Also ASC is a symmetric measure. As SC, it is only invariant against intensity shifts. An additional advantage is that the two sums of $\hat{S}$ in Equation (13) can be precomputed individually for each image column whereas S in Equation (9) is inseparable and therefore has to be computed for all column *pairs*. To avoid division by zero, we add 1 to the denominator in our integer-based implementation of Equation (14). We again test versions with and without scaling of the derivatives; see Section 3.5. The tunable form, TASC, is given by

$$J_{\text{TASC}}(\mathbf{a}, \mathbf{b}) = w J_{\text{ADS}}(\mathbf{a}, \mathbf{b}) + (1 - w) J_{\text{ASC+}}(\mathbf{a}', \mathbf{b}')$$

While ASC is easier to compute, $\hat{d}_i$ is not continuously differentiable, and closed-form derivations involving ASC would be hampered by the necessary distinction of cases. In SC, d_i and $||\mathbf{s}_i||$ are continuously differentiable except at $a'_i = b'_i = 0$, so only this special case has to be handled. So even though SC is computationally more complex, it may still have its value for closed-form derivations, e.g., as a distance measure in gradient or Newton schemes [40,41]. It may then be possible to apply a similar approximation as the one between SC and ASC in the final expressions.

3.8. Correlation Measures on Edge-Filtered Images

It may be helpful to present the equations of all correlation measures on edge-filtered images together and in comparable form:

$$\begin{aligned}
J_{\text{NCC}}(\mathbf{a}', \mathbf{b}') &= \frac{\sum\limits_i a'_i b'_i}{||\mathbf{a}'||\,||\mathbf{b}'||} \\
J_{\text{SC}}(\mathbf{a}', \mathbf{b}') &= \frac{2 \sum\limits_i \frac{a'_i b'_i}{||(a'_i, b'_i)||}}{\sum\limits_i ||(a'_i, b'_i)||} \\
J_{\text{ASC}}(\mathbf{a}', \mathbf{b}') &= \frac{\sum\limits_i (|a'_i + b'_i| - |a'_i - b'_i|)}{\sum\limits_i |a'_i| + \sum\limits_i |b'_i|}
\end{aligned}$$

ASC uses an approximation of the numerator of SC, but a different denominator. There cannot be a direct transformation between NCC and SC since NCC is tolerant against scaling and SC is not.

3.9. Correlation of Edge-Filtered 2D Image Patches

In the present form of min-warping, the relations between neighboring columns in the images are not considered since the neighborhood relation changes with the azimuthal distortion of the images and therefore differs between snapshot and current view. Properly including the horizontal interrelations in the search phase would substantially increase the computational effort since optimization techniques like dynamic programming would be required; whether an approximative solution is possible has to be explored. Edge filtering in TENCC/TEZNCC and SC/ASC is therefore only performed in vertical direction within the image columns. For other applications, however, edges can in addition be detected in horizontal direction. Vertical and horizontal differences are simply stacked in a joint vector and compared by the difference measure. An example closely related to min-warping is the visual compass [25,44]. Here entire images—in this case sets of all horizontal and vertical differences—are compared by the difference measure for varying relative azimuthal orientation between the two images. Note that, for correlation-type measures, the normalization here only has to be applied once at the end if the distance measure is applied to the entire image rather than to image columns (which of course gives up local illumination invariance achieved by applying the distance measure individually to each image column as done in min-warping).

4. Experiments

4.1. Implementation

All methods were implemented in C using SSE intrinsics (mostly SSE2) provided by the gcc compiler. SSE intrinsics allow an access to the SSE SIMD machine instructions of the Intel architecture from C. Using these instructions, operations can be performed in parallel on multiple floating point or integer numbers within 128 bit registers. In the largest portion of the code we use eightfold parallelity based on 8 integer words of 16 bit each. Only the vertical edge filtering for the TENCC, TEZNCC, TSC, and TASC is done on float images (input images are provided as float). In SC, the computation of the square root and the division in Equation (8) is done using packed floats (4 floats processed in parallel); this requires two conversion steps between words and floats.

Input images are provided with floating point pixels in the range $[0, 1]$. When the images (or their edge-filtered versions) are converted to word format, pixels are multiplied by a factor. Another factor is used to convert the correlation quotients (range $[0, 2]$) to words. Both factors are adjusted such that the range of the words (signed or unsigned) is utilized optimally while avoiding overflows.

4.2. Image Databases

Min-warping computes home vectors and compass estimates. All possible applications of min-warping—like following the repeatedly updated home vector to approach a goal location, or triangulating image capture points for navigation in topological-metrical maps [54,55]—depend on the precision of the home vector estimates (note that home vectors depend on both estimated parameters α and ψ). We therefore directly evaluate the precision of home vector estimates using image databases. In each database, panoramic images were captured by placing the robot at known positions on a grid. From this ground truth, true home vector angles can be computed and compared with the home vector angles estimated by the different methods (here min-warping with different distance measures in the first phase); see Section 4.3. The database approach is most suitable for a systematic evaluation of the performance of local visual homing methods since it allows to evaluate thousands of home vector computations (different combinations of snapshots and current views) under controlled conditions such as different conditions of illumination. In our database collection, images were repeatedly captured at the same locations, but under different conditions of illumination. To evaluate illumination tolerance, cross-database experiments can then select snapshot and current view taken under different conditions of illumination.

For the experiments, 6 groups of databases were used, see Table 1 and Figure 10; these databases were collected by Andrew Vardy (B*, D*: [79]) and Sven Kreft (living*: [80]). All databases within a group were collected at the same locations but under different conditions of illumination. In the *B group* of databases, a laboratory room was illuminated from the ceiling lamps, the curtains were closed. In Boriginal, all lamps were on, in Bdoorlit only a row of lamps close to the room's door side was lit, in Bwinlit only a row of lamps close to the room's window side was lit. In the *D group*, the same laboratory room was captured with curtains closed and illumination from the ceiling lamps (Doriginal) and with open curtains at day plus additional illumination from the ceiling lamps (Dday). In the *living group* of databases, 4 different regions (1–4) of one and the same living room were captured, in each pair during the day and with open curtains (day) and during the night with open curtains and artificial illumination (night). As the experiments show, the changes in illumination are substantial enough (particularly in pairings from the B group and pairings from the living group) to seriously affect the navigation performance of min-warping for some of the distance measures.

Table 1. Image databases used in the experiments: size of the grid, grid spacing in (m) (in parentheses), and image size.

Database Group	Grid		Image
Boriginal, Bdoorlit, Bwinlit	10×17	(0.3)	288×30
Doriginal, Dday	10×17	(0.3)	288×30
living1day, living1night	22×4	(0.1)	288×29
living2day, living2night	22×5	(0.1)	288×29
living3day, living3night	21×7	(0.1)	288×29
living4day, living4night	15×3	(0.1)	288×29

For the *cross-database experiments*, snapshots are taken from one database and current views from a different database from the same group, in that way representing strong changes in illumination. All databases within a group provide the snapshots in turn (leading to 6 pairings in the B group and 2 pairings in all other groups, 16 pairings in total). For the *same-database experiments*, snapshots and current views are taken from the same database (13 pairings). Since the collection of an image database takes some time, at least in those databases collected at day with open curtains, moderate changes in illumination are possible. Additional changes are introduced by the camera gain control (living databases) and by histogram equalization (used in all databases). From each database pair, 100 randomly selected pairs of snapshots and current views are selected which gives us 1600 home vectors in the cross-database tests and 1300 in the same-database tests. Before being passed to min-warping, both snapshots and current views are rotated randomly in azimuthal direction.

Figure 10. Image databases used in the experiments. An example image from the center of each database is shown as it would be used as input to min-warping (low-pass filtered, histogram-equalized, resolution 1.25 deg/pixel). All images are shown in the same azimuthal orientation.

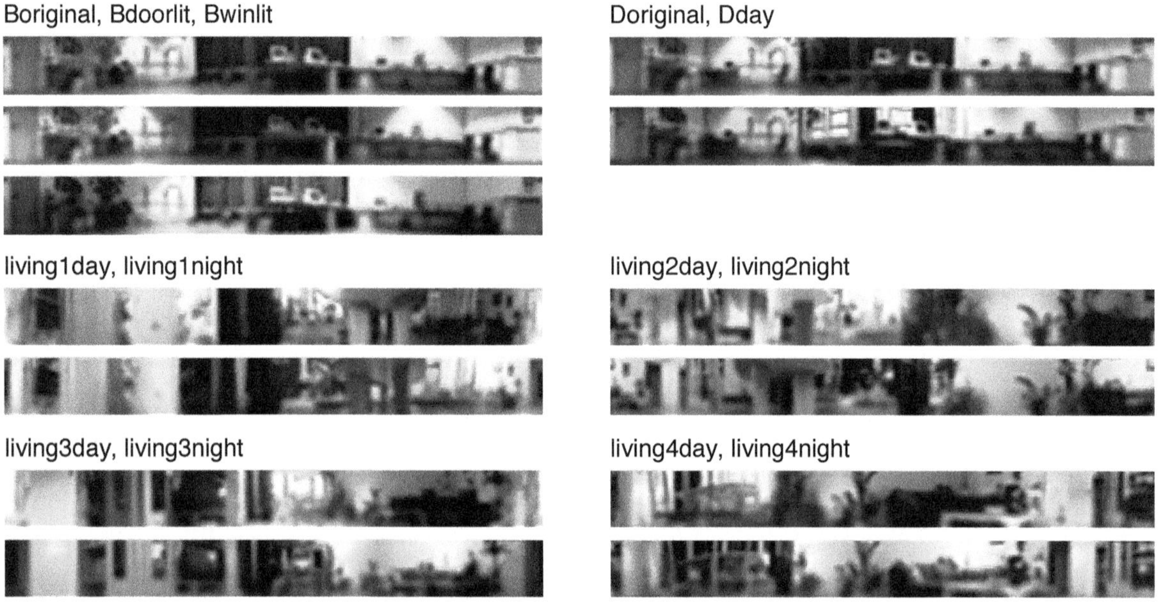

4.3. Evaluation by Angular Error

For the evaluation of the navigation performance, we use the median and the mean of the angular error (AE, from the interval $[0, \pi]$) between the computed and the true home-vector direction, the latter being available for all databases since the capture positions (ground truth) of the views are known. From median and mean together we could derive two parameters called "spread" and "error" indicating the precision of the correct home vectors and the portion of erroneous (and therefore uniformly distributed) home vectors, respectively; see [20]. While we do not analyze spread and error here, we nevertheless present both median and mean (always in radians). In most cases there is an optimal weight w of the tunable distance measures in both median and mean; in the few cases where there is no simultaneous optimum for both median and mean, we treat the data point with the optimal median as optimum.

4.4. Statistical Analysis

As in our previous publication [20] we use bootstrapping with BC_α correction [81] as the statistical test to compare the median and mean angular error of different distance measures. In each bootstrapping test, the data set comprises angular errors for a pair of distance measures (dependent sample) over all snapshot-current view pairs. This data set is re-sampled 10.000 times. We use a two-sided test with $\alpha = 1\%$. In the text we specify whether the differences are significant in both median and mean (*), only in the mean (* mean), only in the median (* median), or not significant (n.s.). The results of statistical tests are only presented for the pooled data in the experiments with strong illumination changes.

4.5. Min-Warping Parameters

In all experiments, we use min-warping, one version of warping on two-dimensional images; for the meaning of the following parameters please refer to [20]. We use 9 scale planes (see Figures 2 and 4) for images in different vertical magnification according to scale factors from the set

$$\{0.50, 0.59, 0.71, 0.83, 1.0, 1.2, 1.4, 1.7, 2.0\}$$

The thresholds separating the scale-factor ranges are

$$\{0.55, 0.65, 0.77, 0.91, 1.1, 1.3, 1.55, 1.85, 3.0\}$$

they define which scale plane is consulted in the search phase. The search range for the two parameters (movement direction α and rotation ψ) spans the entire interval $[0, 2\pi)$ in 72 discrete steps for each parameter. We use a novel version of min-warping (suggested in [82]) with improved performance where the search phase is performed twice, the second time with exchanged images. The stack of scale planes is re-arranged for the second search (a re-computation of the column distances is not required), and the results of the two search phases (match qualities for each parameter combination) are superimposed.

The weight factor w of the tunable distance measures is chosen from the set $\{0.0, 0.01, 0.02, 0.03, 0.04, 0.06, 0.08, 0.1, 0.2, 0.333, 0.4, 0.5, 0.7, 1.0\}$ for the measures TSSD, TZSSD, TNCC, TZNCC, TENCC, and TEZNCC. For the measures TSC and TASC, we use the following weights $\{0.0, 0.03, 0.07, 0.1, 0.13, 0.17, 0.2, 0.23, 0.27, 0.3, 0.5, 0.7, 1.0\}$. The weight sets were chosen such that the optimal values of w are covered with sufficient resolution in the diagrams of the median and mean angular error.

All images were low-pass filtered using a Butterworth filter (3rd order, relative cut-off frequency 0.1); for details see [51].

In contrast to [20], we don't use the approximation of image magnification described there in Equation (5), but solve the magnification equation exactly. The vertical resolution of the image (pixels per angle) is required for the exact solution.

5. Results

5.1. Strong Illumination Changes

Our objective was to develop distance measures which are tolerant against strong changes in illumination. We therefore analyze results obtained from cross-database experiments in more detail. Figure 11 (left) shows the median and mean angular error for TSSD and TZSSD. The plot confirms our expectation that there is an optimal, intermediate weight w for which the best performance is achieved (TSSD: $w = 0.04$). The optimal performance is much better than both the illumination-invariant portion alone (curve ends marked by large black squares) (*) and the illumination-sensitive portion alone (other curve ends). It is also better than the special case of the Euclidean distance at $w = \frac{1}{3}$ (*). Both for $w = 0$ in both measures (large black squares) and for the optimal w in both measures (data points at the bottom left corner), TSSD performs better than TZSSD (*), thus we can conclude that double invariance (against scale *and* shift) indeed impairs the performance.

Figure 11. Cross-database tests: Median and mean angular error (radians) for varying w. The larger black squares mark data points for $w = 0$. **Left**: TSSD *vs*. TZSSD. **Right**: TNCC *vs*. TZNCC (only bottom-left region is shown).

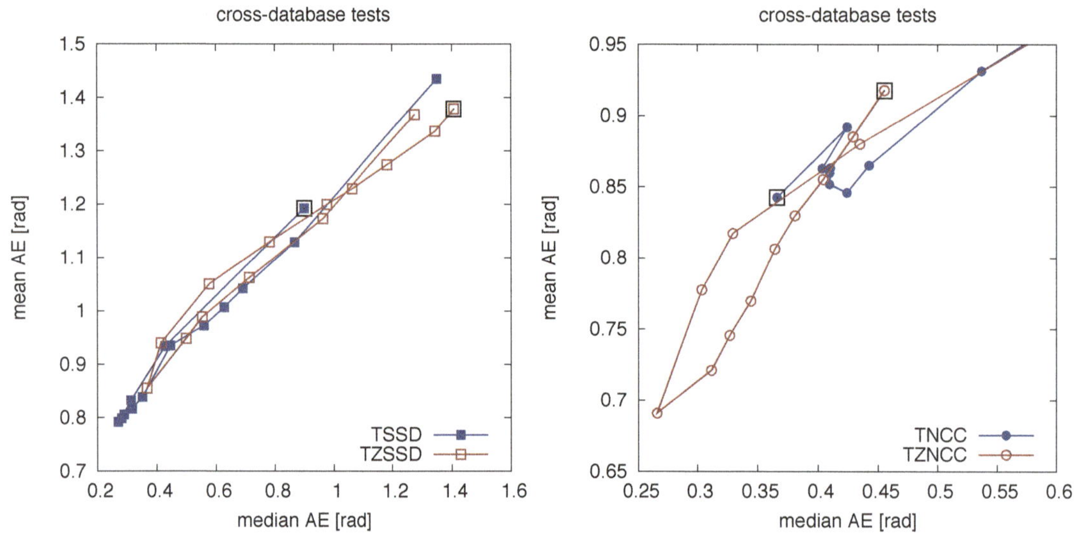

Figure 11 (right) shows a distinct optimum for TZNCC as well ($w = 0.2$); the performance is better than that of the illumination-invariant term alone (*). However, even though NCC outperforms ZNCC ($w = 0$) (*)—invariance to scale beats invariance to both scale and shift, and numerical problems with low-variance vectors may affect ZNCC—NCC seems to be incompatible with the illumination-sensitive term: Increasing w leads to worse angular errors. We are fairly sure that this is not a programming error since the code computing the illumination-sensitive term is shared by both measures. At the moment we have no explanation for this effect.

A comparison of the winners TSSD and TZNCC shows that at the best w, TZNCC mostly improves the mean angular error (* mean). This confirms our prediction that methods with strong scale invariance outperform methods with only weak scale invariance (see Section 2).

Figure 12 compares distances measures applied to edge-filtered columns. The *top left plot* visualizes TENCC. Surprisingly, the correct solution of scaling magnified edge-filtered columns ("+scale") performs worse than the version were the scaling is not applied ("-scale"), both for $w = 0$ and for the optimal w in both measures, respectively (*); see Section 3.5. We again see an optimal w, even though the gain compared to the results obtained for $w = 0$ (large black squares) is relatively small (TENCC+scale: * mean, TENCC-scale: *). Both for $w = 0$ and for the optimal w, the performance of TENCC-scale is much better than the optimum of TZNCC (*) (see Figure 11; note the different axis ranges), so edge filtering is obviously responsible for the good illumination tolerance. The *top right plot* compares TEZNCC with and without scaling of the derivatives. Again, the version without scaling performs better, both for $w = 0$ and for the optimal w (*). The changes over w are comparable to TENCC, and the optima are better than $w = 0$ (*). The *bottom left plot* compares TENCC and TEZNCC for the winning versions (without scaling of magnified edge-filtered columns). We see that TENCC outperforms the zero-mean version TEZNCC, both for $w = 0$ and for the optimal w (*)—again, double invariance doesn't pay off. For TENCC, the optimum is at $w = 0.08$. In the *bottom right plot* it is visible that the best method from the three other plots, TENCC (without scaling), is outperformed by both TASC and TSC (without scaling), even though the differences are small: TASC ($w = 0$) is better than TENCC

($w = 0$) (* mean), but not better than the optimal TENCC (n.s.), and TSC ($w = 0$) is better than TENCC for $w = 0$ (*) but not for the optimal w (n.s.). The original version (SC) and the approximated one (ASC) have a similar performance (n.s.). For both, the gain achieved by adding an illumination-sensitive term ($w > 0$) is so small (TSC: n.s., TASC: * median) that the additional computational effort is wasted. Also for TSC and TASC, the versions without scaling of magnified edge-filtered columns shown in the plot perform better than the versions with scaling, which are not shown (*).

Figure 12. Cross-database tests, see caption of Figure 11. All plots show only the bottom-left regions. *Top left:* TENCC with ("+scale") our without scaling ("-scale") of the magnified edge-filtered columns. *Top right:* TEZNCC with and without scaling. *Bottom left:* TENCC *vs.* TEZNCC (without scaling). *Bottom right:* TENCC *vs.* TSC *vs.* TASC (without scaling).

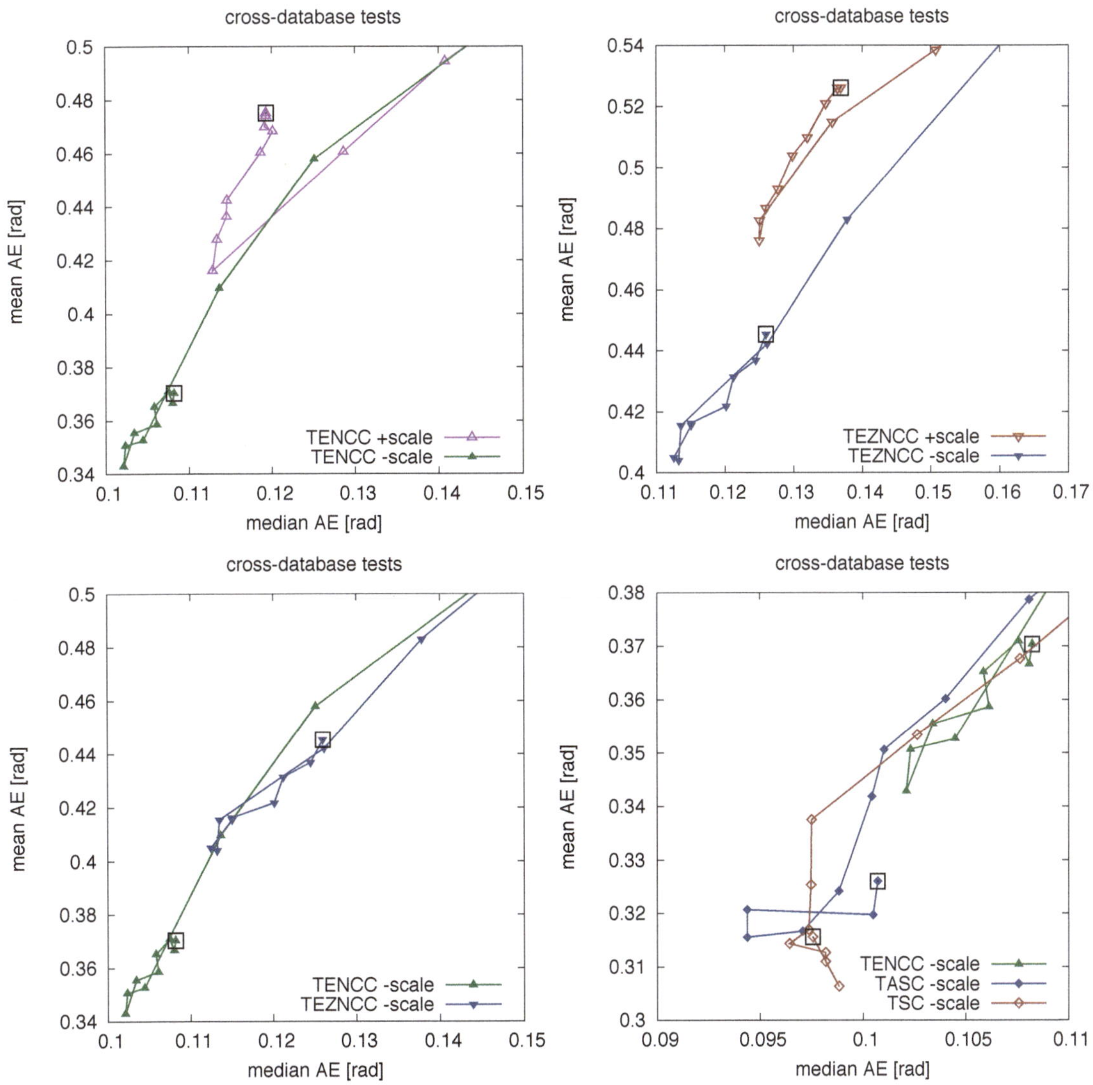

Figure 13 (left) compares selected measures from the previous figures. The best performance for TSSD was achieved for $w = 0.04$ (as expected with a stronger influence of the illumination-invariant portion than in the Euclidean distance, $w = \frac{1}{3}$), the best performance of TZNCC for $w = 0.2$ (this value is not interpretable quantitatively). At the optimal weight, TZNCC performs better than TSSD (* mean). However, even for the best w, TZNCC fares very poorly compared to the TENCC measure ($w = 0$ and optimal w) which is based on edge-filtered columns (*). Without an additional illumination-sensitive portion ($w = 0$), the difference between TENCC and TASC/TSC is noticeable, but for the optimal $w = 0.08$, TENCC comes close to the performance of our novel correlation measures (statistics see previous paragraph). Edge filtering obviously is a prerequisite for good illumination tolerance. For TASC and TSC we only plotted the version with the illumination-invariant term alone ($w = 0$) since the gain of adding an illumination-sensitive term was negligible (TSC: n.s., TASC: * median). This is actually an advantage of these methods since there is no need to compute the additional illumination-sensitive term.

Figure 13. Cross-database tests, see caption of Figure 11. *Left:* Comparison of selected measures (pooled data). *Right:* Comparison of the navigation performance for cross-database experiments on single database pairs. The first database provides the snapshots, the second the current views. Small symbols: TASC, $w = 0$. Large symbols: TSSD, $w = 0.04$.

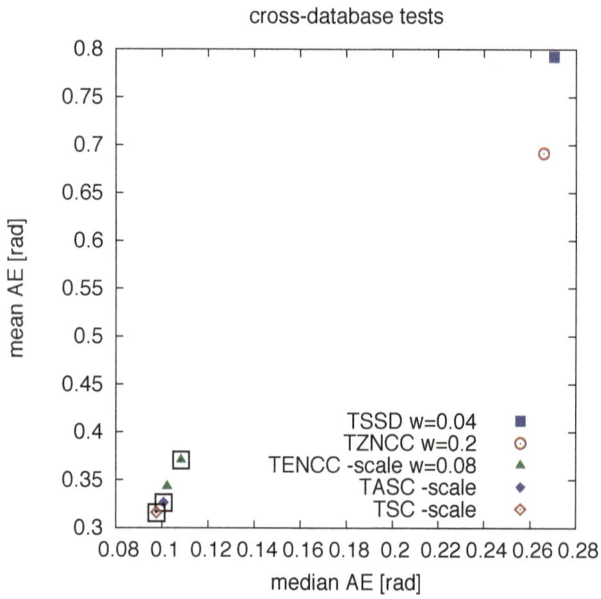

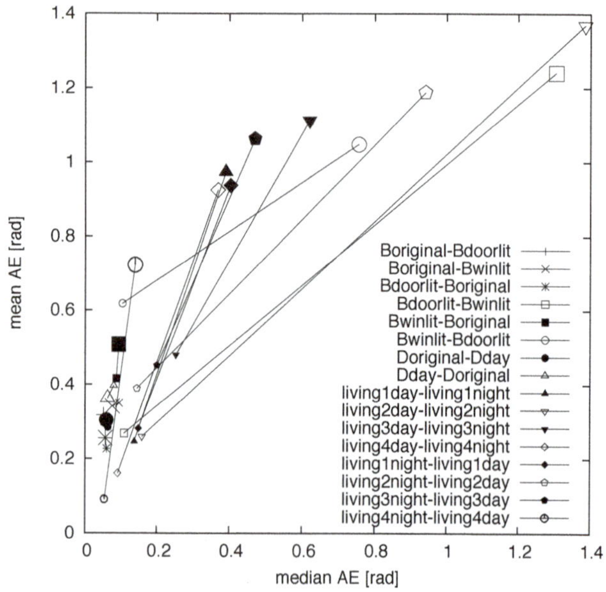

Figure 13 (right) compares the performance of the methods TSSD ($w = 0.04$) and TASC ($w = 0$)—for which the pooled performance is shown in Figure 13 (left)—for the individual database pairings used in the cross-database experiments. We see a large performance advantage of TASC for most of the database pairs. Only for the pairs Boriginal-Bwinlit and Dday-Doriginal, TASC performs slightly worse than TSSD. TSSD has difficulties with all living databases and with the two pairings of Bdoorlit and Bwinlit, whereas the remaining database pairs show a good performance even for TSSD. We can summarize that the performance differences obtained for the pooled data are also visible for the majority and not only for a small number of database pairs. However, since cross-databases were only available for two types of environments (B*/D*: lab room, living*: living room), we cannot make any claims on applicability to a broader range of environments.

For three of the measures in Figure 13, we show examples of home-vector fields for the database pair living1day/living1night in Figure 14. The differences in performance are clearly visible.

Figure 14. Home-vector fields for cross-database tests with the database pair living1day/living1night. The snapshot was taken at the position marked by the square. Gray vectors indicate perfect home vectors computed from the ground truth, black vectors are computed by min-warping using the different column distance measures.

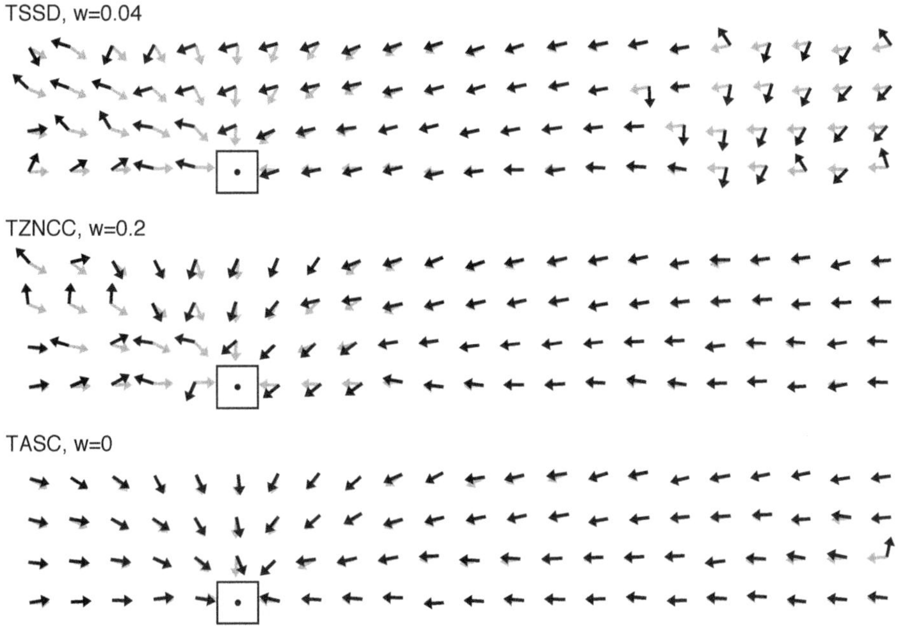

5.2. Moderate Illumination Changes

Measures insensitive to strong illumination changes should of course also perform well if the illumination changes are small. Figure 15 compares the performance of selected methods for tests where snapshots and current views were taken from the same database. For all methods, we see pronounced optima for intermediate values of w. As could be expected, the best performance is better than for the strong illumination changes. TSSD achieves the best performance for $w = 0.08 \ldots 0.2$, again below $w = \frac{1}{3}$ (Euclidean distance). The magnified bottom-left region (right plot) shows that the optima of all methods are close to each other. Even for $w = 0$, TASC and TSC achieve reasonably good performance such that adding the illumination-sensitive term is probably not required. The optima of TASC and TSC are better than all other methods, but they are achieved for high values of $w = 0.3 \ldots 0.5$ which are not suitable for strong illumination changes. Values around $w = 0.15$ appear to be a good compromise working reasonably well both under strong and moderate illumination changes, but we recommend to use $w = 0$—the performance gain obtained by including the illumination-sensitive term ($w > 0$) is so small (for both strong and moderate illumination changes) that the computational effort for this term is not justified.

Figure 15. Same-database tests. Only the best-performing versions (w/o zero mean, w/o scaling) are shown. *Left:* Entire plot. *Right:* Bottom-left region (marked in the left plot).

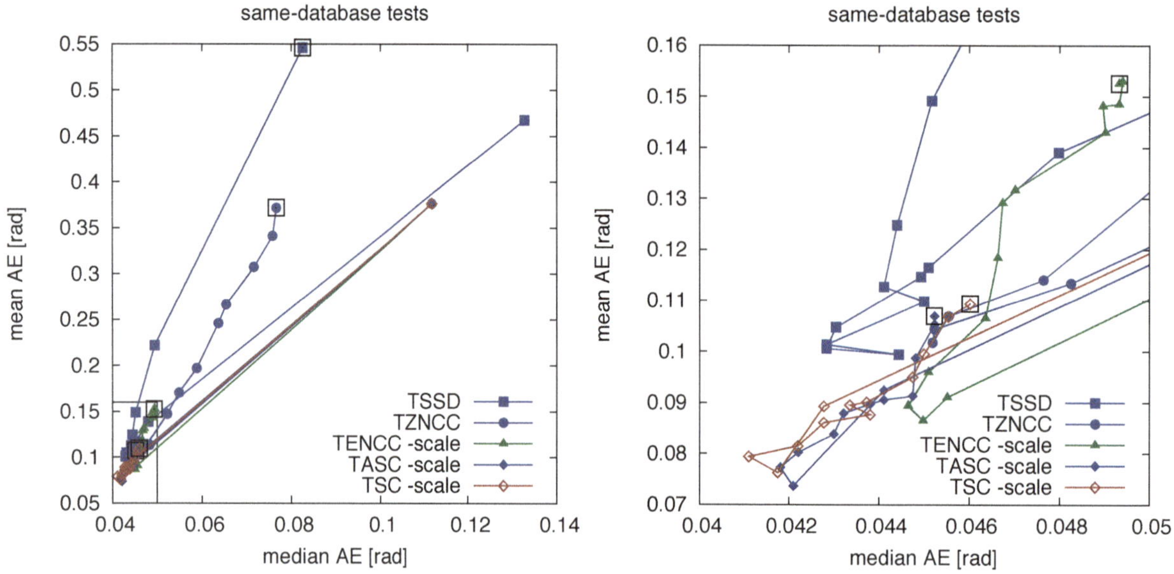

5.3. Computational Effort

Table 2 lists the computation times for the different measures (Intel Core i7-3610QM CPU, 2.30 GHz). The fastest measure is TSSD, but TASC (which exhibits a much better navigation performance in the cross-database tests) is only slightly slower (at least for $w = 0$ where the illumination-sensitive term is not computed).

Table 2. Average computation times for the different methods (Intel Core i7-3610QM CPU, 2.30GHz). Time 1 is the time for the computation of the illumination-invariant term alone ($w = 0$), time 2 is the time for the computation of both terms ($w \neq 0$). In TSSD, both terms are always computed. The second phase of min-warping (search) always takes about 24 ms (full search range, double search; 96 steps of both parameters, used for time measurement only).

Method	**Time 1 (ms)**	**Time 2 (ms)**
TSSD	3.5	3.5
TZSSD	5.4	7.4
TNCC	6.5	8.4
TZNCC	8.0	9.9
TENCC	6.1	8.1
TEZNCC	6.1	8.1
TASC, Equation (11)	4.5	6.5
TASC, Equation (12)	4.3	6.3
TSC	30.9	32.9

TSC is almost an order of magnitude slower than TASC (and that at comparable navigation performance) which is a result of the square root and division operations required for each segment.

All other methods are somewhat slower than TASC (up to a factor of about 2); however, in these cases we did not invest as much effort into speed tuning as in TASC, so there may be more efficient implementations. Computing the illumination-sensitive term (time 2 minus time 1) costs about 2 ms. It may be possible to improve the performance by reducing the number of memory accesses; this could be achieved if the illumination-sensitive term is not computed separately (as done now) but together with the illumination-invariant term. For TASC, we report the computation times for both alternative expressions for $\hat{d}_i$.

6. Discussion

The results show that tunable distance measures which combine an illumination-invariant and an illumination-sensitive term perform best for a certain mixture of the two measures. The illumination-invariant term alone is obviously not sufficient for min-warping which uses all image columns as features without any selection of key points. Whether a tunable distance measure also has an advantage for methods with feature selection has to be explored. For TSSD we can interpret the optimal mixture weight ($w = 0.04$, obtained for strong illumination changes) since we know that $w = \frac{1}{3}$ corresponds to the Euclidean distance; compared to the Euclidean distance, the illumination-invariant portion is weighted more strongly. For the other tunable measures we can only say that an intermediate value of w is optimal. TZNCC shows the same behavior (optimum at $w = 0.2$), but TNCC seems to be incompatible with the chosen illumination-sensitive term. We didn't further explore whether another form of illumination-sensitive term would be better suitable since at $w = 0$ the measure performs much worse than measures working on edge-filtered images. At least for the illumination-invariant portion alone, we always see that introducing additional tolerance against offsets by subtracting the mean leads to impaired performance compared to tolerance against scaling alone. We assume that double invariance introduces false positive matches.

The better performance of TZNCC over TSSD can be explained by the strong *vs.* weak type of scale invariance. Even though we didn't test other methods with only weak scale invariance such as the co-linearity criterion suggested by [76] or the image Euclidean distance by [77], we think that they will also be beaten by measures with strong scale invariance.

Figure 16. Effect of edge filtering (**bottom**) on a 1D intensity image (**top**) where one region became brighter and another region darker (solid to dashed).

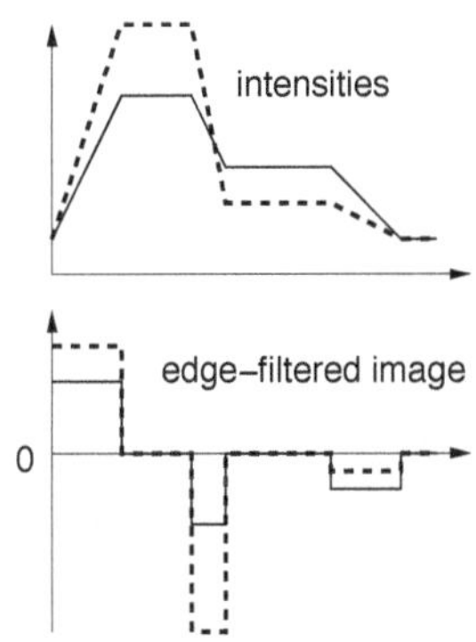

Edge filtering (here only tested for the correlation-type measures TENCC and TEZNCC) seems to be the best way to achieve tolerance against illumination changes. While NCC on intensities can only remove intensity scaling affecting the entire image region (in our case an image column), edge filtering works locally within the image region: In sub-regions with constant intensity, differences in intensity are completely removed from the image region, even if one part of the region gets brighter and another one darker (and thus scale invariance with respect to the entire region doesn't help), and intensity changes only affect the slope of the edges (see Figure 16). Still, the slope changes can in principle be strong, and we currently can't present a mathematical analysis that would show why slope changes should be better tolerated by normalized cross-correlation than changes in the intensity image. Whether the fact that neighborhood relations are considered through edge filtering is an advantage is not clear as well. Note that the good illumination tolerance achieved by edge filtering may be paid for by reduced tolerance against spatial translations in the images [26], e.g., in our example caused by tilt of the robot platform.

If we compare TENCC and TEZNCC, we see that the performance drops if the mean is subtracted, presumably since this step destroys the relation between original and edge-filtered image (very different images can have the same edge-filtered, zero-mean versions, see Figure 6). The optimal weight of TENCC is $w = 0.08$.

Edge filtering is also the first step of our novel "sequential" correlation measures which treat the two intensity vectors as a *sequence* of 2D curve points rather than as a mere collection of points without any order. Here the gain achieved by a tunable form is so small that the additional computational effort is not justified. Sequential correlation alone ($w = 0$) performs best in cross-database experiments (strong illumination changes) and fairly well in same-database experiments (moderate illumination changes). We could show that the performance gain of sequential correlation compared to the best TSSD method which we obtained for the *pooled* data is also visible for the majority of the *individual* databases.

We can only guess why the performance of sequential correlation (SC, ASC) is slightly better than the standard normalized cross-correlation on edge-filtered images (TENCC). Normalized cross-correlation is already (strongly) invariant against scaling, and edge filtering introduces additional invariance against shifts. Double invariance impaired the performance for all distance measures applied to the original images (*i.e.*, without edge filtering), at least for $w = 0$, presumably due to false positive matches introduced by too much tolerance. In contrast, sequential correlation is only tolerant against shifts—as a result of edge filtering—but not against scaling, which may reduce the number of false positive matches.

At the moment we cannot fully explain the following side observation. There are two ways to perform edge filtering in min-warping: Either an image is magnified (in vertical direction around the horizon) first and then edge-filtered, or it is first edge-filtered and then magnified. Since together with nearest-neighbor interpolation we get erroneous derivatives in the first form, we use the second form in our implementation. To achieve the same effect as in the first form, the edge-filtered image should be scaled down if it is magnified. However, leaving out this scaling operation actually improved the performance in all tested methods. We assume that this version of image magnification fits better with the true magnification resulting from decreasing the spatial distance to the feature. Sharp edges may actually have a similar slope regardless of the spatial distance to the feature, but this explanation needs experimental verification.

At least for a fast, integer-based implementation using SIMD instructions (SSE), the original sequential correlation measure requires considerably more computation time than all other distance measures studied here. We suggested an approximated form which can be computed an order of magnitude faster and which does not differ significantly in the navigation performance. We noticed that, compared to normalized cross-correlation, it was easier to chose the float-to-int conversion factors for approximated sequential correlation in accordance with the upper limits imposed by the integer implementation since the method mostly contains additions, subtractions, and the absolute value function but no multiplication.

Invariance against changes in illumination can be established in multiple ways; Table 3 provides a summary. Weak invariance against *scaling* can be achieved by finding the optimal scaling such that the distance (e.g., SSD) between two vectors becomes minimal (parametric SSD measures); strong scale invariance is obtained by normalizing the two vectors to the same length before their distance is computed (NCC measures). Invariance against *shifts* can be obtained by either subtracting the mean from the two vectors before they are compared (zero-mean versions of several measures), or by computing the distance between edge-filtered vectors (NCC on edge-filtered images).

Table 3. Types of invariance of illumination-invariant terms used in the tunable distance measures presented in this work, and navigation performance on databases with strong changes in illumination. Abbreviations: finding the optimal scaling ("scaling"; leads to weak scale invariance), normalizing the vector ("norm"; leads to strong scale invariance), subtracting the mean ("zero-mean"), applying an edge filter ("edge"). The comment "mismatches!" refers to Figure 6.

Measure	Scale Invariance	Shift Invariance	Navigation Performance
PSSD	scaling (weak)		
PZSSD	scaling (weak)	zero-mean	worse than PSSD (*)
NCC+	norm (strong)		better than PSSD (*)
ZNCC+	norm (strong)	zero-mean	worse than NCC+ (*)
ENCC+	norm (strong)	edge	much better than NCC+ (*)
EZNCC+	norm (strong)	edge, zero-mean	worse than ENCC+ (*); mismatches!
SC+		edge	better than ENCC+ (*)
ASC+		edge	close to SC+ (n.s.)

We can provide the following explanations for the results:

Normalization is better than scaling of vectors. Normalization (NCC+) turned out to be the better way to achieve scale invariance than finding the optimal scaling (PSSD), even though normalization amplifies noise for short vectors. This can be explained by the fact that NCC+ exhibits strong but PSSD only weak scale invariance.

Invariance against both scaling and shift may lead to mismatches. When only the invariant term of the tunable measure was used, combining invariance against scale and shift always resulted in worse navigation performance than using scale invariance alone. We assume that this is due to increased false positive matches introduced by the double invariance.

Subtracting the mean in edge-filtered images may lead to mismatches. Introducing two-fold invariance against shifts by first applying an edge filter and then additionally subtracting the mean may lead to mismatches as visualized in Figure 6.

In NCC+, edge filtering is the best way to achieve shift invariance. Even though NCC+ on edge-filtered images is both scale- and shift-invariant, it performs much better than NCC+ on the original intensity images. Edge filtering seems to introduce some kind of local invariance (see Figure 16) against illumination changes which improves the performance. This is in line with results presented in [83] where it was derived, by using a probabilistic method, that edge information (specifically the gradient direction) is the best way to accomplish illumination invariance (for object recognition).

Sequential correlation performs best since it only relies on edge filtering. Our novel correlation measures (SC+, ASC+) *exclusively* rely on shift invariance (and there only on edge filtering); this may explain their small performance advantage over NCC+ on edge-filtered images which is invariant against scale *and* shift. This stands in contrast to feature-based methods such as SIFT or SURF where apparently tolerance against both shift and scaling is beneficial. As NCC+, also SC+ and ASC+ are correlation-type measures with a normalization which may contribute to their good performance. An additional illumination-sensitive term is not required for SC+ and ASC+.

We did not test *rank-based distance measures* such as Spearman's rank correlation coefficient [84] which tolerates non-linear illumination changes. Rank-based correlation is not necessarily much more time-consuming since it is applied to each image individually before the two images are interrelated, the latter being the most time-consuming step. However, as we have seen, even simultaneous invariance against both additive and multiplicative changes in intensity resulted in a loss of home vector precision. We therefore expect that the additional tolerance introduced by a rank-based method would further impair the performance.

We also did not test whether a *binary representation* of each column vector and a distance computation by the Hamming distance as in the BRIEF, ORB, or FREAK descriptors would improve the performance. This would require an optimization process for the selection of pixels pairs within a column from which the binarized intensity difference are computed [9]. It is difficult to predict whether such an approach would improve the performance. Binarizing edge-filtered images (*i.e.*, neighboring pixels form pairs) may impair the performance: Small spatial shifts of the columns introduced by camera tilt would lead to small changes in smooth first derivatives but to larger changes in binarized first derivatives. Binarizing differences between more distant pixels from the column may also not be beneficial: If one part of the column is increased in intensity, local edge filters still recognize the structure in each part whereas differences between more distant pixels may have changed in sign.

All tests described in this paper were performed in two indoor environments, a lab room and a living room, for which cross-illumination databases were available. To the best of our knowledge, there is presently no outdoor image database available where images from the same locations were collected under different conditions of illumination and with sufficiently precise ground truth. We will therefore collect additional cross-illumination databases in indoor and outdoor environments and repeat our tests for these data. For outdoor environments, we have suggested methods where illumination tolerance is achieved by computing contrast measures between UV and visual channels [85,86]; it may improve the performance if the distance measures suggested in this paper are applied to data preprocessed by such contrast methods.

The question whether the results are transferable to other holistic methods, e.g., methods from the DID framework, cannot be answered without systematically investigating these methods. We performed preliminary experiments where we compared the performance of a holistic visual compass method [25] for the TSSD measure ($w = 0.04$) and for the ASC measure. Due to similarities between the first phase of min-warping and the visual compass (see Section 3.3 in [20]), we computed a single scale plane using our min-warping implementation and averaged column distances with the same difference angle $\Theta' - \Theta$; the minimal average provides the rotation estimate. Each randomly rotated snapshot was compared to each randomly rotated current view from the same database or from a cross database, using the same databases as for the min-warping experiments. The mean and median angular error between true and estimated azimuthal rotation were computed.

Surprisingly, here the TSSD measure performed *better* than ASC, both for cross-database and for same-database experiments (see Table 4, first and second row). We can provide the following explanation: Min-warping *distorts and rotates* one of the input images according to the simulated movement. For the best match, a large portion of pixels belonging to the same point in space will share the same image position with the other image. In contrast, the visual compass only *rotates* one of the images. Therefore the image distortions caused by the spatial distance between the vantage points are not compensated for, and a larger portion of pixels belonging to the same landmark will not appear at the same image position. Since ASC uses edge-filtered images where low image frequencies have been removed, small distances between pixel locations are sufficient to miss a match between edges. TSSD works on intensities and will therefore tolerate larger distortions. This explanation is in accordance with the observation that rotational image difference functions have very sharp minima if low image frequencies are removed [26]. We expect a similar effect for the DID method where image distortions are not compensated for as well. As a side observation we see from the bottom row in Table 4 that the rotation estimate obtained from min-warping (which considers image distortions) is much better than the rotation estimate obtained from the visual compass (which doesn't).

Table 4. Mean and median angular error [rad] of the compass estimate (ψ) for cross-database and same-database experiments. The first and second row show the results of the visual compass. For comparison, the bottom row presents the error of min-warping's compass estimate.

	Cross-Database		**Same-Database**	
Method	**Mean AE**	**Median AE**	**Mean AE**	**Median AE**
compass, TSSD ($w = 0.04$)	0.77	0.41	0.49	0.24
compass, ASC	0.88	0.48	0.74	0.33
min-warping, ASC	0.27	0.04	0.12	0.02

7. Application: Cleaning Robot Control

In this section, our intention is not to provide a further test of illumination tolerance, but rather to demonstrate that min-warping with the novel approximated sequential correlation measure can successfully be applied to the control of cleaning robots in unmodified indoor environments. With this we hope to dispel doubts that proponents of feature-based methods may harbor on the general applicability of the holistic min-warping method. This is also one of the first robot experiments we performed where we used approximated sequential correlation in the first phase of min-warping instead of TSSD which was used in [55].

The navigation method used by the cleaning robot is described in detail in [55]. The robot attempts to cover the entire reachable space by adjoining meander parts. On each meander lane, it collects panoramic snapshot images every 10 cm. It then computes home vectors and compass estimates with respect to snapshot images collected on the previous lane of the same meander part or on a previous meander part. Based on these measurements, a particle filter updates the robot's position estimate, and a controller corrects the movement direction of the robot such that a lane distance of 30 cm is maintained. Each particle cloud associated with a panoramic view can later be used as landmark position estimate. Views and clouds form nodes of a graph-based, topological-metrical map of the environment which is constructed during the cleaning run.

Figure 17. Internal representation obtained in a cleaning run covering three rooms of an apartment (schematic layout is shown in the inset). Target lanes are shown as red lines. Position estimates (averages of particle clouds) are visualized by light blue circles. Parts of the trajectory where the robot traversed existing meander parts to reach the starting point of a new part are omitted for clarity. Black dots are range sensor measurements. The final position of the robot (close to the starting point of the very first lane) is indicated.

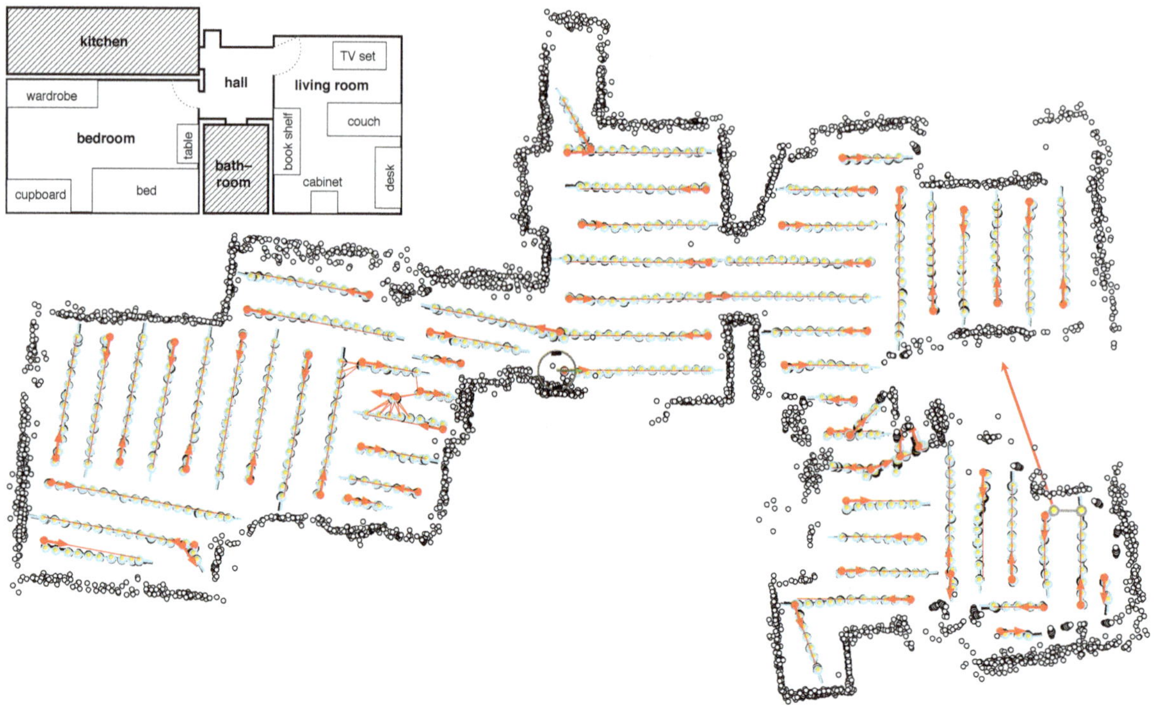

Figure 17 shows the internal representation of the robot's trajectory for a cleaning run that covers multiple rooms of an apartment. Red lines are target lanes, light blue circles visualize position estimates at all map nodes (averages of particle clouds). It can be seen that a wide portion of the reachable space was covered. (Note that the gap in the range sensor measurements (black dots) close to the robot is actually closed, but the range sensor measurements which originally filled the gap were discarded and could not be replaced by fresh sensor measurements since the robot returned to the start location immediately afterwards and finished the run.) Black dots are range sensor measurements. These are only used to identify uncleaned free space, not for updating the position estimates. Range sensor measurements are attached to the nodes of the topological-metrical map and therefore depend on the position estimates of the corresponding map nodes. When looking at the range sensor measurements it is apparent that even though the map is not fully metrically consistent over several parts, there is not much positional drift within each part—otherwise walls would appear multiple times in the plot. The robot successfully returned to the starting point after finishing the last part (the final position is indicated). Light blue lines depict the snapshot pairings for which min-warping was applied. For each visual correction of the position estimate, typically three home vectors are computed from the current snapshot and three snapshots on the previous lane. The first one, shown in red, points to a snapshot approximately 45 deg from the forward direction, the second one, shown in green, points to the closest snapshot on the previous lane, and the third one, shown in blue, points to the snapshot approximately 45 deg from the backward direction. The quality of the estimates provided by min-warping is apparent from the observation that the home vectors produced by min-warping with ASC almost never switch order (red, green, blue) with respect to their direction and usually correspond well with the estimated direction to the node on the previous lane (light blue). Note, however, that the plots show position estimates, not true positions, and that these position estimates are updated by the same home vectors.

Figure 18 shows all home vectors computed during the cleaning run (compass estimates are omitted for clarity). Figure 19 magnifies the part indicated by the rectangle in Figure 18.

Figure 18. Home vectors (red, green, blue) computed for the update of the position estimate. Light blue lines indicate snapshot pairings from which home vectors (and compass estimates) are computed. The black rectangle indicates the part magnified in Figure 19.

Figure 19. Magnification of the part enclosed by the black rectangle in Figure 18. Each robot position estimate is shown as a black dot with a black orientation line.

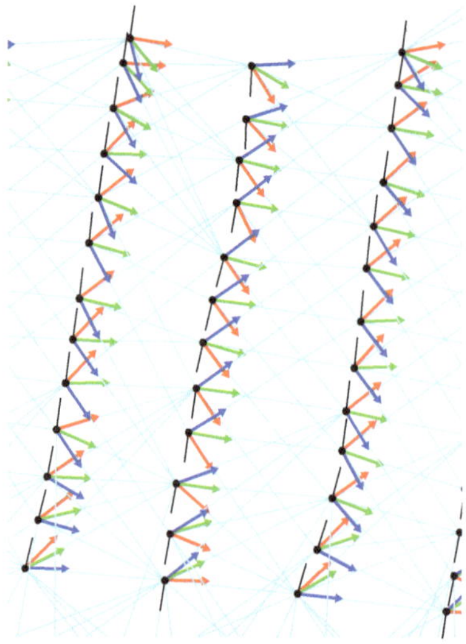

A quantitative evaluation of the home vectors is not possible without ground truth (which is not available for this run); our intention is just to give a visual impression. We know, however, that metrical consistency within each meander part is usually good enough to judge homing performance from such a visual inspection [55].

Figure 20. Images captured on the center lane and the right lane shown in Figure 19. The images are ordered spatially according to where their corresponding map nodes appear in Figure 19.

Figure 20 gives a visual impression of the panoramic images (size 320×48) attached to the map nodes of the center lane and the right lane shown in Figure 19. A red line through the center of each image was added which corresponds to the backward viewing direction of the panoramic camera. Visual features underneath the line shift only by small amounts which indicates that each lane was approximately straight; this corresponds with the position estimates in Figure 19.

8. Conclusions

The distance measures TSSD, TZSSD, and TZNCC, which work on original (not on edge-filtered) images, profit from a tunable combination of an illumination-invariant and an illumination-sensitive term. In TSSD, this combination is elegant since it contains the Euclidean distance as special case. However, applying distance measures to edge-filtered images leads to a much larger improvement in homing performance, and for these methods, the gain by the weighed superposition with an illumination-sensitive term is relatively small. For TSC and TASC methods introduced in this paper—which perform best among all tested methods—an additional illumination-sensitive term is not required since the resulting performance gain is negligible.

Applying distance methods to edge-filtered images generally leads to good homing performance of the min-warping method. Correlation-based methods exhibit the strong type of scale invariance (TENCC) or are shift-invariant (SC, ASC) and should be preferred. Among these, we can recommend our novel approximated sequential correlation (ASC) measure. The small advantage of ASC over normalized cross-correlation on edge-filtered images can possibly be explained by the fact that ASC is only tolerant against shifts whereas NCC on edge-filtered images is tolerant against both scaling and shifts which may introduce too much tolerance and therefore false positive matches. How ASC performs for holistic methods other than min-warping or for feature-based methods which select high-variance image regions (such as corners) for matching has yet to be explored.

Acknowledgements

We are grateful to Andrew Vardy for his helpful comments on an earlier version of this manuscript.

Author Contributions

Ralf Möller developed and implemented the distances measures and performed the experiments evaluating the home vector quality. Michael Horst and David Fleer integrated min-warping into their framework of cleaning robot navigation and performed the cleaning robot experiments. The paper was written by Ralf Möller.

Conflicts of Interest

The authors declare no conflicts of interest.

References

1. Mikolajczyk, K.; Schmid, C. A performance evaluation of local descriptors. *IEEE Trans. Pattern Anal. Mach. Intell.* **2005**, *27*, 1615–1630.

2. Mikolajczyk, K.; Tuytelaars, T.; Schmid, C.; Zisserman, A.; Matas, J.; Schaffelitzky, F.; Kadir, T.; van Gool, L. A comparison of affine region detectors. *Int. J. Comput. Vis.* **2005**, *65*, 43–72.
3. Tuytelaars, T.; Mikolajczyk, K. Local invariant feature detectors: A survey. *Found. Trends. Comp. Graphics Vis.* **2007**, *3*, 177–280.
4. Li, J.; Allinson, N.M. A comprehensive review of current local features for computer vision. *Neurocomputing* **2008**, *71*, 1771–1787.
5. Gauglitz, S.; Höllerer, T.; Turk, M. Evaluation of interest point detectors and feature descriptors for visual tracking. *Int. J. Comput. Vis.* **2011**, *94*, 335–360.
6. Rosten, E.; Porter, R.; Drummond, T. Faster and better: A machine learning approach to corner detection. *IEEE Trans. Pattern Anal. Mach. Intell.* **2010**, *32*, 105–119.
7. Lowe, D.G. Distinctive image features from scale-invariant keypoints. *Int. J. Comput. Vis.* **2004**, *60*, 91–110.
8. Bay, H.; Ess, A.; Tuytelaars, T.; van Goal, L. Speeded-up robust features SURF. *Comp. Vis. Image Und.* **2008**, *110*, 346–359.
9. Calonder, M.; Lepetit, V.; Strecha, C.; Fua, P. BRIEF: Binary Robust Independent Elementary Features. In Proceedings of the European Conference Computer Vision (ECCV 10), Crete, Greece, 5–11 September 2010; pp. 778–792.
10. Calonder, M.; Lepetit, V.; Özuysal, M.; Trzcinski, T.; Strecha, C.; Fua, P. BRIEF: Computing a local binary descriptor very fast. *IEEE Trans. Pattern Anal. Mach. Intell.* **2012**, *34*, 1281–1298.
11. Rublee, E.; Rabaud, V.; Konolige, K.; Bradski, G. ORB: An Efficient Alternative to SIFT or SURF. In Proceedings of the IEEE International Conference Computer Vision (ICCV 11), Barcelona, Spain, 5–11 November 2011; pp. 2564–2571.
12. Alahi, A.; Ortiz, R.; Vandergheynst, P. FREAK: Fast Retina Keypoint. In Proceedings of the IEEE Conference Computer Vision and Pattern Recognition (CVPR 12), Providence, RI, USA, 16–21 June 2012; pp. 510–517.
13. Scaramuzza, D.; Fraundorfer, F. Visual odometry. Part I: The first 30 years and fundamentals. *IEEE Robot. Autom. Mag.* **2011**, *18*, 80–92.
14. Fraundorfer, F.; Scaramuzza, D. Visual odometry. Part II: Matching, robustness, optimization, and applications. *IEEE Robot. Autom. Mag.* **2012**, *19*, 78–90.
15. Lemaire, T.; Lacroix, S. SLAM with panoramic vision. *J. Field Rob.* **2007**, *24*, 91–111.
16. Gamallo, C.; Mucientes, M.; Regueiro, C. Visual FastSLAM through Omnivision. In Proceedings of the Towards Autonomous Robotic Systems (TAROS 09), Derry, UK, 12–14 September 2009; pp. 128–135.
17. Gil, A.; Martinez Mozos, O.; Ballesta, M.; Reinoso, O. A comparative evaluation of interest point detectors and local descriptors for visual SLAM. *Mach. Vis. Appl.* **2010**, *21*, 905–920.
18. Schmidt, A.; Kraft, M.; Fularz, M.; Domagała, Z. Comparative assessment of point feature detectors and descriptors in the context of robot navigation. *J. Autom. Mob. Rob. Intell. Syst.* **2013**, *7*, 11–20.
19. Valgren, C.; Lilienthal, A.J. SIFT, SURF & seasons: Appearance-based long-term localization in outdoor environments. *Rob. Auton. Syst.* **2010**, *58*, 149–156.

20. Möller, R.; Krzykawski, M.; Gerstmayr, L. Three 2D-warping schemes for visual robot navigation. *Auton. Robot.* **2010**, *29*, 253–291.
21. Möller, R. A model of ant navigation based on visual prediction. *J. Theor. Biol.* **2012**, *305*, 118–130.
22. Zeil, J. Visual homing: An insect perspective. *Curr. Opin. Neurobiol.* **2012**, *22*, 285–293.
23. Collett, M.; Chittka, L.; Collett, T.S. Spatial memory in insect navigation. *Curr. Biol.* **2013**, *23*, R789–R800.
24. Möller, R. Do insects use templates or parameters for landmark navigation? *J. Theor. Biol.* **2001**, *210*, 33–45.
25. Zeil, J.; Hoffmann, M.I.; Chahl, J.S. Catchment areas of panoramic images in outdoor scenes. *J. Opt. Soc. Am. A* **2003**, *20*, 450–469.
26. Stürzl, W.; Zeil, J. Depth, contrast and view-based homing in outdoor scenes. *Biol. Cybern.* **2007**, *96*, 519–531.
27. Basten, K.; Mallot, H.A. Simulated visual homing in desert ant natural environments: Efficiency of skyline cues. *Biol. Cybern.* **2010**, *102*, 413–425.
28. Graham, P.; Philippides, A.; Baddeley, B. Animal cognition: Multi-modal interactions in ant learning. *Curr. Biol.* **2010**, *20*, R639–R640.
29. Baddeley, B.; Graham, P.; Phillipides, A.; Husbands, P. Holistic visual encoding of ant-like routes: Navigation without waypoints. *Adapt. Behav.* **2011**, *19*, 3–15.
30. Baddeley, B.; Graham, P.; Husbands, P.; Phillipides, A. A model of ant route navigation driven by scene familiarity. *PLoS Comput. Biol.* **2012**, doi:10.1371/journal.pcbi.1002336.
31. Lambrinos, D. Navigation in Biorobotic Agents. Ph.D. Thesis, Department of Computer Science, University of Zurich, Zurich, Switzerland, 1999.
32. Lambrinos, D.; Möller, R.; Labhart, T.; Pfeifer, R.; Wehner, R. A mobile robot employing insect strategies for navigation. *Rob. Auton. Syst. Spec. Issue: Biomim. Robot.* **2000**, *30*, 39–64.
33. Möller, R. Insect visual homing strategies in a robot with analog processing. *Biol. Cybern.* **2000**, *83*, 231–243.
34. Mangan, M.; Webb, B. Modelling place memory in crickets. *Biol. Cybern.* **2009**, *101*, 307–323.
35. Gerstmayr-Hillen, L.; Schlüter, O.; Krzykawski, M.; Möller, R. Parsimonious Loop-Closure Detection Based on Global Image-Descriptors of Panoramic Images. In Proceedings of the IEEE Xplore 15th International Conference Advanced Robotics (ICAR), Sarajevo, Bosnia, 20–23 June 2011; pp. 576–581.
36. Hillen, L. From Local Visual Homing Towards Navigation of Autonomous Cleaning Robots. Ph.D. Thesis, Bielefeld University, Bielefeld, Germany, 2013.
37. Stürzl, W.; Cheung, A.; Cheng, K.; Zeil, J. The information content of panoramic images I: The rotational errors and the similarity of views in rectangular experimental arenas. *J. Exp. Psychol. Anim. B* **2008**, *34*, 1–14.
38. Cheung, A.; Stürzl, W.; Zeil, J.; Cheng, K. The information content of panoramic images II: View-based navigation in nonrectangular experimental arenas. *J. Exp. Psychol. Anim. B.* **2008**, *34*, 15–30.

39. Arena, P.; de Fiore, S.; Fortuna, L.; Nicolosi, L.; Patené, L.; Vagliasindi, G. Visual homing: Eperimental Results on an Autonomous Robot. In Proceedings of the IEEE Xplore 18th European Conference on Circuit Theory and Design, Sevilla, Spain, 26–30 August 2007; pp. 304–307.
40. Möller, R.; Vardy, A. Local visual homing by matched-filter descent in image distances. *Biol. Cybern.* **2006**, *95*, 413–430.
41. Möller, R.; Vardy, A.; Kreft, S.; Ruwisch, S. Visual homing in environments with anisotropic landmark distribution. *Auton. Robot.* **2007**, *23*, 231–245.
42. Labrosse, F. Short and long-range visual navigation using warped panoramic images. *Rob. Auton. Syst.* **2007**, *55*, 675–684.
43. Pajdla, T.; Hlaváč, V. Zero Phase Representations of Panoramic Image for Image Based Localization. In Proceedings of the 8th International Conference Computer Analysis of Images and Patterns, Ljubljana, Slovenia, 2–5 September 1999; pp. 550–557.
44. Labrosse, F. The visual compass: Performance and limitations of an appearance-based method. *J. Field Rob.* **2006**, *23*, 913–941.
45. Stürzl, W.; Möller, R. An Insect-Inspired Active Vision Approach for Orientation Estimation with Panoramic Images. In *Bio-inspired Modeling of Cognitive Tasks*; Springer: Berlin, Germany, 2007; Volume 4527, pp. 61–70.
46. Saez Pons, J.; Hübner, W.; Dahmen, H.; Mallot, H.A. Vision-Based Robotic Homing in Dynamic Environments. In Proceedings of the 13th IASTED International Conference Robotics and Applications, Wuerzburg, Germany, 15–17 June 2007; pp. 293–298.
47. Zhang, A.M.; Kleeman, L. Robust appearance based visual route following for navigation in large-scale outdoor environments. *Int. J. Rob. Res.* **2009**, *28*, 331–356.
48. Franz, M.O.; Schölkopf, B.; Mallot, H.A.; Bülthoff, H.H. Where did I take that snapshot? Scene-based homing by image matching. *Biol. Cybern.* **1998**, *79*, 191–202.
49. Stürzl, W.; Mallot, H.A. Efficient visual homing based on Fourier transformed panoramic images. *Rob. Auton. Syst.* **2006**, *54*, 300–313.
50. Franz, M.O.; Stürzl, W.; Hübner, W.; Mallot, H.A. A Robot System for Biomimetic Navigation—From Snapshots to Metric Embeddings of View Graphs. In *Robotics and Cognitive Approaches to Spatial Mapping*; Jefferies, M.E., Yeap, W.K., Eds.; Springer: Berlin, Germany, 2008; Chapter 14, pp. 297–314.
51. Möller, R. Local visual homing by warping of two-dimensional images. *Rob. Auton. Syst.* **2009**, *57*, 87–101.
52. Franz, M.O.; Schölkopf, B.; Mallot, H.A.; Bülthoff, H.H. Learning view graphs for robot navigation. *Auton. Robot.* **1998**, *5*, 111–125.
53. Hübner, W.; Mallot, H.A. Metric embedding of view-graphs—a vision and odometry-based approach to cognitive mapping. *Auton. Robot.* **2007**, *23*, 183–196.
54. Gerstmayr-Hillen, L.; Röben, F.; Krzykawski, M.; Kreft, S.; Venjakob, D.; Möller, R. Dense topological maps and partial pose estimation for visual control of an autonomous cleaning robot. *Rob. Auton. Syst.* **2013**, *61*, 497–516.

55. Möller, R.; Krzykawski, M.; Gerstmayr-Hillen, L.; Horst, M.; Fleer, D.; de Jong, J. Cleaning robot navigation using panoramic views and particle clouds as landmarks. *Rob. Auton. Syst.* **2013**, *61*, 1415–1439.
56. Narendra, A.; Gourmaud, S.; Zeil, J. Mapping the navigation knowledge of individually foraging ants, *Myrmecia croslandi. Proc. R. Soc. B* **2013**, *280*, 20130683.
57. Scaramuzza, D.; Fraundorfer, F.; Pollefeys, M. Closing the loop in appearance-guided omnidirectional visual odometry by using vocabulary trees. *Rob. Auton. Syst.* **2010**, *58*, 820–827.
58. Milford, M. Vision-based place recognition: How low can you go? *Int. J. Rob. Res.* **2013**, *32*, 766–789.
59. Schatz, A. Visuelle Navigation mit "Scale Invariant Feature Transform". Diploma Thesis, Faculty of Technology, Bielefeld University, Bielefeld, Germany, 2006.
60. Aschwanden, P.; Guggenbühl, W. Experimental Results from A Comparative Study on Correlation-Type Registration Algorithms. In *Robust Computer Vision: Quality of Vision Algorithms*; Förstner, W., Ruwiedel, S., Eds.; Wichmann: Karlsruhe, Germany, 1992; pp. 268–288.
61. Chambon, S.; Crouzil, A. Dense Matching Using Correlation: New Measures That are Robust near Occlusions. In Proceedings of the British Machine Vision Conference, Norwich, UK, 8–11 September 2003; pp. 15.1–15.10.
62. Chambon, S.; Crouzil, A. Similarity measures for image matching despite occlusions in stereo vision. *Pattern Recognit.* **2011**, *44*, 2063–2075.
63. Giachetti, A. Matching techniques to compute image motion. *Image Vis. Comput.* **2000**, *18*, 247–260.
64. Pan, B. Recent progress in digital image correlation. *Exp. Mech.* **2011**, *51*, 1223–1235.
65. Pan, B.; Xie, H.; Wang, Z. Equivalence of digital image correlation criteria for pattern matching. *Appl. Opt.* **2010**, *49*, 5501–5509.
66. Hirschmüller, H. Evaluation of stereo matching costs on images with radiometric differences. *IEEE Trans. Pattern Anal. Mach. Intell.* **2009**, *31*, 1582–1599.
67. Tombari, F.; di Stefano, L.; Mattoccia, S.; Galanti, A. Performance Evaluation of Robust Matching Measures. VISAPP 2008: In Proceedings of the 3rd International Conference Computer Vision Theory and Applications, Madeira, Portugal, 22–28 January 2008; pp. 473–478.
68. Rubner, Y.; Tomasi, C.; Guibas, L.J. The earth mover's distance as a metric for image retrieval. *Int. J. Comput. Vis.* **2000**, *40*, 99–121.
69. Viola, P.; Wells, W.M. Alignment by maximization of mutual information. *Int. J. Comput. Vis.* **1997**, *24*, 137–154.
70. Corke, P.; Paul, R.; Churchill, W.; Newman, P. Dealing with Shadows: Capturing Intrinsic Scene Appearance for Image-Based Outdoor Localisation. In Proceedings of the IEEE/RSJ International Conference Intelligent Robots and Systems (IROS), Tokyo, Japan, 3–8 November 2013; pp. 2085–2092.
71. Sünderhauf, N.; Neubert, P.; Protzel, P. Predicting the Change—A Step towards Life-Long Operation in Everyday Environments. In Proceedings of the Robotics Challenges and Vision Workshop, Berlin, Germany, 24–28 June 2013.

72. Johns, E.; Yang, G.Z. Dynamic Scene Models for Incremental, Long-Term, Appearance-Based Navigation. In Proceedings of the IEEE International Conference Robotics and Automation (ICRA), Karlsruhe, Germany, 2013; pp. 2731–2736.
73. Vonikakis, V.; Chrysostomou, D.; Kouskouridas, R.; Gasteratos, A. A biologically inspired scale-space for illumination-invariant feature detection. *Meas. Sci. Technol.* **2013**, *24*, 074024.
74. Milford, M.; Vig, E.; Scheirer, W.; Cox, D. Towards Condition-Invariant, Top-Down Visual Place Recognition. In Proceedings of the Australasian Conference Robotics and Automation, Sydney, Australia, 2–4 December 2013.
75. Rodgers, J.L.; Nicewander, W.A. Thirteen ways to look at the correlation coefficient. *Am. Stat.* **1988**, *42*, 59–66.
76. Mester, R.; Aach, T.; Dümbgen, L. Illumination-Invariant Change Detection Using a Statistical Colinearity Criterion. In Proceedings of the Pattern Recognition, 23rd DAGM-Symposium, Munich, Germany, 12–14 September 2001; Volume 2191, pp. 170–177.
77. Wang, L.; Zhang, Y.; Feng, J. On the Euclidean distance of images. *IEEE Trans. Pattern Anal. Mach. Intell.* **2005**, *27*, 1334–1339.
78. Li, J.; Lu, B.L. An adaptive image Euclidean distance. *Pattern Recognit.* **2009**, *42*, 349–357.
79. Vardy, A. Biologically Plausible Methods for Robot Visual Homing. Ph.D. Thesis, Carleton University, Carleton, Canada, 2005.
80. Kreft, S. Reinigungstrajektorien Mobiler Roboter unter Visueller Steuerung. Diploma Thesis, Faculty of Technology, Bielefeld University, Bielefeld, Germany, 2007.
81. Efron, B.; Tibshirani, R.J. *An Introduction to the Bootstrap*; CRC: New York, NY, USA, 1998.
82. Gedicke, T. Warping for 3D Laser Scans. Bachelor's Thesis, University of Osnabrück, Osnabrück, Germany, 2012.
83. Chen, H.F.; Belhumeur, P.N.; Jacobs, D.W. In search of Illumination Invariants. In Proceedings of the IEEE Conference Computer Vision and Pattern Recognition CVPR'00, Graz, Austria, 7–13 May 2000; Volume 1, pp. 254–261.
84. Spearman, C. The proof and measurement of association between two things. *Am. J. Psychol.* **1904**, *15*, 72–101.
85. Möller, R. Insects could exploit UV-green contrast for landmark navigation. *J. Theor. Biol.* **2002**, *214*, 619–631.
86. Kollmeier, T.; Röben, F.; Schenck, W.; Möller, R. Spectral contrasts for landmark navigation. *J. Opt. Soc. Am. A* **2007**, *24*, 1–10.

4

Psychophysiological Methods to Evaluate User's Response in Human Robot Interaction: A Review and Feasibility Study

Lorenza Tiberio [1,2,*], Amedeo Cesta [1] and Marta Olivetti Belardinelli [2,3]

[1] CNR—Italian National Research Council, ISTC, Via San Martino Della Battaglia 44, Rome 00185, Italy; E-Mail: amedeo.cesta@istc.cnr.it

[2] ECONA (Interuniversity Centre for the Research on Cognitive Processing in Natural and Artificial Systems), Rome 00185, Italy; E-Mail: marta.olivetti@uniroma1.it

[3] Department of Psychology, Università degli Studi di Roma "La Sapienza", Rome 00185, Italy

* Author to whom correspondence should be addressed; E-Mail: lorenza.tiberio@istc.cnr.it

---

Abstract: Implementing psychophysiological measures is a worthwhile approach for understanding human reaction to robot presence in terms of individual emotional state. This paper reviews the suitability of using psychophysiological assessment in human-robot interaction (HRI) research. A review of most common psychophysiological parameters used in a controlled laboratory setting is provided and advantages and challenges of their utilization in HRI experiments are described. Exemplar studies focused on the implementation of psychophysiological measures for the evaluation of the emotional responses of the participants to the robots' presence are described. Based on the reviewed literature, the paper also describes the results of our own research experience to make the most of the emerged recommendations. We planned and performed a study aimed at implementing psychophysiological measurements for assessing the human response of two groups of older adults (Healthy *vs.* Mild Cognitive Impairment subjects) towards a telepresence robot. Finally, the paper provides a summary of lessons learned across the field in using psychophysiological measures in HRI studies.

Keywords: psychophysiological measurements; human-robot interaction; human response; socially assistive robots

---

1. Introduction

Humans generally interact with robots in the same way they might interact with other people, establishing social relationships and emotional ties with them [1–4]. One of the main characteristics relative to assistive social robotics is that a robot proactively engages users in a social manner creating an interaction with the person for the purpose of giving assistance and support in certain activities of daily living and care [4–6]. Gross *et al.* [7] report that the application of assistive social robots for domestic use has gone one step further towards the introduction of robotics technologies in real homes.

As assistive social robots are being designed to become a part of our life, an extremely important issue regards the development of methods to assess user's experience with a robot, while understanding how humans feel during their interaction with it [8]. A robot operating at home, in nursing care or in a rehabilitation unit is required to have determined social capabilities, like natural language communication skills or emotion expression and/or perceptual skills. Such features, in fact, maintain and facilitate the interaction with human users [9,10]. How people respond and react to the robot's social skills, and in general, to other features such as its functionality, role and appearance, is an important aspect of social interaction, which affects the acceptance and use of the robot. In addition, it is difficult to imagine social interactions occurring without emotional responses and therefore the analysis of the interactions between a human and a robot should be based also on a clear understanding of the human's reaction to the robot in terms of emotions. An emotional response against a stimulus, such as the daily interaction at home with a robot, implies an involvement of the central nervous system (CNS) through multiple cortical and subcortical regions and an activation (in terms of arousal) of the autonomic nervous system (ANS), which can be registered by physiological measurements [11–14]. In brief, the emotional states can be categorized along two main dimensions, one that explains the valence of the emotion (in a continuum of pleasantness-unpleasantness) and another that refers to the corresponding level of arousal/physiological activation. According to the "circumplex model of affect" [15], every emotion can be explained as a linear combination of valence (positive or negative) and intensity of activation (arousal). An additional dimension (dominance/control) has been frequently suggested for emotions with the same level of arousal and valence (e.g., fear and anger), but that are distinguishable in everyday life [16].

In this paper, the main focus is on the psychophysiological measurements of human response to a robot. Psychophysiological measures are considered one of the main methods of assessment used for human studies in human robot interaction (HRI) together with self-report, behavioral measures and task performance analysis, as Bethel *et al.* report in [8]. In this survey, the authors provide a detailed overview of the process of designing and planning user evaluation based on a psychophysiological approach applied to detecting and/or identifying users' emotions, modifying in real-time robot control and behavior, as well as evaluating the user's reactions to technology. The aim pursued in this review is to expand the knowledge on the assessment users' reactions to technology by means of psychophysiological measures as reported in [8]. In the scope, we have taken into account recent researches aimed at determining how humans react in terms of physiological response during an interaction with a robot in order to establish the real effectiveness of human-robot interaction. The comprehension of these aspects will encourage users to collaborate and share the environment with a robot designed, for instance, to provide them with social, health-care and rehabilitative services.

The paper is organized as follows: Section 2 briefly presents a description of psychophysiological terminology along with a survey of applications in the human-computer interaction research. Advantages and disadvantages of using psychophysiological measures are also reported. Section 3 focuses on the most common psychophysiological measures used in human-robot interaction studies in a controlled laboratory setting. Section 4 provides an overview of the most recent studies in the field of HRI that apply psychophysiological methods to evaluate the human's reaction to a robot in terms of affective-emotional response. Section 5 presents our feasibility study focused on the assessment of the human response to a telepresence robot in a group of healthy older adults and in a group of older patients suffering from mild cognitive impairment (MCI). Section 6 summarizes the results of both the reviewed literature and of our study, leading to the final discussion.

2. Psychophysiology for Human Studies

Citing from [17]: "Psychophysiology concerns the study of cognitive, emotional and behavioral phenomena as related to and revealed through physiological principles and events". Through the monitoring and interpreting of physiological signals during the interaction between a human user and an assistive robot, it is possible to obtain information about the user's inner experience in terms of affective and cognitive states [8,18,19].

Methods based on psychophysiological assessment have been previously used in human computer interaction (HCI) to evaluate users' cognitive workload [20], emotional responses [21], and cognitive-emotional states [22] during interaction with a technology. A significant increase of electromyography activity and hearth rate, and a significant decrease in blood volume pulse were found in research focused on individual's response to audio and video degradations in videoconferencing software [23]. Scheirer *et al.* [24] used electromyography activity and blood volume pulse to distinguish subjects' frustration during the utilization of a pre-programmed mouse delay. Particularly, the psychophysiological game evaluation measures the effects that games have on the players [25], how people respond to game features [26], users' experience and emotional engagement [27]. Mandryk *et al.* [28] analyze user experience with entertainment technology. Participants in their experiment report higher electromyography and electrodermal activity when playing against a friend than against a computer. In another study, Mandryk and Atkins [22] evaluate different methods for modeling emotion using physiological data through two fuzzy logic approaches. The first fuzzy logic model transformed electrodermal (GSR), electromyography (smiling and frowning) and cardiovascular activity (HR) into arousal and valence. The second model transformed arousal and valence into five emotions. Their results prove the reliability of physiological metrics to model emotional user's experience with interactive play technologies.

Psychophysiological measures have been more recently used in HRI human studies [8] in combination with subjective research methods, as behavioral and self-report measurements, in order to achieve a more complete and reliable knowledge of the person's experience. Generally, when a person communicates his/her own personal opinion about a research issue, as in our case about the interaction with a robot, this may be affected by mood and/or social desirability. In addition, observer bias may have an effect on participants' responses and consequently reactions or behaviors may not reflect the exact feeling of the persons involved in an evaluation study of HRI [8,29].

The psychophysiological measures offer numerous advantages and disadvantages that are well identified in the literature [8,18,30,31]. One of the main benefits of using these techniques is the objectivity of the assessment since participants generally cannot manipulate their own physiological reactions. Moreover, psychophysiological monitoring is a non-invasive method that is able to detect continually and in real time different cognitive and affective states when they occurs in response to a stimulus. However, the use of psychophysiological techniques poses also different challenges related to the modality of data acquisition and interpretation. Moreover, the quality of the recording may be influenced by both confounding environmental factors, such as noise or lightning, and by individual psychological internal state during the evaluation that might undermine the reliability and the correct interpretation of the gathered data. Physiological monitoring requires the use and management of specialized equipment. The accurate placement of electrodes on the body of the participant is essential for a correct acquisition of the parameters. Generally the monitoring of these measures occurs in a laboratory setting where the naturalness of the task or the interaction is lacking and the participants' response may be biased.

Wherever designing a study aimed at the evaluation of any human-system interaction by means of psychophysiological measurements, specific guidelines should be taken into account in order to gather useful data for a precise analysis of the user's experience [32]. Every experimental design should consider a set of phases. First of all, a researcher should select the appropriate psychophysiological parameters to be employed in order to identify definite cognitive and affective user's states during the interaction with a robot. It is necessary to determine the sample size and characteristics. After determining the experimental setting (laboratory or real-world) and equipment, a pre-testing assessment should be performed with the purpose of verifying potential confounding factors. Participants should be informed about the aim and the procedure of the study through a consent form to be signed. Prior to the experimental evaluation, the researcher should gather some health and psychological information of the participants. Prior and following the assessment, the experimental design should also consider the administering of other tools, such as questionnaires/interviews or video-recording, to acquire additional information that might help the interpretation of the psychophysiological signals and of the users' experience. The gradual adaptation to the stimulus, or in this case to the interaction with a robot, could modify participants' answers. Finally, a debriefing phase should be conducted to reflect on the experience and make it meaningful by identifying any issue that may have had an effect on the participant's response and behavior.

These guidelines are well discussed in Bethel *et al.* [32], and although the psychophysiological approach can be useful in human-robot interaction analysis in order to assess participants' reactions and responses in terms of affective and cognitive states, in the experimental design, a researcher should always bear in mind advantages and disadvantages of the use of psychophysiological measures. The evaluation of the participants' experience should focus on different methods assessment in order to gain a more complete understanding of the aspects that affect the quality of the interaction between a service robot and a human.

The next section illustrates the psychophysiological measures mostly applied in HRI.

3. Common Psychophysiological Measures in HRI

The nervous system can be divided into two parts: the central nervous system and the peripheral nervous system. The peripheral nervous system is subdivided into the sensory-somatic nervous system and the autonomic nervous system. As mentioned in the introduction, the central and autonomic nervous system responses can indicate possible modifications in our emotions, motivation, attention, and preferences. Without going into details, it can be stated that a system of multiple cortical (e.g., frontal, temporal, and parietal) and subcortical (e.g., basal ganglia, thalamus, amygdala, and hippocampus) regions contributes to the experience of emotion. Each cerebral area has a different role and a specific contribution to the elaboration of the stimulus [12]. According to right-hemisphere and valence hypothesis, this cerebral hemisphere is dominant for emotional processing, and in particular, for decoding negative emotions differently from left-hemisphere specialized for positive emotions [33,34]. The ANS is part of the peripheral nervous system (PNS) containing all the nerves that lie outside the central nervous system (CNS). The key role of the PNS is to connect the CNS to the organs, limbs and skin and to adapt organisms' internal state to the internal and environmental demands maintaining homeostasis. The ANS is further divided into two branches: the sympathetic system regulates the "flight-or-fight" responses, while the parasympathetic system helps to maintain normal body functions and conserves physical resources. An activation of ANS is also recognized as consequence of emotional responses [35]; negative emotions seem to activate a stronger autonomic reaction than the positive ones. The functioning and activation of the ANS can be assessed through psychophysiological parameters. Recently, a psychophysiological approach has been adopted to assess the internal state of subjects interacting with the technology. In this section, we briefly introduce some widely used psychophysiological measures applied in human-robot interaction studies.

3.1. Cardiovascular Activity

The main measures of cardiovascular activity are heart rate (HR), blood pressure (BP), blood volume pulse (BVP) and heart rate variability (HRV). HR is the number of heartbeats per unit of time-typically expressed as beats per min (bpm). In terms of autonomic responding, increases in sympathetic activity are associated with heart rate rise, and relative increases in parasympathetic activity are associated with heart rate reduction [36,37]. The force exerted by blood on the walls of blood vessels is the blood pressure. Systolic blood pressure (SBP) is a measure of blood pressure when the ventricles are contracting and the heart beats. Diastolic blood pressure (DBP) is a measure of blood pressure while the heart is relaxed. Sympathetic stimulation causes an increase in blood pressure, while on the contrary, the parasympathetic system decreases cardiac activity reducing blood pressure [38]. HRV is the sequence of time intervals between heart beats. This interbeat interval time series is used to calculate the variability in the timing of the heart beat in response to changing physiological conditions and factors such as respiration rate or emotional states. Consequently, HRV is a clear effect of regulation by the sympathetic and parasympathetic nervous system. Sympathetic activity is associated with the low frequency components (LF) while parasympathetic activity is associated with the higher frequency components (HF) of modulation frequencies of the HR [39].

Other cardiovascular measures include respiratory sinus arrhythmia (RSA) defined as the natural variation in heart rate that is primarily driven by breathing patterns and its regulating influence of sympathetic activity on the heart. Although the role of RSA in emotional response and regulation is not yet clear, according to the literature, its response measures might be a reliable index of emotional valence in situations demanding emotion regulation [40,41]. Inhaling inhibits the activity of the sympathetic system, increases heart rate, and decreases HRV. Exhaling activates the sympathetic branch, decreases heart rate, and increases HRV.

Several studies have investigated the relationship of cardiovascular indicators with arousal and cognitive load [42,43]. Cardiovascular response is a good marker of emotional states of the subjects [35,44–46]. An increase in HR, SBP, DBP and in respiratory activity, but not in HRV, has been reported in anger, anxiety and happiness responses [40,47,48].

3.2. Electrodermal Activity

Electrodermal activity (or galvanic skin response or skin conductance) is based on the variation of the skin electrical resistance in response to various emotional stimuli. This variation is mainly due to the level of sweat of the skin as a result of the action of the eccrine glands [49–51]. Sweat glands are primarily driven by sympathetic innervations. There are two main components of electrodermal activity:

(1) the tonic component (Skin Conductance Level) is the absolute value of the skin conductance at a given moment in absence of a specific stimulus onset. So-called nonspecific skin conductance responses also occur in absence of specific (individual) emotional stimuli, with their frequency proportional to arousal.
(2) the phasic component (Skin Conductance Response) is the response to an emotional stimulus.

Electrodermal activity is measured placing two electrodes on the skin surface of two adjacent fingers and passing a tiny electric charge. When the subject responds to a stimulus, the skin immediately becomes a slightly better conductor of electricity. External emotional stimuli produce a drop in the skin electrical resistance and in particular at the level of the palm and soles of the feet.

Electrodermal activity is related to the level of arousal elicited by a wide range of psychological and emotional states with either positive or negative valence. Different studies investigating anxiety, anger, fear and also joy experience report increased electrodermal activity [52,53]. Electrodermal activity is also an indicator of the cognitive load, stress and arousal [42,51,54,55].

3.3. Electromyography Activity

Electromyographic signals (EMG) are produced by the activation of the muscle motor unit and provide an index of the muscle activity. A muscle is composed of a multitude of motor units that are activated by electrical signals elicited by the nervous system. When a motor unit is activated, an electrochemical impulse is fired and propagates through the nerve fibers. Along the nerve fibers, the impulse stimulates the skeletal muscle creating muscle contraction, and consequently, the movement of human limbs [56]. Surface EMG is a method of recording muscle activity by electrodes mounted directly on the skin. Surface EMG is evaluated by means of the analysis of the signal amplitude in microvolt. The muscles more frequently monitored are those reflecting the human emotional state and that can serve to assess the emotional regulation (frontalis, corrugator supercilii, zygmaticus major, levator labii superioris, orbicularis oculi) [11,57]. The facial electromyography gives information related to the affective state of subjects. A rise of zygomatic major muscle activity results in a happy facial stimuli, whereas people tend to contract corrugators supercilii in presence of angry facial stimuli [35,58,59]. The zygomatic muscle lifts up the lips to produce a smile, whereas the corrugator muscle knits the eyebrows forming a frown facial expression. Dimberg *et al.* [60] show that electromyographic responses to positive and negative stimuli can be generated without any conscious awareness of the emotional eliciting stimuli.

3.4. Brain Activity

Psychophysiological reactions based on the central nervous system activity can provide information about the brain responses to emotional stimuli. Electroencephalography (EEG) is a technique that records the brain electrical activity through the placement of electrodes over the surface of the scalp. The electrodes record the rapidly fluctuating of brain wave patterns, which are a sign of continuous electrical changes in the brain. EEGs show a different involvement of the two hemispheres of the brain (cerebral asymmetry) in responding to positive and negative emotional stimuli. A higher activation of the left hemisphere indicates a positive reaction, whereas the right hemisphere is sensitive to negative affect [35,61]. Specifically, prefrontal cortical regions of the brain are involved in emotional responding. Several studies show that frontal EEG asymmetry is considered a mediator of emotional processing [62]. Some of the earliest studies about the frontal EEG asymmetry and emotions were done by Davidson *et al.* [61], and concerned differences in hemispheric activity during the viewing of emotional films. Results showed that more active right hemisphere activity was related to negative content in films while more left hemisphere activity was related to happy films. Recently, evidence suggests a relationship between brain activity and anxiety [63]. Distinct patterns of brain activity are associated with different types of anxiety. For example, anxious apprehension involves an increased left-hemisphere activity and anxious arousal is related to a greater right-hemisphere activity [64]. Electroencephalography activity can also be used as a measure of mental workload. Chaouachi *et al.* [65] present a workload index based on features extracted from EEG signals. Wilson [66] classifies operators' workload level by taking EEG as an input and reports results of 90% accuracy.

Functional magnetic resonance imaging (fMRI) is a further method used to evaluate human response during the interaction with a robot. This technique displays the hemodynamic response (changes in blood flow and oxygenation) related to neural activity in the brain or spinal cord. When nerve cells are active, they consume oxygen carried by hemoglobin. The effect of this consumption is an increase in blood flow in regions where there is a greater brain activity. When measuring changes in blood flow associated with increasing neuronal activity, it is possible to map the activation of brain areas during an activity or in response to a stimulus. Specific brain areas (amygdala, medial prefontral cortex, anterior cingulated cortex, insula) are supposed to be involved for emotional processes [67]. Saygin *et al.* underline the value of an interdisciplinary collaboration between robotics and cognitive neuroscience to improve the understanding of how robots' actions, behavior and embodiment are perceived and processed in the human brain [68]. Some relevant research has focused on Theory of Mind (ToM) [69]. When we interact with each other, it is possible to infer the interlocutor's internal state, observing behavior, ways of speaking and facial expressions. The other's internal state refers to their intentions, goals and beliefs, on the basis of which, it is possible to interpret the behavior of others. Brain imaging methods, such as fMRI have been widely applied also to identify brain areas for ToM in order to explore how human users respond, perceive and represent robots' actions, behavior and embodiment during an interaction [70].

4. Psychophysiology Applied in HRI Studies: An Overview

As mentioned in the introduction, Bethel *et al.* identify three main categories of research studies implementing psychophysiological measurements in HRI [8]. The first is focused on detecting emotions expressed by human users, based on the development of control architecture, behavior system or online emotion detection algorithms able to recognize, classify and interpret participants' affective state [19,70–80]. The second category is related to the analysis of subjects' psychophysiological responses for development and implementation of real-time robot control architectures, and adaptation of robot behaviors in order to, for instance, maximize users' motivation, engagement and performance during rehabilitation therapy [64,81–86]. The third category is the objective of this review, and it comprises research studies aimed to evaluate human experience and emotional reaction to the technology. The intent in the next subsection is to review the most relevant and recent works specifically oriented to using psychophysiology measures in understanding humans' emotional response and reactions towards a robot. A separate subsection describes a set of recent studies implementing functional magnetic resonance imaging (fMRI) as a method to evaluate how the human brain reacts to the robot's features and behavior.

4.1. Psychophysiological Studies Evaluating Humans' Response to Robots

In the field of research that implement psychophysiological measures to assess how humans react to a robot, Kulić and Croft evaluated the feasibility of physiological measures as indicators of human subjective response towards motions of a robotic manipulator arm [72]. Electrodermal (GSR), electromyography (corrugators muscle) and cardiovascular activity (HR) of 36 participants were measured during an experiment in which the human response to two different robot motions was assessed. The robot could be attracted to the target location and be rejected avoiding obstacles (potential field planned robot motions) and it could have the capability to avoid obstacles (safe planned robot motions). Two different tasks were performed by the manipulator robotic arm: (1) pick and place (2) reach and retract. Three different speeds for the manipulator arm were provided for each task.

Psychophysiological reactions were analyzed using a fuzzy inference engine as an estimation method for the relationship between physiological responses and emotional categories. Fuzzification is the process of converting the input variables' values into a set of membership values in the interval {0,1} in the corresponding fuzzy sets [87]. Physiological signals are not measured directly via the fuzzy inference engine but they are preprocessed. The relevant features extracted are used for hypothesis and for establishing a set of rules base to estimate the human emotional state. In this research, the fuzzification was derived using data from psychophysiological research [17].

Psychophysiological data were compared with subjective responses based on a Likert scale developed to assess anxiety, calm and surprise levels of the participants. Results showed a correlation between anxiety and speed, and surprise and speed of the robotic manipulator arm. In addition, subjects tended to show a stronger response in electrodermal, electromyography and cardiovascular activity (in terms of higher levels of anxiety and surprise and lower levels of calm) for potential field planned motions than safe planned motions. An estimate of arousal and valence of human emotional state was obtained from the fuzzy inference engine, but it has not proven a capability to successfully estimate valence.

In a second study, researchers developed a new human affective state estimator based on the hidden markov model (HMM). They performed the same experimental method and procedure in [72] and compared the HMM with the fuzzy inference engine. Their results showed that HMM is a better affective state estimator, able to identify the relationship between physiological signals and valence and arousal [19].

Dehais *et al.* [88] evaluate human response to different types of robot motions. They compare different ways to move a robotic arm close to the "human space". Twelve participants were recruited in order to evaluate the effects to three different robot motion types (*Motion 1.* with planner, with grasp detection, medium velocity; *Motion 2.* no planner, no grasp detection, high velocity; *Motion 3.* with planner, no grasp detection, low velocity) on physiological parameters, such as electrodermal activity, deltoid and ocular activities. The task consisted of taking a bottle from the robot's grip and participants were then asked to report their opinion on a nine-point visual analog scale, reporting the legibility of the different types of robot motions, safety and physical comfort perceived during the experiment. Their results, similar to those of Kulić and Croft in [72], show a higher electrodermal activity for high velocity robot motion compared to the other two motions characterized by medium or low velocity. The participants interpreted the electrodermal activity with a higher level of arousal and anxiety in Motion 2. The lowest deltoid and ocular activity (in terms of fixation time) was found for Motion 1. with planner, grasp detection and medium velocity.

In the results of the above studies, a common aspect lies in the belief that by combining different measurements (physiological parameters and self-report measures), it is possible to obtain information on the nature of the interaction between a human and a robot, and on the specific human reaction. The results show that the robot's behavior has an important effect on the user's reactions and it can influence acceptance toward a robot and the user's perception of interaction safety.

Human physiological responses are also involved in the assessment of specific user's responses to different tasks performed through a robot. Goljar *et al.* [89] recruited three different groups of subjects: a subacute stroke (N = 23 subjects), a chronic stroke (N = 10 subjects) and a control group (N = 23). Electrodermal, cardiovascular activity and respiratory rate were monitored during a reaching and grasping exercise task performed through a manipulator robotic arm and a haptic robotic interface. The aim of the study was focused on understanding if stroke patients' physiological responses were different from the responses of healthy subjects in relation to a physically undemanding task, and which physiological responses were most reliable in evaluating psychological states in motor rehabilitation training. Two different types of each were presented to the subjects (virtual task *vs.* harder virtual task), having different complexity, hence being different in their mental demand. The virtual task required to catch a ball before it reached the lower end of a virtual table. The manipulator guided the subjects' arms in order to move left or right and reach the ball. In the harder virtual task, subjects performed the same exercises but the robot had inverted left-right controls. Physiological data was also compared with 9-point arousal and valence scales from the Self-Assessment Manikin, in order to have an index for the participants of their level of emotional valence and arousal during the experiment. Results showed that the electrodermal signal appears to be the most valuable for obtaining useful information about participants' arousal state. Specifically, electrodermal activity of the control group was higher than the subacute and chronic stroke groups during both types of task. A significant correlation between electrodermal activity and Self-Assessment Manikin was also found in all three groups. Beside some limitations of the study discussed in the paper, results confirmed the utility of selecting appropriate psychophysiological measures to get supplementary information about users' psychological state and the efficacy of robots in clinical practice.

Also in this study, feedback obtained from the user's physiological parameters is useful for evaluating the emotional status of a human who performs a rehabilitation task through a robot. Individual reactions in rehabilitation practices may have an effect on the patient's performance that could be affected by the robot's behavior. The efficacy of human-robot interaction may depend also on individual motivation and engagement so it is important to ensure that the robot's behavior is able to adapt to a user's state, needs and capabilities.

Recently, Swangnetr and Kaber [90] recruited 24 participants at a senior center in order to evaluate the relationship between physiological response and emotional states, and to develop a machine-learning algorithm for their accurate classification. Heart Rate (HR) and electrodermal activity were collected during a simulated medicine delivery experiment in which participants interacted with a PeopleBot robot. The experimental design counted three independent variables: robot facial feature (abstract *vs.* android); speech capability (synthesized *vs.* digitized voice; and different modes of user interaction (visual messages *vs.* physical confirmation of receipt of medication with a touch screen). All participants were exposed to each variable. At the end of each experimental session (14 test trials), subjects were asked to complete the Self-Assessment Manikin in order to indicate their emotional response to the different robot patterns. In general, respect to the variables on the robot feature type, the results showed a significant difference in terms of an increased HR and electrodermal activity to the condition in which the robot required physical confirmation of receipt of medication with a touch screen. Referring to the development of a machine-learning algorithm, for HR and electrodermal activity features extraction, a wavelet analysis was used with the purpose of classifying human emotional state in real time.

Results showed that HR was a stronger index of valence compared to electrodermal activity, while arousal states had a strong relationship with electrodermal activity. The wavelet analysis was confirmed as a valuable method for identifying associations between physiological response and arousal and valence. A correct classification of human emotional state in response to a robot in a healthcare setting could contribute to ensure efficacy and quality of the human-robot interaction.

Similar results in Zhang and colleagues [91] showed that interactivity features of a service robot stimulated higher cardiovascular and electrodermal response in 24 older participants during a set of experimental trials in which a robot executed a medicine delivery task. Participants' cardiovascular (HR) and electrodermal activity was collected during the experimental trials during the interaction with the robot. A fuzzy logic model was adopted in order to gather physiological data in terms of arousal and valence. The objective of the study was to understand how different service robot interfaces (face, voice messaging and user interactivity) affect elderly people's perceptions and emotional response.

The robot had an abstract or human-like face and a synthesized or digitized human voice. User interactivity consisted in two levels of interaction based on the reading of a visual message and the request to confirm medicine delivery. Participants indicated a more positive emotional experience interacting with a service robot with anthropomorphic features based on interactivity aspects.

Broadbent and colleagues [92] explored if mental schemas about robots' humanness were associated with their psychophysiological reactions to a robot. They recruited 57 adult participants and asked them to draw their idea of a healthcare robot before seeing a real robot (PeopleBot), explaining how to attach the blood pressure cuff for measuring it. Blood pressure was monitored at baseline and then during the interaction with a robot. Participants completed also the positive and negative affect scale (PANAS) [93]. An increase in blood pressure, as an index of the ANS, was registered in participants who had drawn the robot more human-like compared to those that imagined a more machine-like robot when they interacted with the real robot. In addition, participants who drew a robot more human-like reported also more negative emotion after the interaction with the real robot compared to those who drew a machine-like robot.

A possible interpretation of these results may refer to the high responsiveness to stress and arousal by the ANS. Assessment and interpretation of human physiological responses towards features and behavior of a robot provide valuable elements for understanding the degree of applicability of a robot based on the context of use, on the type of task, on interaction modality and on users' needs and preferences.

4.2. fMRI Studies Evaluating Humans' Responses to Robots

A further technique to assess the human reaction during the interaction with a robot is found in the study of brain response. The assessment of brain activity is an interdisciplinary area involving different disciplines, as cognitive neuroscience, psychology, robotics engineering and social science, and gaining a great interest in the field of HRI studies. Saygin *et al.* [68] recruited 20 participants in a fMRI experiment in which they viewed video clips of three agents carrying out actions, such as drinking water from a cup or waving their hand. The three agents differed in form: human; robot with human form; robot with nonhuman form. The aim of the study was to explore whether the human brain would show a different specialization area for human form and motion. Researchers expected similar brain responses for human and robot with human form, and differentials for robot with nonhuman form, while for motion they supposed similar responses for robot with human form and robot with nonhuman form, and differentials for human. Results of this study demonstrated that robot with human form was processed in a different brain area compared to both human and robot with nonhuman form. Specifically they found an activation in the parietal cortex for the robot with nonhuman form. The researchers supposed a prediction error based on the incongruity between the human-like form and the mechanical movements that the brain is not able to integrate. Chaminade *et al.* [94] used fMRI as a method to confirm the hypothesis of a preference in autistic patients for a small humanoid robot during a computerized computer game. They recruited two experimental groups of subjects: 18 healthy students and 12 participants with Autism diagnoses. All subjects participated in the rock-paper-scissors game in a computerized version. Participants were told that they would play against three different opponents: the experimenter, an artificial agent and a generator embedded in a computer. The subjects knew that the first two opponents (active opponent) playing strategies used to win. The fMRI data were collected during the three game sessions. Results confirmed that autistic patients considered the artificial agent as a human opponent since the same level of brain activity in the temporal area of autistic patients was registered by means of fMRI when they interacted with the robot and with the human. In another study [95], fMRI was used to test the hypothesis of a reduced brain activity in areas involved in motor and emotional processing during the interaction with a robotic agent compared to human agent. The experimental paradigm consisted of 13 healthy young participants and 16 experimental conditions in which human and robot agents impersonated four facial expressions (anger, joy, disgust and speech). Participants observed the human and robot agents performing the gestures by way of short videos while their brain activity was scanned. An increased activity of occipital and posterior temporal regions was found in response to the gesture of robotic agent compared to the human agent, while a response in insula and orbitofrontal cortex respectively for the perception of disgust and anger emerged only for human gesture.

In order to determine the brain human response to the robot's embodiment during an interaction, Miura *et al.* [96] analyzed fMRI data of 39 students during an interaction with bipedal or wheel drive humanoid robots. A human like embodiment of the robot would involve a greater emotional response in terms of brain activity compared to a robot with a more machine-like embodiment. The experimental procedure consisted in six conditions in which a human, a bipedal robot (more human-like) and a wheel drive humanoid robot (more machine-like) performed emotionally positive and neutral daily life actions (e.g., walking happily). The fMRI scanning was along the six conditions. Results revealed that the left orbi-frontal cortex was more active during the interaction with the bipedal humanoid robot performing emotionally positive actions compared to the wheel drive humanoid robot. It is worth noting how the same brain response was found during the interaction with the human.

The studies described above confirm that the psychophysiological approach is well applicable to assess emotional response and quality of experience during human robot interaction. The benefits from the use of physiological parameters in HRI research are well documented by [8].

5. Using Psycho-Physiological Measures to Evaluate Older Adults' Response to a Telepresence Robot: A Feasibility Study

To ensure an effective human-robot interaction, it is necessary to understand also how end users react to the presence of a robot and to its features (e.g., abilities, appearance, behavior). Given the mentioned benefits of psychophysiological approach and drawing inspiration from the reviewed literature, we designed a study aimed to implement psychophysiological measurements for assessing the human response towards a telepresence robot (results in this section partially appeared in [97]). Telepresence robots have been more and more often proposed to be used to enhance older adults' social and daily functioning and to supplement human care-giving. Robots, as in Giraff [98], Texai [99], and Kompai [100], have been put forward as a proposed form of assistive and communicative device, able to deliver health and social services. To date, literature has not focused on empirical studies that employ a psychophysiological approach specifically for human-telepresence robot interaction. Our intent was to explore the impact of a telepresence robot in terms of emotional human response in older adults. We have relied on the use of psychophysiological measures presuming also that this approach may offer valuable information also about potential problems in the human-robot interaction, that need to be taken into account in order to make the proposed technology more usable, satisfactory and adaptable to the user's needs. The objective of the study was to examine the effectiveness of a remotely controlled telepresence robot, called Giraff [101], as a tool for health and social services. Subjective and psychophysiological measures were collected in order to evaluate older users' responses towards robots' presence and to obtain initial feedback about the suitability of a telepresence robot to support older adults.

5.1. Apparatus

Giraff is a social robotic telepresence system manufactured by the Giraff Technologies AB (Figure 1). It is a human-height physical robot integrated with a videoconferencing system mounted on a mobile robotic base. A user (called *pilot*) can remotely connect through an Internet connection to the robot via a computer interface and can move around in the environment where the robot is located and interact with other humans (called *local users*). Giraff is part of the ExCITE European Project [102] dedicated to study the use of robotic telepresence as a means to promote social interaction and inclusion of older adults [98].

Figure 1. The Giraff robot.

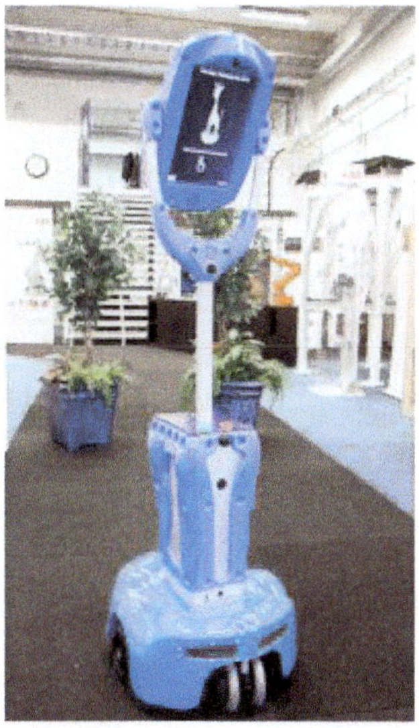

The experimental setting was composed of two rooms. In the first room participants interacted directly with the human experimenter, whereas the second room was used by the experimenter to interact through the telepresence robot. The experimenter maneuvered the robot through the pilot from the second room and met the subject remotely.

5.2. Physiological Parameters

The physiological monitoring during the study was performed through a Holter monitor. Cardiovascular activity in terms of heart rate (HR) and heart rate variability (HRV) was measured. As described in the previous sections, HR and HRV result from the activation of the ANS, which consists of two components: the sympathetic and the parasympathetic systems. Based on this assumption, an increase in sympathetic activity was expected, associated with heart rate rise, as well as relative increases in parasympathetic activity, associated with slower heart rate [4]. HRV is a consequence of regulation by the sympathetic and parasympathetic nervous system. The most frequently used HRV components are spectral parameters such as low-frequency (LF) band and high-frequency (HF) band percentual powers. Respectively, the LF band is considered as markers of the sympathetic nervous system, while the HF band reflects parasympathetic (vagal) activity A same pattern of activation of ANS was expected both in the interaction between human experimenter and participants and in the interaction with the robot. The protocol should have included an analysis of the combination of several physiological parameters, but some issues, related to the timing of the collaboration between the parties involved in the research, have limited the study only to the analysis of HR and HRV. We do not exclude the possibility of expanding the results by analyzing other physiological measures in later stages of the study.

5.3. Subjective Measures

The administered self-report measures were:

(1) The *State-Trait Anxiety Inventory* [103] (STAI). It is a self-report questionnaire consisting of two scales (Y1 for state anxiety and Y2 for trait anxiety) containing 20 items each. All items are rated on a 4-point scale (*i.e.*, from 1 = almost never, to 4 = almost always).
(2) The *Positive and Negative Affect Schedule* [93] (PANAS). It consists of two scales containing 10 items each, one measuring positive affect and the other measuring negative affect. Each item is rated on a 5-point scale (*i.e.*, from 1 = very slightly or not at all, to 5 = extremely).
(3) A concluding interview on the robot experience was administered in order to get information about: **social presence** ("*I had the feeling that the investigator was in the same place where I was*"), **utility and advantages of the robot** ("*I think that Giraff could be useful in the home*"), **engagement** ("*I felt physically involved in the experience with Giraff*"), **physical aspect of the robot** ("*Giraff is pleasing to see*"), **privacy issues** ("*It may be annoying to receive virtual visits through the Giraff at home*").

5.4. Participants

Seventeen participants were recruited at the rehabilitation department of the Fondazione "Don Gnocchi" of Rome. Given the intent to validate the use of the telepresence robot as an aid for health and social services, it was decided to focus on a specific type of older users, recruiting a group of elderly with a diagnosis of mild cognitive impairment (MCI) and a group of healthy older adults. The elderly group with a diagnosis of MCI was or had been engaged in cognitive rehabilitation interventions at the Fondazione. The characteristics of the sample are as follows:

(1) Normal control group (NC), which consisted of 9 participants (age 70 ± 4.3 years, age range 65–75, 6 males, 3 females) with no cognitive impairment. Subjects were tested with the Mini-Mental State Examination (MMSE) reporting a score of 30. Mean score on Trait Anxiety scale (Y2) was 32.3 (with *Spielberger's recommended cut-off of 39/40* [82]).

(2) MCI group, which consisted of 8 participants (age 73.5 ± 5.2 years, age range 65–79, 5 males, 3 females). MCIs were assessed by a battery of standardized neuropsychological tests and their score on MMSE was between 25 and 30 (mean 27 ± 1.18). Mean score on Trait Anxiety scale (Y2) was 34.5 (with *Spielberger's recommended cut-off of 39/40* [82]).

5.5. Procedure

The study has received the consent by the Ethical Committee. On the basis of agreements with representative health professionals of the Fondazione, experimental procedure took place over the course of four-day laboratory sessions performed at the Fondazione "Don Gnocchi". Upon arrival to the first-day laboratory session, participants were asked to read the consent form and to sign it. They were also asked to fill the STAI-Y2 and a demographics form. Participants were asked to avoid nicotine, caffeine, alcohol, other substances, and strenuous exercise at least three hours before their scheduled laboratory session. The experimental procedure is show in Figure 2 and it was articulated into four main sessions in which the participants interacted:

Figure 2. Experimental procedure.

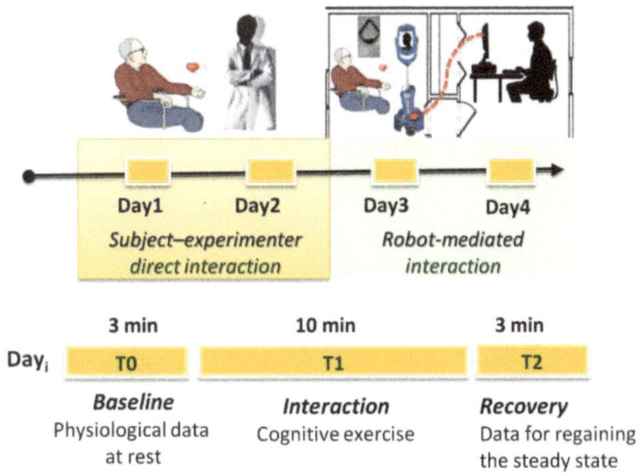

(1) directly with the human experimenter during the first two sessions;

(2) through the telepresence robot during the subsequent two sessions.

Two sessions for each type of interaction were scheduled in order to allow participants to know the experimenter and the robot (the sequence of the sessions was established according to the agreement with the limitations imposed by the Foundation for the patient's safeguard). Each interaction session had three main phases. During the *Baseline* (T0), HR and HRV assessment at rest was performed. Baseline was followed by an *Interaction Phase* (T1) in which the experimenter administered cognitive stimulation tasks to the participant. Finally a Recovery Phase (T2) was monitored for regaining HR and HRV steady state. Physiological assessment was continuous during each phase of each session. During the laboratory session the participant was seated in front of the experimenter and the Holter electrode was applied to the subject's chest.

As indicated by the data literature of human studies in HRI measures, a timing between 2 to 5 min is reported to compute cardiovascular parameters [64,84,91]. In line with this indication, and according to the healthcare professionals of the Fondazione, we chose a duration of 3 min for the *baseline* phase, during which participants were asked to rest. Following the Baseline Phase (T0), participants were engaged in 10 min of *Interaction Phase* with cognitive stimulation (T1). Following the interaction phase, participants were asked to remain seated for 3 min (*recovery* phase, T2). STAI-Y1 and PANAS scale were administered at the end of the recovery phase of the first and third session, in order to gain feedback in terms of anxiety and positive/negative affects for both types of interaction (human and robot). At the end of the fourth day laboratory session, participants met the experimenter again for the final interview focused on their experience with the telepresence robot.

Cognitive Stimulation Tasks

Cognitive stimulation tasks were usual activities for the participants of our experiment. We have decided to employ some tasks similar to the cognitive exercises of the cognitive rehabilitation interventions of the Fondazione to which patients had been or were engaged previously. This choice was purposeful also for examining the human/robot interaction in a rehabilitation context.

Particularly, in accordance to the cognitive therapist of the Fondazione, we have chosen to administer the following cognitive exercises:

(1). Word list memory: a list of 10 words is read to the subject at a constant rate of 1 word every 2 s. The word list is presented 3 times to the subject. At the end of each of the 3 presentations, the subject is asked to recall the list of words.
(2). Verbal fluency test: subject is asked to name in one minute a list of words within a specific semantic (e.g., vegetables or professions) or phonetic category (words that begin with letter F, A, and S).
(3). Story recall: the participants are asked to memorize the story once and then to complete an immediate recall test, at the end of the Interaction Phase.
(4). Digit span Forward / Backward: in the forward digit-span task a series of numbers (e.g., 7–3–9) are presented to the subjects and they have to immediately repeat them back in the given order. If they do this successfully, they are given a longer list. In the backward digit-span task the participant needs to reverse the order of the numbers.
(5). Numerical and abstract reasoning: in the first case participants are asked to resolve simple operations in addition, subtraction, multiplication or division; for the abstract reasoning subjects have to interpret some popular proverbs.

5.6. Results

For statistical significance, we used a multivariate analysis of variance (MANOVA) with one *between* subjects factor (*Group*: NC or MCI) and three *within* subjects factor (*Presence_Type*: human or robot; *Session*: first or second observation; *Phase*: baseline, interaction and recovery). A comparison of the HR variables in all phases of the experiment showed no significant difference in the baseline values.

5.6.1. Heart Rate

As we supposed before, an increase in sympathetic activity associated with heart rate rise was expected in response to the cognitive stimulation activity. Results showed a main effect for the variable *Phase,* $F_{(2.28)} = 35.7$, $p < 0.05$, $\eta^2 = 0.71$, indicating that heart rate was significantly higher during the *interaction* (T1) with respect to the *baseline* (T0) for both groups. A significant interaction *Phase* x *Group* (NC or MCI), $F_{(2.28)} = 5.26$, $p < 0.05$, $\eta^2 = 0.27$, indicated a higher increase in HR at T1 in NCs compared to MCIs (means are reported in Figure 3). A tendency to significant interaction was found for *Presence_Type* and *Group* $F_{(1.15)} = 3.79$, $p < 0.07$, $\eta^2 = 0.20$, indicating a higher HR in NC group compared to MCI during participants-experimenter direct interaction. There were no significant *Presence_Type* and *Session* effects in HR response during the various phases suggesting that there is a comparable cardiovascular reaction in terms of HR both in the participants-experimenter direct interaction and in the robot mediated interactions. A possible explanation of this result might lead to consider the change in Heart Rate in NCs as an index of a higher arousal in response to the cognitive stimulation activity independently on the type of presence (human or robot).

Figure 3. Mean for HR for phase (T1, T2, T3) for both groups during the interaction with human and telepresence robot.

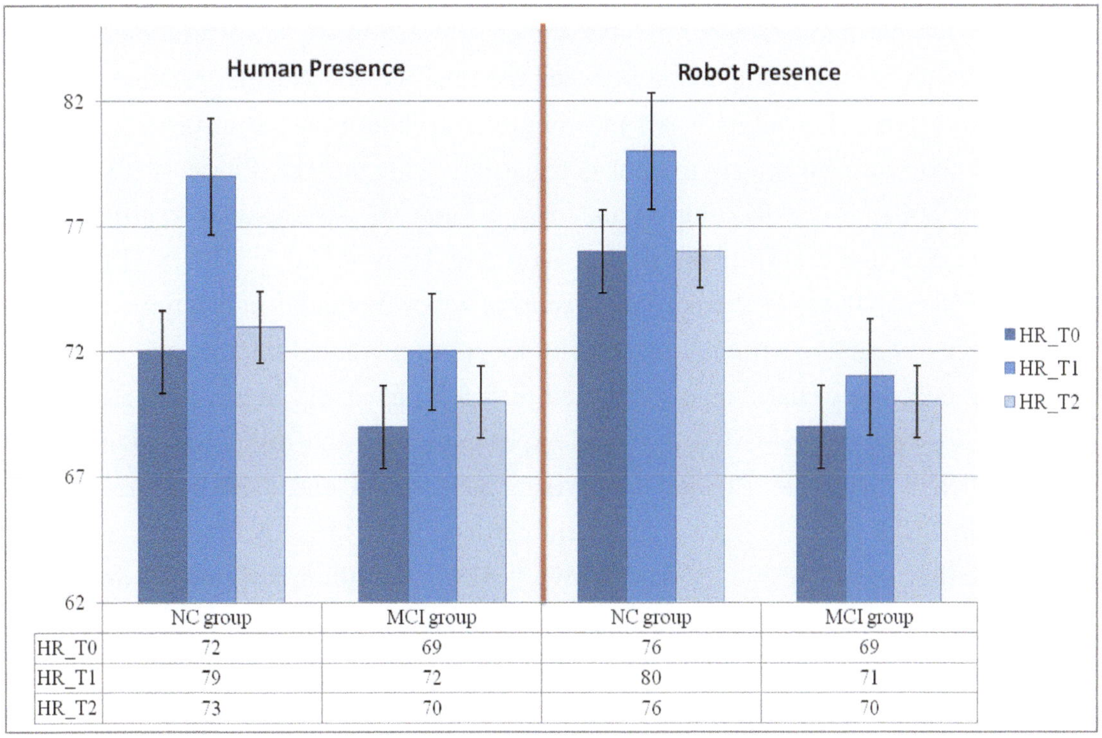

5.6.2. Heart Rate Variability

The mean of low frequency (LF) and high frequency (HF) were considered for each phase of each experimental session. We expected a predominance of the LF band as a marker of sympathetic nervous system activation and a decrease in HF band as result of parasympathetic in response to cognitive stimulation activity during the interaction both in the participants-experimenter direct interaction and in the robot-mediated interactions. A significant interaction *Phase* x *Presence_Type* (human or robot), $F_{(2.28)} = 4{,}18$, $p < 0.05$, $\eta^2 = 0.21$ was found. This result indicated that both groups showed a higher increase in LF percentual power at T1 during the interaction with the robot. There was a significant interaction *Phase* × *Presence_Type* × *Group*, $F_{(2.28)} = 4{,}6$, $p < 0.05$, $\eta^2 = 0.23$. Both groups showed a decrease for HF percentage power at T1. A significant effect for *Presence_Type* was found in both groups but the MCIs showed a higher decrease of HF percentage power at T1 during the robot-mediated interaction with respect to the NC group. A possible explanation might lead to consider these results as an indicator of a higher stress level in MCIs in response to a cognitive stimulation activity during the robot-mediated interaction.

5.6.3. Subjective Measures

State Trait Anxiety Inventory (STAI): no significant differences were observed between groups in terms of state anxiety ($p = 0.33$). In addition, no major effects were observed with respect to the *Presence_Type* (human or robot). No significant correlation was found between state anxiety and physiological parameters.

Positive and Negative Affect Scale (PANAS): mean scores were calculated for participants-experimenter direct interaction and robot-mediated interaction. Mean higher scores on Positive Affect subscale were found in both groups for both *Presence_Type*. NC group's mean score for participants-experimenter direct interaction = 37.3, while mean score for robot-mediated interaction = 37.5; similarly MCI group's mean score for participants-experimenter direct interaction = 32, while mean score for robot-mediated interaction = 31.4. Comparison between the Positive and Negative Affect Scales score and physiological parameters collected indicates meaningful correlation for Negative Affect. We found that Negative Affect indicated higher significant correlations with the HR Phase at T1 ($r = 0.545$, $p < 0.05$) e T2 ($r = 0.503$, $p < 0.05$) during the first session of participants-experimenter direct interaction and T3 ($r=.573$, $p < 0.05$) of the first session and T0 ($r= .560$, $p < 0.05$) T1 ($r = 0.571$, $p < 0.05$) T2 ($r = 0.558$, $p < 0.05$) of the second session of robot-mediated interaction. These results indicated an experience of negative emotions, especially for robot-mediated interaction.

Final interview: in general our participants affirmed they felt physically involved in the experience with the robot (%NC = 100 and %MCI = 87). The experience with the robot was judged as rather pleasant (%NC = 89 and %MCI = 75) and involving (%NC = 78 and %MCI = 62) and totally spontaneous. NCs and MCIs subjects have a sufficient perception of the experimenter, as she was in the same place where they were (%NC = 56 and %MCI = 62). In addition, 78% of NC and 62% MCI participants asserted they would feel safer with the robot at home. However, only 50% of MCI group in case of need would be willing to use the robot compared with 79% of NC group. In case of need, a telepresence robot could be useful at home for 89% of NCs compared with only 25% of MCIs. Most of the participants stated that the presence of the robot would not disturb their life (%NC = 78 and %MCI = 75) but 75% of MCI subjects, compared with 44% of the NC group, considered the robot as annoying as a possible visit from a person. Lastly, the robot was judged pleasant to see (%NC = 100 and %MCI = 75) but all the participants of NC group, compared with 25% of MCI, judged the robot too big.

5.6.4. Summary and Discussion of Results of the Study

The feasibility study demonstrated that the robot presence did not cause adverse effects in terms of cardiovascular response and these results support the possibility for using a robotic mediated interaction by a telepresence system in clinical practice and also for a specific target patient as older adults with MCI diagnosis. During the participants-experimenter direct and robot-mediated interactions, HR data showed a pattern similar for the NCs and the MCIs group as a response to cognitive stimulation tasks. HR was higher during the *interaction* (T1) with respect to the *baseline* (T0) for both groups. However, the significant main effect of the Group on heart rate response showed higher mean heart rate for NC group than MCI group. The NC group's heart rate increases from baseline during the interaction period and this could be considered as an indicator of higher arousal. The increase of arousal was slightly greater in the NC group during participants-experimenter direct interaction.

In terms of ANS response, the MCI group showed a higher mean sympathetic activity (predominance of LF band) and a reduced parasympathetic activity (decreasing in HF band) compared to the NC group. The higher sympathetic activity dominance found in the MCI group could be an index of the MCIs higher mental stress during interactions mediated by the telepresence robot. Despite the MCI group exhibited significant more mental stress, they reported subjective anxiety scores similar to the NC group for participants-experimenter direct and robot-mediated interactions. A general positive attitude of the telepresence robot was reported by participants, and this aspect might influence older adults' perceptions of the usefulness and ease of use of robots [104].

Some limitations of our study are consistent with other studies mentioned in this review [68,88,90,91,94] and are worth being highlighted. First, the small sample size of this study could preclude the generalizability of our results. The recruitment of MCI patients has encountered several difficulties. For instance, patients were asked to come to the Fondazione, and this request represented more effort for participants' relatives. Second, a weak association between physiological responses and self-reported measures in the case of the MCI participants would require further investigation. Although there is a clear difference between the groups in HR response during the interaction phase, we may affirm that the presence of the robot has not caused anxiety in the participants. However, based on the correlation between the Negative Affect scale of PANAS, the robot's presence may have had a slight negative connotation in terms of affects. Third, the MCI's medications might have altered some physiological responses (especially for HR) despite the two experimental groups showing the same cardiovascular clinical pattern. Fourth, the indications for experimental procedure suggested by the representative healthcare staff of the Fondazione did not allow us to control possible habituation effects. Finally, the laboratory setting/experiment may have increased the task engagement in the NC group because they could be more motivated to achieve a satisfactory performance.

Notwithstanding the limitations of the study, our results confirm the efficacy of the psychophysiological approach in studying human robot interaction as a valuable method also for comparing the reactions of different user groups. Telepresence robots, such as Giraff, have the potential to guarantee continuity of care, allowing users to be video-monitored by caregivers or healthcare providers remotely. Before the implementation of a robot in a user's life environment, it is recommended to evaluate users' characteristics and their emotional reaction to the robot. A positive reaction will affect the quality of experience and the effectiveness of the interaction. Psychophysiological assessment is a viable method to understand how a patient with specific diseases can react to the robot's presence and to its features, offering valuable feedback to understand the feasibility of an assistance program based on a robotic device.

6. Some Lessons Learned from the Literature

Starting with the assumption that assistive social robotics has been increasingly proposed as a tool for health care and social services, especially for older adults and their caregivers [105], a very important issue must be to expand our knowledge about how humans may react and interpret the robot's behavior, robot-performed human-like gestures, the robot's ability to perform specific tasks, and how the robot's physical embodiment may affect an interaction. The psychophysiological approach is confirmed as a valuable method to investigate the human's response towards an interacting robot. This section is intended to provide a summary of the guidelines underscored above, re-examined in the light of the reviewed literature and of our study on the older adults' responses to a telepresence robot.

Psychophysiological measures and their detection: Cardiovascular, electrodermal, electromyography and brain activity are widely employed in the studies of this review and their results are in line with previous studies on physiology of emotion [11,35], confirming that electrodermal (GSR) and cardiovascular (HR) measures are the most commonly assessed indices of ANS activation and of stimulus valence and arousal activation [14,90,106,107]. Consequently, it is recommended to refer mainly to cardiovascular and electrodermal variables in order to obtain an autonomic assessment. Psychophysiological parameters should be suitable for the experimental environment's features (laboratory, virtual or field setting), for the participant's profile (children-young and old), and as already mentioned, for the specific emotional reaction to measures. Similarly, the selection of the most appropriate equipment for detecting variables of research interest have to be considered one of the strategies for the overall control of the research plan. In a psychophysiological assessment, it is important that the experimenter learn to manage the equipment for the detection of the physiological parameters. In addition, it is suggested that the experimenter be supported by neuropsychopathology staff for the collection and interpretation of data. Physiological equipment or devices are not always easy to wear, and some of them may be invasive, affecting the participant's behavior and reaction. It is recommended, whenever it is possible, to choose non-invasive devices that do not exceedingly obstruct the person's physical body movements. Regarding this aspect, a phase of equipment functioning pre-assessment, events timing, and acquisition of the data to be recorded, should be taken into account.

Robots and participants as part of the experiment: The participants involved in the studies of this review interacted and reacted to different robot's features regarding the behavior, the level of anthropomorphism and the type of task performed. These variables can affect the robot's safety and effectiveness of the interaction perceived in relation to the user needs and context of use. In [22,68,94] participants reported positive emotions in response to human-like robots, confirming the results of Minato *et al.* [108] based on the Uncanny Valley hypothesis [109]. Human-like robot's features in terms of gestures and movements generate a positive response among the participants and consequently affect their attitude and acceptance towards the robot. The most used robots are generally robotic manipulator arms [19,72,89] social assistive robots, such as PeopleBot [91,92], mechanical humanoid robots [68,94,96], and robotic haptic interfaces [89], with which participants do not have sufficient familiarity. As reported in the literature, user's previous experience and familiarity contribute to modulate acceptance, especially in older adults [110,104]. It is likely that a lack of familiarity may have an adverse effect on the user's reaction. In the recruitment of the sample it is useful to consider all the external variables (such as technology experience) as well as the users' characteristics that may generally affect their attitude and performance [104]. Among users' characteristics, healthy status and psychological information should be evaluated at the beginning of the experiment. Clearly, the cardiovascular pattern will affect the participant's reaction and it will be taken into account for possible confounds. One point is worth being underscored: the subjects recruited for sampling in many of the HRI studies described in this review are university students or young adults (e.g., [84,89,111]). Furthermore, stroke victim subjects tend to be younger adults with an average age of 50 years. The average age of an elderly subject enrolled in several studies was maximum 69 years. Just two of the mentioned studies refer to a sample composed exclusively of older adults [90,91] with an average of 80 years. The sample size varies from a minimum of 12 [88] to a maximum of 57 participants [92]. In the research of the category of study considered in this review, the issue of sample size identified by Bethel persists [8]. However, in [8] Kulic *et al.* [72], they had the largest sample (with 36 individuals), and in this review Broadbent *et al.* research had the largest sample recruited (with 57 participants). Determining the sample size before collecting data is important in order to achieve sufficient significance from the results [32]. This is not always easy, especially when the sample, such as ours, includes people with a specific disease whose recruitment and participation in the experiment can be demanding. As suggested by Bethel [32], a power analysis would always be advisable to determine the sample size required to detect an effect of a given size.

Combining objective and subjective measures: The experimental method of the studies is based on the main recommendations in HRI research when using a psychophysiological approach [32,112,113]: (1) combining multiple psychophysiology measures in order to find correlations in the results; (2) employing psychophysiology parameters as a valuable complement of self-report measures. Among the most commonly used self-report measures administered, the Self-Assessment Manikin [16] is a very common assessment technique to measure the pleasure, arousal, and dominance associated with participants' emotional reaction to robot stimuli as reported in [89,90]. Other research includes self-report assessment developed ad hoc [19,72,88] or a scale for evaluating affective response as positive and negative affect schedule (PANAS) [93]. A multimodal approach can be convenient to facilitate the understanding of the human response based on the evaluation of the correlation between behavioral feedback/physiological indicators in response to the stimulus (*i.e.*, the robot) and what the participant reports about the experience with the robot itself. In addition, objectivity of psychophysiological methods is a resource to avoid social desirability effects because we know that self-report assessment is subject to sources of error and bias; the participant usually wishes to provide a "socially acceptable" answer rather than to report his/her true feelings [114]. A study protocol should also consider the habituation effect that is very common in psychophysiological studies. Habituation has several implications for the reliability and stability assessment of results produced by studies based on the psychophysiological approach. Administering self-report measures or asking participants to evaluate their experience in the presentation of each task or stimulus of the experiment [8] is a consolidation strategy to avoid or reduce the negative effects of habituation, as reported in [72,88–91].

Briefing and Debriefing: In the planning phase of a study designed to evaluate the effectiveness of human-robot interaction through emotional response, it is advisable to define a written study protocol including, in addition to the different phases of the research, a phase of initial briefing to give all the necessary instructions and to obtain consent forms from the participants, as well as a final debriefing phase. A debriefing phase is considered, for example, in [90] by means of a final interview performed after the experiment. Debriefing is a worthwhile means to gather information about more significant aspects of the user's experience and to ensure that the experience did not have a negative impact on the participants. In this way it is possible to reduce the effect of possible confounds and to get more information for a correct understanding of the physiological data collected during the experiment.

Experimental setting: The laboratory setting remains the primary experimental environment, because the methods for physiological parameters assessment require a controlled setting where it is easier to use technical equipment and to prevent unexpected events that could affect participants' reaction. In addition, the laboratory setting allows experimenters to operate faster on any malfunction of the robot and on the equipment used for the experiment. One of the main disadvantages of a laboratory setting consists in the minor ecological validity of the results. Field experiments take place in the participant's natural environment and consequently their results can be more easily generalized. Therefore, whenever it is possible, the experimental setups should be as close as possible to the participants' real life environment or even so familiar, using known contexts that are representative of real life situations (as a clinic for cognitive rehabilitation) in which the participant is already used to environmental and organizational features.

The lessons learned from this review confirm the value of developing guidelines for both planning and conducting the experiments that implement psychophysiological assessment in HRI. We believe that pulling out guidelines from different experiences in this research field is the best strategy for defining standardized methods to maximize reliability and generalizability of the results.

7. Conclusions

This paper presented a literature review of studies conducted in the HRI field research focused on the psychophysiological assessment as an evaluation method for measuring human emotional reactions during the interaction with a robot. The interest lies in expanding our knowledge of how humans react and what their affective and emotional responses are when they interact and collaborate with a robot. The skills and features of assistive social robots represent an important aspect of human robot interaction which, involving human emotional response, may contribute to affect both the quality and effectiveness of the interaction itself. From the psychophysiological feedback, it is possible to assess both the valence and levels of arousal associated to the interaction experience between a human and a social assistive robot. According to a "circumplex model of affect", emotional responses occur from two fundamental neurophysiological systems, one related to valence (a pleasure-displeasure continuum) and the other to arousal [15,115]. An emotional response, such as stress, is modeled as high arousal and low valence, while joy would be modeled as high arousal and high valence. This kind of information may be helpful to clarify if the user interprets the interaction with a robot as a negative or a positive experience. Positive experience will facilitate human-robot interaction and encourage collaboration between robot and user. In a context of real utilization, the assessment of individual emotional responses, as well as the analysis of user needs, is essential to the implementation of a robotic device.

We reported significant studies in this research field, and we also included studies implementing fMRI technique to assess the feedback from brain activity relative to how human users respond, perceive and represent robots' actions, behavior and embodiment during an interaction. From the analysis of this literature we derived the elements useful to contribute to the reinforcement and development of guidelines to standardized methods based on the psychophysiological approach.

Overall, HRI studies based on the psychophysiological approach can be demanding, but following a careful planning of the research and taking into account some useful recommendations, it is possible to obtain encouraging results. The value of the psychophysiological feedback provides a key factor in increasing the knowledge about how the human being reacts during interactions with robots. Using these type of measures, it is possible to explore issues like understanding: (i) how to facilitate human-robot interaction; (ii) which users are more suitable to interact with a particular type of social assistive robot; (iii) what is the adequate level of anthropomorphism; and (iv) what is the best behavior that the robot should adopt relative to its features, the user's characteristics and the task to be performed for the human. Based on this information, we believe that it will be possible to render the human-robot interaction increasingly effective. A successful interaction will have a positive impact on individual motivation, attitude and performance, and will encourage the user to collaborate with the robot. We believe that the emotional response assessment based on psychophysiological measures could be a helpful approach to match individuals with the most appropriate robotic device depending on their needs and use preferences.

Acknowledgments

Authors would like to thank: the Fondazione Don C. Gnocchi of Rome for the field environment for this research and for the precious support of the neuropsychopathology staff, under the direction of Luca Padua, in the data collection and management. CNR authors would like to acknowledge Gabriella Cortellessa for joint work on the topic and the ExCITE Project (EU AAL JP) for partially supporting this work.

References

1. Breazeal, C. Towards sociable robots. *Robot. Auton. Syst.* **2003**, *42*, 167–175.
2. Forlizzi, J.; di Salvo, C. Service Robots in the Domestic Environment: A Study of the Roomba Vacuum in the Home. In Proceedings of the 1st Annual Conference on Human-Robot Interaction, Salt Lake City, UT, USA, March 2006; pp. 258–265.
3. Reeves, B.; Nass, C. *The Media Equation*; CSLI Publications: Cambridge, UK, 1996.
4. Sung, J.Y.; Guo, L.; Grinter, R.E.; Christensen, H.I. My roomba is rambo: Intimate home appliances. *Ubiquitous Comp.* **2007**, *4717*, 145–162.
5. Broekens, J.; Heerink, M.; Rosendal, H. Assistive social robots in elderly care: A review. *Gerontechnology* **2009**, *8*, 94–103.
6. Feil-Seifer, D.; Mataric, M.J. Defining Socially Assistive Robotics. In Proceedings of the IEEE International Conference on Rehabilitation Robotics, Chicago, IL, USA, 28 June–1 July 2005; pp. 465–468.
7. Gross, H.-M.; Schröter, C.H.; Müller, S.; Volkhardt, M.; Einhorn, E.; Bley, A.; Langner, T.; Merten, M.; Huijnen, C.; van den Heuvel, H.; *et al.* Further Progress towards a Home Robot Companion for People with Mild Cognitive Impairment. In Proceedings of the IEEE International Conference on Systems, Man, and Cybernetics (IEEE-SMC 2012), Seoul, Korea, 14–17 October 2012; pp. 637–644.

8. Bethel, C.L.; Salomon, K.; Murphy, R.R.; Burke, J.L. Survey of Psychophysiology Measurements Applied to Human-Robot Interaction. In Proceedings of the 16th IEEE International Symposium on Robot & Human Interactive Communication, Jeju Island, Korea, 26–29 August 2007.
9. Dautenhahn, K. Socially intelligent robots: Dimensions of human-robot interaction. *Philos. Trans. Roy. Soc. Biol. Sci.* **2007**, *362*, 679–704.
10. Fong, T.; Nourbakhsh, I.; Dautenhahn, K. A survey of socially interactive robots. *Rob. Auton. Syst.* **2003**, *42*, 143–166.
11. Cacioppo, J.T.; Berntson, G.G.; Klein, D.J.; Poehlmann, K.M. The psychophysiology of emotion across the lifespan. *Annu. Rev. Gerontol. Geriatr.* **1997**, *17*, 27–74.
12. Hagemann, D.; Waldstein, S.; Thayer, J.F. Central and autonomic nervous system integration in emotion. *Brain Cogn.* **2003**, *52*, 79–87.
13. Oatley, K.; Keltner, D.; Jenkins, J.M. *Understanding Emotions*, 2nd ed.; Wiley-Blackwell Publishing: Oxford, UK, 2006.
14. Swangnetr, M. Analysis of Patient-Robot Interaction Using Statistical and Signal Processing Methods. Ph.D. Thesis, North Carolina State University, Raleigh, NC, USA, 2010.
15. Russell, J.A. A circumplex model of affect. *J. Pers. Soc. Psychol.* **1980**, *39*, 1151–1178.
16. Morris, J.D. SAM: The self-assessment manikin. An efficient cross-cultural measurement of emotional response. *J. Advert. Res.* **1995**, *35*, 63–68.
17. Cacioppo, J.T.; Tassinary, L.G. Psychophysiology and psychophysiological inference. In *Principles of Psychophysiology*; Cacioppo, J.T., Tassinary, L.G., Eds.; Cambridge University Press: Cambridge, UK, 1990; pp. 3–33.
18. Dirican, A.C.; Göktürk, M. Psychophysiological measures of human cognitive states applied in human computer interaction. *Procedia Comput. Sci.* **2011**, *3*, 1361–1367.
19. Kulić, D.; Croft, E. Affective state estimation for human-robot interaction *IEEE Trans. Robot.* **2007**, *23*, 991–1000.
20. Gevins, A.; Smith, M.E. Neurophysiological measures of cognitive workload during human-computer interaction. *Theor. Issues Ergon. Sci.* **2003**, *4*, 113–131.
21. Van Reekum, C.M.; Johnstone, T.; Banse, R.; Etter, A.; Wehrle, T.; Scherer, K.R. Psychophysiological responses to appraisal dimensions in a computer game. *Cogn. Emotion* **2004**, *18*, 663–668.
22. Mandryk, R.L.; Atkins, M.S. A fuzzy physiological approach for continuously modeling emotion during interaction with play technologies. *Int. J. Hum. Comp. Stud.* **2007**, *65*, 329–347.
23. Wilson, G.M.; Sasse, M.A. Investigating the Impact of Audio Degradations on Users: Subjective *vs.* Objective Assessment Methods. In Proceedings of OZCHI 2000: Interfacing Reality in the New Millennium, Sydney, Australia, 4–8 December 2000; pp. 135–142.
24. Scheirer, J.; Fernandez, R.; Klein, J.; Picard, R. Frustrating the user on purpose: A step toward building an affective computer. *Interact. Comp.* **2002**, *14*, 93–118.
25. Barlett, C.P.; Anderson, C.A.; Swing, E.L. Video game effects—Confirmed, suspected, and speculative. A review of the evidence. *Simul. Gam.* **2009**, *40*, 377–403.
26. Hebert, S.; Beland, R.; Dionne-Fournelle, O.; Crete, M.; Lupien, S.J. Physiological stress response to video-game playing: The contribution of built-in music. *Life Sci.* **2005**, *76*, 2371–2380.

27. Ravaja, N.; Saari, T.; Salminen, M.; Laarni, J.; Holopainen, J.; Järvinen, A. Emotional Response Patterns and Sense of Presence during Video Games: Potential Criterion Variables for Game Design. In Proceedings of the NordiCHI 2004, Tampere, Finland, 26–27 October 2004; pp. 339–347.
28. Mandryk, R.L.; Atkins, M.S.; Inkpen, K.M. A Continuous and Objective Evaluation of Emotional Experience with Interactive Play Environments. In Proceedings of the Conference on Human Factors in Computing Systems (CHI 2006), Montreal, QC, Canada, 24–27 April 2006; pp. 1027–1036.
29. Kivikangas, M.; Chanel, G.; Cowley, B.; Ekman, I.; Salminen, M.; Järvelä, S.; Ravaja, N. A review of the use of psychophysiological methods in game research. *J. Gam. Virt. Worlds* **2011**, *3*, 181–199.
30. Kidd, C.D.; Breazeal, C. Human-Robot Interaction Experiments: Lessons Learned. In Proceedings of the AISB'05 Symposium Robot Companions: Hard Problems and Open Challenges in Robot-Human Interaction, Hertfordshire, UK, 14–15 April 2005; pp. 141–142.
31. Steinfeld, A.; Fong, T.; Kaber, D.; Lewis, M.; Scholtz, J.; Schultz, A.; Goodrich, M. Common Metrics for Human-Robot Interaction. In Proceedings of the 1st ACM SIGCHI/SIGART Conference on Human-Robot Interaction, Salt Lake City, UT, USA, 2–3 March 2006.
32. Bethel, C.L.; Salomon, K.; Burke, J.L.; Murphy, R.R. Psychophysiological Experimental Design for Use in Human-Robot Interaction Studies. In Proceedings of the 2007 International Symposium on Collaborative Technologies and Systems (CTS 2007), Orlando, FL, USA, 25–26 May 2007.
33. Borod, J.C.; Bloom, R.L.; Brickman, A.M.; Nakhutina, L.; Curko, E.A. Emotional processing deficits in individuals with unilateral brain damage. Emotional processing deficits in individuals with unilateral brain damage. *Appl. Neuropsychol. Adult* **2002**, *9*, 23–36.
34. Demaree, H.A.; Everhart, D.E.; Youngstrom, E.A.; Harrison, D.W. Brain lateralization of emotional processing: Histrical roots and a future incorporating dominance. *Behav. Cognit. Neurosci. Rev.* **2005**, *4*, 3–20.
35. Cacioppo, J.T.; Berntson, G.G.; Larsen, J.T.; Poehlmann, K.M.; Ito, T.A. The psychophysiology of emotion. In *The Handbook of Emotion*, 2nd ed.; Lewis, R., Haviland-Jones, J.M., Eds.; Guilford Press: New York, NY, USA, 2000; pp. 173–191.
36. Ruediger, H.; Seibt, R.; Scheuch, K.; Krause, M.; Alam, S.S. Sympathetic and parasympathetic activation in heart rate variability in male hypertensive patients under mental stress. *J. Hum. Hypert.* **2004**, *18*, 307–315.
37. Thayer, J.F.; Ahs, F.; Fredrickson, M.; Sollers, J.J.; Wager, T.D. A meta-analysis of heart rate variability and neuroimaging studies: Implications for heart rate variability as a marker of stress and health. *Neurosci. Biobehav. Rev.* **2012**, *36*, 747–756.
38. Joyner, J.M.; Charkoudian, N.; Wallin, B.G. A sympathetic view of the sympathetic nervous system and human blood pressure regulation. *Exp. Physiol.* **2008**, *93*, 715–724.
39. Acharya, R.U.; Joseph, P.K.; Kannathal, N.; Lim, C.M.; Suri, J.S. Heart rate variability: A review. *Med. Biol. Eng. Comp.* **2006**, *44*, 1031–1051.
40. Rainville, P.; Bechara, A.; Naqvi, N.; Damasio, A.R. Basic emotions are associated with distinct patterns of cardiorespiratory activity. *Int. J. Psychophysiol.* **2006**, *61*, 5–18.

41. Thérèse, J.M.O.; van de Boxtel, A.; Westerink, D.M. Respiratory sinus arrhythmia responses to induced emotional states: Effects of RSA indices, emotion induction method, age, and sex. *Biol. Psychol.* **2012**, *91*, 128–141.
42. Park, B. Psychophysiology as a Tool for HCI Research: Promises and Pitfalls. In Proceedings of the 13th International Conference on Human-Computer Interaction Part I: New Trends, San Diego, CA, USA, 19–24 July 2009; pp. 141–148.
43. Wilson, G. Operator functional state assessment for adaptive automation implementation. In *Biomonitoring for Physiological and Cognitive Performance during Military Operations*; Caldwell, J.A., Wesensten, N.J., Eds.; SPIE: Orlando, FL, USA, pp. 100–104.
44. Blechert, J.; Lajtman, M.; Michael, T.; Margraf, J.; Wilhelm, F.H. Identifying anxiety states using broad sampling and advanced processing of peripheral physiological information. *Biomed. Sci. Instrum.* **2006**, *42*, 136–141.
45. Etzel, J.A.; Johnsen, E.L.; Dickerson, J.A.; Tranel, D.; Adolphs, R. Cardiovascular and respiratory responses during musical mood induction. *Int. J. Psychophysiol.* **2006**, *61*, 57–69.
46. Khalfa, S.; Roy, M.; Rainville, P.; Bella, S.D.; Peretz, I. Role of tempo entrainment in psychophysiological differentiation of happy and sad music? *Int. J. Psychophysiol.* **2008**, *68*, 17–26.
47. Murakami, H.; Ohira, H. Influence of attention manipulation on emotion and autonomic responses. *Percept. Motor Skills* **2007**, *105*, 299–308.
48. Pauls, C.A.; Stemmler, G. Repressive and defensive coping during fear and anger. *Emotion* **2003**, *3*, 284–302.
49. Stern, R.M.; Ray, W.J.; Quigley, K.S. *Psychophysiological Recording*, 2nd ed.; Oxford University Press: New York, NY, USA, 2001.
50. Benedek, M.; Kaerbach, C. A continuous measure of phasic electrodermal activity. *J. Neurosci. Method* **2010**, *190*, 80–91.
51. Dawson, M.E.; Schell, A.M.; Filion, D.L. The electrodermal system. In *Handbook of Psychophysiology*, 2nd ed.; Cacippo, J.T., Tassinary, L.G., Berntson, G.G., Eds.; Cambridge University Press: Cambridge, UK, 2000; pp. 200–223.
52. Ritz, T.; Steptoe, A.; Wilde, S.D.; Costa, M. Emotions and stress increase respiratory resistance in asthma. *Psychos. Med.* **2000**, *62*, 401–412.
53. Stemmler, G.; Heldmann, M.; Pauls, C.A.; Scherer, T. Constraints for emotion specificity in fear and anger: The context counts. *Psychophysiology* **2001**, *38*, 275–291.
54. Haapalainen, E.; Kim, S.J.; Forlizzi, J.F.; Dey, A.K. Psycho-Physiological Measures for Assessing Cognitive Load. In Proceedings of the 12th ACM International Conference on Ubiquitous Computing, Copenhagen, Denmark, 26–29 September 2010; pp. 301–310.
55. Shi, Y.; Ruiz, N.; Taib, R.; Choi, E.; Chen, F. *Galvanic Skin Response* (*GSR*) *as an Index of Cognitive Load. Ext. Abstracts CHI 2007*; ACM Press: New York, NY, USA, 2007; pp. 2651–2656.
56. Raez, M.B.I.; Hussain, M.S.; Mohd-Yasin, F. Techniques of EMG signal analysis: Detection, processing, classification and applications. *Biol. Proced. Online* **2006**, *8*, 11–35.
57. Neumann, D.L.; Westbury, H.R. The psychophysiological measurement of empathy. In *Psychology of Empathy*; Scapaletti, D.J., Ed.; Nova Science Publishers Inc.: Hauppauge, NY, USA, 2011; pp. 119–142.

58. Dimberg, U.; Thunberg, M.; Grunedal, S. Facial reactions to emotional stimuli: Automatically controlled emotional responses. *Cogn. Emotion* **2002**, *16*, 449–471.
59. Wied, M.A.; van de Boxtel, A.; Zaalberg, R.; Goudena, P.P.; Matthys, W.C.H.J. Facial EMG responses to dynamic emotional facial expressions in boys with disruptive behavior disorders. *J. Psychiatr. Res.* **2006**, *40*, 112–121.
60. Dimberg, U.; Thunberg, M.; Elmehed, K. Unconscious facial reactions to emotional facial expressions. *Psychol. Sci.* **2000**, *11*, 86–89.
61. Davidson, R.J.; Pizzagalli, D.; Nitschke, J.B.; Putnam, K.M. Depression: Perspectives from affective neuroscience. *Annu. Rev. Psychol.* **2002**, *53*, 545–574.
62. Coan, J.A.; Allen, J.J.B. Frontal EEG asymmetry as a moderator and mediator of emotion. *Biol. Psychol.* **2004**, *67*, 7–49.
63. Engels, A.S.; Heller, W.; Mohanty, A.; Herrington, J.D.; Banich, M.T.; Webb, A.G.; Miller, G.A. Specificity of regional brain activity in anxiety types during emotion processing. *Psychophysiology* **2007**, *44*, 352–363.
64. Nitschke, J.B.; Heller, W.; Miller, G.A. Anxiety, Stress, and Cortical Brain Function. In *The Neuropsychology of Emotions*; Borod, J.C., Ed.; Oxford University Press: New York, NY, USA, 2000; pp. 298–319.
65. Chaouachi, M.; Jraidi, I.; Frasson, C. Modeling mental workload using eeg features for intelligent systems. User modeling, adaption and personalization. *Lect. Note. Comput. Sci.* **2011**, *6787*, 50–60.
66. Wilson, G.F. An analysis of mental workload in pilots during flight using multiple psychophysiological measures. *Int. J. Aviat. Psychol.* **2002**, *12*, 3–18.
67. Phan, K.L.; Wager, T.D.; Taylor, S.F.; Liberzon, I. Functional brain imaging studies of human emotion. *CNS Spectr.* **2004**, *9*, 258–266.
68. Saygin, A.P.; Chaminade, T.; Urgen, B.A.; Ishiguro, H. Cognitive neuroscience and robotics: A mutually beneficial joining of forces. In *Robotics: Systems and Science*; Takayama, L., Ed.; University of Southern California: Los Angeles, CA, USA, 2011.
69. Dennett, D. *The Intentional Stance*; MIT Press: Cambridge, MA, USA, 1987.
70. Kim, K.J.; Lipson, H. Towards a Simple Robotic Theory of Mind. In Proceedings of the 9th Workshop on Performance Metrics for Intelligent Systems, PerMIS 2009, Gaithersburg, MD, USA, 21–23 September 2009; pp. 131–138.
71. Bonarini, A.; Mainardi, L.; Matteucci, M.; Tognetti, S.; Colombo, R. Stress Recognition in a Robotic Rehabilitation Task. In Proceedings of the Robotic Helpers: User Interaction, Interfaces and Companions in Assistive and Therapy Robotics, Amsterdam, the Netherlands, 12–15 March 2008.
72. Kulić, D.; Croft, E.A. Anxiety Detection during Human-Robot Interaction. In Proceedings of the IEEE International Conference on Intelligent Robots and Systems, Edmonton, AB, Canada, 2–6 August 2005; pp. 616–621.
73. Kulic, D.; Croft, E. Estimating Intent for Human-Robot Interaction. In Proceedings of the 11th International Conference on Advanced Robotics, Coimbra, Portugal, 30 June–3 July 2003; pp. 810–815.

74. Mower, E.; Feil-Seifer, D.J.; Mataric, M.J.; Narayanan, S. Investigating Implicit Cues for User State Estimation in Human-Robot Interaction Using Physiological Measurements. In Proceedings of the 16th IEEE International Symposium on Robot and Human Interactive Communication, RO-MAN 2007, Jeju Island, Korea, 26–29 August 2007; pp. 1125–1130.
75. Nasoz, F.; Alvarez, K.; Lisetti, C.L.; Finkelstein, N. Emotion recognition from physiological signals for presence technologies. *Int. J. Cogn. Technol. Work* **2003**, *6*, 1–14.
76. Picard, R.W.; Vyzas, E.; Healy, J. Toward machine emotional intelligence: Analysis of affective psychological states. *IEEE Trans. Pattern Anal. Mach. Intell.* **2001**, *23*, 1175–1191.
77. Rani, P.; Sims, J.; Brackin, R.; Sarkar, N. Online stress detection using psychophysiological signals for implicit human-robot cooperation. *Robotica* **2002**, *20*, 672–685.
78. Rani, P.; Sarkar, N. A new approach to implicit human-robot interaction using affective cues. In *Mobile Robots, towards New Applications*; Lazinica, A., Ed.; InTech: New York, NY, USA, 2006; pp. 233–252.
79. Rani, P.; Sarkar, N.; Smith, C.A.; Adams, J.A. Affective Communication for Implicit Human-Machine Interaction. In Proceedings of the IEEE International Conference on Systems, Man, and Cybernetics, Washington, DC, USA, 5–8 October 2003; Volume 5, pp. 4896–4903.
80. Zhai, J.; Barreto, A.B.; Chin, C.; Chao, Li. Realization of Stress Detection Using Psychophysiological Signals for Improvement of Human-Computer Interactions. In Proceedings of the IEEE SoutheastCon 2005, Fort Lauderdale, FL, USA, 8–10 April 2005; pp. 415–420.
81. Itoh, K.; Miwa, H.; Nukariya, Y.; Zecca, M.; Takanobu, H.; Roccella, S.; Carrozza, M.C.; Dario, P.; Atsuo, T. Development of a Bioinstrumentation System in the Interaction between a Human and a Robot. In Proceedings of the International Conference of Intelligent Robots and Systems, Beijing, China, 9–15 October 2006; pp. 2620–2625.
82. Badesa, F.J.; Morales, R.; Garcia-Aracil, N.; Sabater, J.M.; Perez-Vidal, C.; Fernandez, E. Multimodal interfaces to improve therapeutic outcomes in robot-assisted rehabilitation. *IEEE Trans. Syst. Man Cybern. Appl. Rev.* **2012**, *42*, 1152–1158.
83. Guerrero, C.R.; Marinero, J.F.; Turiel, J.P.; Farina, P.R. Using Psychophysiological Feedback to Enhance Physical Human Robot Interaction in a Cooperative Scenario. In Proceedings of the 4th IEEE RAS & EMBS International Conference on Biomedical Robotics and Biomechatronics (BioRob), Roma, Italy, 24–27 June 2012; pp. 1176–1181.
84. Koenig, A.; Novak, D.; Omlin, X.; Pulfer, M.; Perreault, E.; Zimmerli, L.; Mihelj, M.; Riener, R. Real-time closed-loop control of cognitive load in neurological patients during robot-assisted gait training. *IEEE Trans. Neural Syst. Rehabil. Eng.* **2011**, *19*, 453–464.
85. Liu, C.; Conn, K.; Sarkar, N.; Stone, W. Online affect detection and robot behavior adaptation for intervention of children with autism. *IEEE Trans. Robot.* **2008**, *24*, 883–896,
86. Rani, P.; Sarkar, N.; Smith, C.A.; Kirby, L.D. Anxiety detecting robotic system towards implicit human-robot collaboration. *Robotica* **2004**, *22*, 85–95.
87. Ganesh, M. *Introduction to Fuzzy Sets and Fuzzy Logic*; Prentice Hall of India: New Delhi, India, 2006.
88. Dehais, F.; Sisbot, E-A.; Alami, R.; Causse, M. Physiological and subjective evaluation of a human-robot object hand over task. *Appl. Ergon.* **2011**, *42*, 785–791.

89. Goljar, N.; Javh, M.; Poje, J.; Ocepek, J.; Novak, D.; Ziherl, J.; Olenšek, A.; Mihelj, M.; Munih, M. Psychophysiological Responses to Robot Training in Different Recovery Phases after Stroke. In Proceedings of the IEEE International Conference on Rehabilitation Robotics (ICORR 2011), Zurich, Switzerland, 29 June–1 July 2011.
90. Swangnetr, M.; Kaber, D.B. Emotional state classification in patient-robot interaction using wavelet analysis and statistics-based feature selection. *IEEE Trans. Hum. Mach. Syst.* **2013**, *43*, 63–75.
91. Zhang, T.; Kaber, D.B.; Zhu, B.; Swangnetr, M.; Mosaly, P.; Hodge, L. Service robot feature design effects on user perceptions and emotional responses. *Intell. Serv. Robot.* **2010**, *3*, 73–88.
92. Broadbent, E.; Lee, Y.I.; Stafford, R.Q.; Kuo, Y.H.; MacDonald, B.A. Mental schemas of robots as more human-like are associated with higher blood pressure and negative emotions in a human-robot interaction. *Int. J. Soc. Robot* **2011**, *3*, 291–297.
93. Watson, D.; Clark, L.A.; Tellegen, A. Development and validation of brief measures of positive and negative affect: The PANAS scales. *J. Pers. Soc. Psychol.* **1988**, *54*, 1063–1070.
94. Chaminade, T.; da Fonseca, D.; Rosset, D.; Lutcher, E.; Cheng, G.; Deruelle, C. FMRI Study of Young Adults with Autism Interacting with a Humanoid Robot. In Proceedings of the 21st IEEE International Symposium on Robot and Human Interactive Communication, Paris, France, 9–13 September 2012; pp. 380–385.
95. Chaminade, T.; Zecca, M.; Blakemore, S.-J.; Takanishi, A.; Frith, C.D.; Micera, S.; Dario, P.; Rizzolatti, G.; Gallese, V.; Umità, M.A. Brain response to a humanoid robot in areas implicated in the perception of human emotional gestures. *PLoS One* **2010**, *5*, doi: 10.1371/journal.pone.0011577.
96. Miura, N.; Sugiura, M.; Takahashi, M.; Moridaira, T.; Miyamoto, A.; Kuroki, Y.; Kawashima, R. An Advantage of Bipedal Humanoid Robot on the Empathy Generation: A Neuroimaging Study. In Proceedings of the IEEE/RSJ International Conference on Intelligent Robots and Systems, Nice, France, 22–26 September 2008; pp. 2465–2470.
97. Tiberio, L.; Cesta, A.; Cortellessa, G.; Padua, L.; Pellegrino, A.R. Assessing Affective Response of Older Users to a Telepresence Robot Using a Combination of Psychophysiological Measures. In Proceedings of the IEEE RO-MAN: The 21st IEEE International Symposium on Robot and Human Interactive Communication, Paris, France, 9–13 September 2012.
98. Cesta, A.; Cortellessa, G.; Orlandini, A.; Tiberio, L. *Addressing the Long-Term Evaluation of a Telepresence Robot for the Elderly*; Joaquim, F., Ana, L.N., Eds.; SciTePress: Rome, Italy, 2012; pp. 652–663.
99. Beer, J.; Takayama, L. Mobile Remote Presence Systems for Older Adults: Acceptance, Benefits, and Concerns. In Proceedings of the 6th International Conference on Human-Robot Interaction, Lausanne, Switzerland, 6–9 March 2011; pp. 19–26.
100. Sharkey, A.; Sharkey, N. Children, the elderly, and interactive robots. *IEEE Robot. Autom. Mag.* **2011**, *18*, 32–38.
101. Giraff. Available online: http://www.giraff.org (accessed on 17 May 2013).
102. ExCITE (Enabling SoCial Interaction Through Embodiment). Available online: http://www.oru.se/excite (accessed on 17 May 2013).

103. Spielberger, C.D.; Gorsuch, R.L.; Lushene, R.E. *Manual for the State-Trait Anxiety Inventory*; Consulting Psychologists Press: Palo Alto, CA, USA, 1970.
104. Davis, F.D. Perceived usefulness, perceived ease of use, and user acceptance of information technology. *MIS Quart.* **1989**, *13*, 319–340.
105. Tapus, A.; Matarić, M.J.; Scassellati, S. The grand challenges in socially assistive robotics. *IEEE Robot. Autom. Mag.* **2007**, *14*, 35–42.
106. Kim, K.H.; Bang, S.W.; Kim, S.R. Emotion recognition system using short-term monitoring of physiological signals. *Med. Biol. Eng. Comp.* **2004**, *42*, 419–427.
107. Kreibig, S.D. Autonomic nervous system activity in emotion: A review. *Biol. Psychol.* **2010**, *84*, 394–421.
108. Minato, T.; Shimada, M.; Itakura, S.; Lee, K.; Ishiguro, H. Evaluating the human likeness of an android by comparing gaze behaviors elicited by the android and a person. *Adv. Robot.* **2006**, *20*, 1147–1163.
109. Mori, M. The uncanny valley. *Energy* **1970**, *74*, 33–35.
110. Czaja, S.; Charness, N.; Fisk, A.; Hertzog, C.; Nair, S.; Rogers, W.; Sharit, J. Factors predicting the use of technology: Findings from the center for research and education on aging and technology enhancement. *Psychol. Aging* **2006**, *21*, 333–352.
111. Novak, D.; Mihelj, M.; Ziherl, J.; Olensek, A.; Munih, M. Psychophysiological measurements in a biocooperative feedback loop for upper extremity rehabilitation. *IEEE Trans. Neural Syst. Rehabil. Eng.* **2011**, *19*, 400–410.
112. Fairclough, S.H. Fundamentals of physiological computing. *Interact. Comp.* **2009**, *21*, 133–145.
113. Novak, D.; Mihelj, M.; Munih, M. A survey of methods for data fusion and system adaptation using autonomic nervous system responses in physiological computing. *Interact. Comput.* **2012**, *24*, 154–172.
114. Fisher, R.J. Social desirability bias and the validity of indirect questioning. *J. Consum. Res.* **1993**, *2*, 303–315.
115. Posner, J.; Russell, J.A.; Peterson, B.S. The circumplex model of affect: An integrative approach to affective neuroscience, cognitive development, and psychopathology. *Dev. Psychopathol.* **2005**, *17*, 715–734.

5

Robust Nonlinear Regulation of Limit Cycle Oscillations in UAVs Using Synthetic Jet Actuators

Natalie Ramos Pedroza [1,*], William MacKunis [1,*] and Vladimir V. Golubev [2]

[1] Department of Physical Sciences, Embry-Riddle Aeronautical University, 600 S Clyde Morris Blvd, Daytona Beach, FL 32114, USA

[2] Department of Aerospace Engineering, Embry-Riddle Aeronautical University, 600 S Clyde Morris Blvd, Daytona Beach, FL 32114, USA; E-Mail: golubd1b@erau.edu

* Authors to whom correspondence should be addressed; E-Mails: ramosa66@my.erau.edu (N.R.P.); mackuniw@erau.edu (W.M.)

External Editor: Ikuo Yamamoto

Abstract: In this paper, a synthetic jet actuators (SJA)-based nonlinear robust controller is developed, which is capable of completely suppressing limit cycle oscillations (LCO) in UAV systems with parametric uncertainty in the SJA dynamics and unmodeled external disturbances. Specifically, the control law compensates for uncertainty in an input gain matrix, which results from the unknown airflow dynamics generated by the SJA. Challenges in the control design include compensation for input-multiplicative parametric uncertainty in the actuator dynamic model. The result was achieved via innovative algebraic manipulation in the error system development, along with a Lyapunov-based robust control law. A rigorous Lyapunov-based stability analysis is utilized to prove asymptotic LCO suppression, considering a detailed dynamic model of the pitching and plunging dynamics. Numerical simulation results are provided to demonstrate the robustness and practical performance of the proposed control law.

Keywords: nonlinear; robust; LCO; synthetic jet actuator

---

1. Introduction

There has recently been a surge of interest in the design and application of unmanned aerial vehicles (UAVs). These UAVs can be used in numerous civilian applications, such as urban reconnaissance, package delivery, and area mapping. Besides, UAVs are utilized in various military applications as well. One of the biggest challenges involved in the autonomous operation of UAVs is in the design of flight tracking controllers for UAVs operating in uncertain and possibly adverse conditions. In particular, suppression of limit cycle oscillations (LCO) (or flutter) is an important concern in UAV tracking control design. This is especially true for applications involving smaller, lightweight UAV systems, where the aircraft wings are more susceptible to LCO. These engineering challenges necessitate the utilization of UAV flight controllers, which achieve accurate flight tracking in the presence of dynamic uncertainty while simultaneously suppressing LCO. Moreover, as practical considerations motivate the implementation of smaller UAVs, there is a growing need for UAV flight control designs that do not require heavy mechanical deflection surfaces.

LCO refer to "flutter" behaviors in UAV wings that manifest themselves as constant-amplitude oscillations [1], which result from nonlinearities inherent in the aeroelastic dynamics of the UAV system [2]. Due to these behaviors, the LCO would surpass the limiting safe flight boundaries of an aircraft [3] and could potentially lead to structural damage and catastrophes. Control applications for LCO suppression are often developed (e.g., see [4–6]) using mechanical deflection surfaces (e.g. flaps, ailerons, rudders, and elevators). However, when dealing with small UAVs, practical engineering considerations and physical constraints can preclude the addition of the large, heavy moving parts that are required for the installation of deflection surfaces. To address this challenge, the use of synthetic jet actuators (SJA) in UAV flight control systems is becoming popular as a practical alternative to mechanical deflection surfaces.

The design and application of SJA has recently increased by virtue of their capability to achieve momentum transfer with zero-net mass-flux. This beneficial feature eliminates the need for an external fuel supply, since the working substance is simply the gas (*i.e.*, air) that is already present in the environment of operation [7]. This makes SJA an attractive option in UAV applications, because of the significant reduction in the size of the required equipment. The SJA synthesize the jet flow through the alternating suction and ejection of fluid through an aperture, which is produced via pressure oscillations in a cavity [8] as shown in Figure 1. The pressure oscillations can be generated using various methods, including pistons in the SJA orifices [3] or piezoelectric diaphragms [7,9]. SJA can achieve boundary-layer flow control near the surface of a UAV wing [10], since they can provide instant actuation, unlike conventional mechanical control surfaces. In addition, SJA can expand the usable range of angle of attack, resulting in improved UAV maneuverability [11].

Recently developed nonlinear control methods using SJA typically utilize neural networks and/or complex fluid dynamics computations in the feedback loop (e.g., see [9,10,12–21]). While such approaches have been shown to yield good SJA-based control performance, they can require increased computational resources, which might not be available in small UAV applications. Adaptive control approaches have been applied to linear time-invariant (LTI) dynamic models to compensate for SJA nonlinearities and external disturbances [13]. Adaptive inverse control schemes are another popularly utilized method to compensate for the actuator nonlinearity inherent in SJA [9,14–17]. A robust tracking control method is proposed in [7], which builds on results similar to [9,14–17], to compensate for the SJA nonlinearity at a reduced computational cost. The aforementioned approaches have been shown to achieve good flight tracking performance using SJA; however, nonlinear control approaches have not as often been applied to SJA-based LCO suppression.

Figure 1. Schematic layout of a synthetic jet actuator.

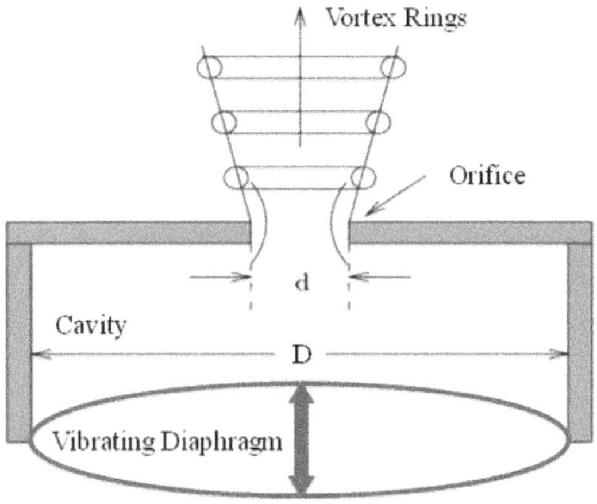

In this paper, an SJA-based nonlinear robust controller is developed, which is capable of completely suppressing LCO in UAV systems with uncertain dynamics. This difficulty was handled via innovative algebraic manipulation in the error system development, along with a Lyapunov-based robust control law. A salient feature about this robust control law is that its structure is continuous, and so bounded disturbances can be asymptotically rejected without the need for infinite bandwidth. A rigorous Lyapunov-based stability analysis is utilized to prove asymptotic pitching and plunging regulation, considering a detailed dynamic model of the pitching and plunging dynamics. Numerical simulation results are provided to demonstrate the performance of the proposed control law.

2. Dynamic Model and Properties

The equation describing LCO in an airfoil approximated as a 2-dimensional thin plate can be expressed as

$$M_s\ddot{p} + C_s\dot{p} + F(p)p = \begin{bmatrix} -Lift \\ Moment \end{bmatrix} \tag{1}$$

where the coefficients M_s, $C_s \in \mathbb{R}^{2\times 2}$ denote the structural mass and damping matrices, $F(p) \in \mathbb{R}^{2\times 2}$ is a nonlinear stiffness matrix, and $p(t) \in \mathbb{R}^2$ denotes the state vector. In Equation (1), $p(t)$ is explicitly defined as

$$p = \begin{bmatrix} h \\ \alpha \end{bmatrix} \tag{2}$$

where $h(t)$, $\alpha(t) \in \mathbb{R}$ denote the plunging [meters] and pitching [radians] displacements describing the LCO effects. Also in Equation (1), the structural linear mass matrix M_s [22]

$$M_s = \begin{bmatrix} m & S_\alpha \\ S_\alpha & I_\alpha \end{bmatrix} \tag{3}$$

where the parameters S_α, $I_\alpha \in \mathbb{R}$ are the static moment and moment of inertia, respectively. The structural linear damping matrix is described as

$$C_s = 2 \begin{bmatrix} \zeta_h \sqrt{k_h m} & 0 \\ 0 & \zeta_\alpha \sqrt{k_\alpha I_\alpha} \end{bmatrix} \tag{4}$$

where the parameters ζ_h, $\zeta_\alpha \in \mathbb{R}$ are the damping logarithmic decrements for plunging and pitching, and $m \in \mathbb{R}$ is the mass of the wing, or in this case, a flat plate. The nonlinear stiffness matrix utilized in this study is

$$F(p) = \begin{bmatrix} k_h & 0 \\ 0 & k_\alpha + k_{\alpha^3}\alpha^2 \end{bmatrix} \tag{5}$$

where k_α, $k_{\alpha^3} \in \mathbb{R}$ denote structural resistances to pitching (linear and nonlinear) and $k_h \in \mathbb{R}$ is the structural resistance to plunging.

In Equation (1), the total lift and moment are explicitly defined as

$$\begin{bmatrix} -Lift \\ Moment \end{bmatrix} = \begin{bmatrix} -(L + L_{v_j}) \\ (M + M_{v_j}) \end{bmatrix} \tag{6}$$
$$= M_a \ddot{p} + C_a \dot{p} + K_a p + L_\eta \eta + Bu$$

where $L_{vj}(t)$, $M_{vj}(t) \in \mathbb{R}$ denote the equivalent control force and moment generated by the jth SJA, and $L(t)$, $M(t) \in \mathbb{R}$ are the aerodynamic lift and moment due to the 2 degree-of-freedom motion [10]. In Equation (6), $\eta(t) \in \mathbb{R}^2$ denotes the aerodynamic state vector that relates the moment and lift to the structural modes. Also in Equation (6), $u(t) \in \mathbb{R}^2$ denotes the SJA-based control input (e.g., the SJA air velocity or acceleration), and $B \in \mathbb{R}^{2\times2}$ is an uncertain constant input gain matrix that relates the control input $u(t)$ to the equivalent force and moment generated by the SJA. Also in Equation (6), the aerodynamic and mode matrices M_a, C_a, K_a, $L_\eta \in \mathbb{R}^{2\times2}$ are described as

$$M_a = \pi\rho b^2 \begin{bmatrix} -1 & ba \\ ba & -b^2\left(\frac{1}{8} + a^2\right) \end{bmatrix} \tag{7}$$

$$C_a = \pi\rho b^2 \begin{bmatrix} 0 & -U \\ 0 & -Ub\left(\frac{1}{2} - a\right) \end{bmatrix} \tag{8}$$
$$+2\pi\rho U b\phi(0) \begin{bmatrix} -1 & -b\left(\frac{1}{2} - a\right) \\ b\left(\frac{1}{2} + a\right) & b^2\left(\frac{1}{2} + a\right)\left(\frac{1}{2} - a\right) \end{bmatrix}$$

$$K_a = 2\pi\rho U b\phi(0) \begin{bmatrix} 0 & -U \\ 0 & b\left(\frac{1}{2} + a\right)U \end{bmatrix} \tag{9}$$

$$L_\eta = 2\pi\rho U b \begin{bmatrix} a_1 b_1 & a_2 b_2 \\ -b\left(\dfrac{1}{2} + a\right) a_1 b_1 & -b\left(\dfrac{1}{2} + a\right) a_2 b_2 \end{bmatrix} \tag{10}$$

where $\phi(0)$ is the Wagner solution function at 0, and the parameters a_1, b_1, a_2, $b_2 \in \mathbb{R}$ are the Wagner coefficients. In addition, $a, b \in \mathbb{R}$ denote the relative locations of the rotational axis from the mid-chord and the semi-chord, respectively. The aerodynamic state variables are governed by [22]

$$\dot{\eta} = C_\eta \dot{p} + K_\eta p + S_\eta \eta \tag{11}$$

The aerodynamic state matrices in Equation (11), C_η, K_η, $S_\eta \in \mathbb{R}^{2\times 2}$, are explicitly defined as

$$C_\eta = \frac{U}{b}\begin{bmatrix} -1 & -b\left(\frac{1}{2} - a\right) \\ -1 & -b\left(\frac{1}{2} - a\right) \end{bmatrix} \tag{12}$$

$$K_\eta = \frac{U}{b}\begin{bmatrix} 0 & -U \\ 0 & -U \end{bmatrix} \tag{13}$$

$$S_\eta = \frac{U}{b}\begin{bmatrix} -b_1 & 0 \\ 0 & -b_2 \end{bmatrix}. \tag{14}$$

By substituting Equation (6) into Equation (1), the LCO dynamics can be expressed as

$$\ddot{p} = -M^{-1}C\dot{p} - M^{-1}Kp + M^{-1}L_\eta \eta + M^{-1}Bu \tag{15}$$

where $C = C_s - C_a$, $K = F(p) - K_a$, and $M = M_s - M_a$. By making the definitions $x_1(t) = h(t)$, $x_2(t) = \alpha(t)$, $x_3(t) = \dot{h}(t)$, $x_4(t) = \dot{\alpha}(t)$, $x_5(t) = \eta_1(t)$, and $x_6(t) = \eta_2(t)$, the dynamic equation in Equation (15) can be expressed in state form as

$$\dot{x} = A(x)x + \bar{B}u \tag{16}$$

where $x(t) \in \mathbb{R}^6$ is the state vector, $A(x) \in \mathbb{R}^{6\times 6}$ is the state matrix (state-dependent). In Equation (16), the input gain matrix $\bar{B} \in \mathbb{R}^{6\times 2}$ is defined as

$$\bar{B} = \begin{bmatrix} 0_{2\times 2} \\ M^{-1}B \\ 0_{2\times 2} \end{bmatrix} \tag{17}$$

where $0_{2\times 2}$ denotes a 2×2 matrix of zeros. The structure of the input gain matrix in Equation (17) results from the fact that the control input $u(t)$ only directly affects $\ddot{h}(t)$ and $\ddot{\alpha}(t)$.

3. Control Development

The objective is to design the control signal $u(t)$ to regulate the plunge and pitching dynamics (*i.e.*, $h(t)$, $\alpha(t)$) to zero. To facilitate the control design, the expression in Equation (15) is rewritten as

$$M\ddot{p} = g(h, \alpha, \eta) + Bu \tag{18}$$

where $g(h, \alpha, \eta)$ is an unknown, unmeasurable auxiliary function.

Remark 1. *Based on the open-loop error dynamics in Equation (18), one of the control design challenges is that the control input $u(t)$ is premultiplied by the uncertain matrix B. In the following control development and stability analysis, it will be assumed that the matrix B is uncertain, and the robust control law will be designed with a constant feedforward estimate of the uncertain matrix. The simulation results demonstrate the capability of the robust control law to compensate for the input matrix uncertainty without the need for online parameter estimation or function approximators.*

To quantify the control objective, a regulation error $e_1(t) \in \mathbb{R}^2$ and auxiliary tracking error variables $e_2(t), r(t) \in \mathbb{R}^2$ are defined as

$$e_1 = p - p_d \quad (19)$$

$$e_2 = \dot{e}_1 + \alpha_1 e_1 \quad (20)$$

$$r = \dot{e}_2 + \alpha_2 e_2 \quad (21)$$

where $\alpha_1, \alpha_2 > 0 \in \mathbb{R}^+$ are user-defined control gains, and the desired plunging and pitching states $p_d = [h, a]^T = [0, 0]^T$ for the plunging and pitching suppression objective. To facilitate the following analysis, Equation (21) is premultiplied by M and the time derivative is calculated as

$$M\dot{r} = M\ddot{e}_2 + \alpha_2 M\dot{e}_2 \quad (22)$$

After using Equations (18)–(21), the open-loop error dynamics are obtained as

$$M\dot{r} = \tilde{N} + N_d + B\dot{u} - e_2 \quad (23)$$

where the unknown, unmeasurable auxiliary functions $\tilde{N}(e_1, e_2, r)$, $N_d(p_d, \dddot{p}_d) \in \mathbb{R}^2$ are defined as

$$\begin{aligned}\tilde{N} \triangleq\ & \dot{g}(p,\eta) - \dot{g}(p_d,\eta) + \alpha_1\left(r - \alpha_2 e_2 - \alpha_1 e_2 + \alpha_1^2 e_1\right) \\ & + \alpha_2 M(r - \alpha_2 e_2) + e_2\end{aligned} \quad (24)$$

$$N_d \triangleq -\dddot{p}_d + \dot{g}(p_d, \eta) \quad (25)$$

The motivation for defining the auxiliary functions in Equations (24) and (25) is based on the fact that the following inequalities can be developed:

$$\|\tilde{N}\| \leq \rho_0\|z\|, \quad \|N_d\| \leq \zeta_{N_d}, \quad \|\dot{N}_d\| \leq \zeta_{\dot{N}_d} \quad (26)$$

where $\rho_0, \zeta_{N_d}, \zeta_{\dot{N}_d} \in \mathbb{R}^+$ are known bounding constants, and $z(t) \in \mathbb{R}^6$ is defined as

$$z \triangleq \begin{bmatrix} e_1^T & e_2^T & r^T \end{bmatrix}^T \quad (27)$$

Based on the open-loop error dynamics in Equation (23), the control input is designed via

$$\dot{u} = \hat{B}^{-1}(-(k_s + I_{2\times 2})r - \beta sgn(e_2(t))) \quad (28)$$

where k_s, $\beta \in \mathbb{R}^{2\times 2}$ denote constant, positive definite, diagonal control gain matrices, and $I_{2\times 2}$ denotes a 2×2 identity matrix. In Equation (28), $\hat{B} \in \mathbb{R}^{2\times 2}$ denotes a constant, feedforward "best guess" estimate of the uncertain input gain matrix B. Note that the control input $u(t)$ does not depend on the unmeasurable acceleration term $r(t)$, since Equation (28) can be directly integrated to show that $u(t)$ requires measurements of $e_1(t)$ and $e_2(t)$ only.

To facilitate the following stability proof, the control gain matrix β in Equation (28) is selected to satisfy the sufficient condition

$$\lambda_{\min}(\beta) > \zeta_{N_d} + \frac{1}{\alpha_2}\zeta_{\dot{N}_d} \tag{29}$$

where $\lambda_{\min}(\cdot)$ denotes the minimum eigenvalue of the argument. After substituting Equation (28) into Equation (23), the closed-loop error dynamics are obtained as

$$M\dot{r} = \tilde{N} + N_d - (k_s + I_{n\times n})r + \beta sgn(e_2(t)) - e_2 \tag{30}$$

To reduce the complexity of the following stability analysis, it is assumed that the product $B\hat{B}^{-1}$ is equal to identity. It can be proven that asymptotic regulation can be achieved for the case where the feedforward estimate $\hat{B}$ is within some prescribed finite range of the actual matrix B. The proof including the uncertainty in B is lengthy and is omitted here for brevity. The complete proof can be found in [23,24]. The following simulation results demonstrate the performance of the controller in the presence of uncertainty in the input gain matrix B.

3.1. Stability Analysis

Theorem 1. *The controller given in Equation (28) ensures asymptotic regulation of pitching and plunging displacements in the sense that*

$$\|e_1(t)\| \to 0 \qquad as \qquad t \to \infty \tag{31}$$

provided the control gain k_s is selected sufficiently large, and β is selected according to the sufficient condition in Equation (29).

Lemma 1. *To facilitate the following proof, let $\mathcal{D} \subset \mathbb{R}^7$ be a domain containing $w(t) = 0$, where $w(t) \in \mathbb{R}^7$ is defined as*

$$w(t) \triangleq \begin{bmatrix} z^T & \sqrt{P(t)} \end{bmatrix}^T \tag{32}$$

In Equation (32), the auxiliary function $P(t) \in \mathbb{R}$ is the generalized solution to the differential equation

$$\dot{P}(t) = -L(t) \tag{33}$$

$$P(0) = \beta\|e_2(0)\| - N_d^T(0)\,e_2(0) \tag{34}$$

where the auxiliary function $L(t) \in \mathbb{R}$ is defined as

$$L(t) = r^T(N_d(t) - \beta sgn(e_2)) \tag{35}$$

Provided the sufficient condition in Equation (29) is satisfied, the following inequality can be obtained:

$$\int_0^t L(\tau)d\tau \le \beta\|e_2(0)\| - N_d^T(0)\,e_2(0) \tag{36}$$

Hence, Equation (36) can be used to conclude that $P(t) \ge 0$.

Proof 1. (See Theorem 1) *Let $V(w,t) : \mathcal{D} \times [0,\infty) \to \mathbb{R}$ be defined as the nonnegative function*

$$V(w,t) \triangleq \frac{1}{2}e_1^T e_1 + \frac{1}{2}e_2^T e_2 + \frac{1}{2}r^T M r + P \tag{37}$$

where $e_1(t)$, $e_2(t)$, and $r(t)$ are defined in Equations (19)–(21), respectively; and the positive definite function $P(t)$ is defined in Equation (33). The function $V(w,t)$ satisfies the inequality

$$U_1(w) \leq V(w,t) \leq U_2(w) \tag{38}$$

provided the sufficient condition introduced in Equation (29) is satisfied, where $U_1(w)$, $U_2(w) \in \mathbb{R}$ denote the positive definite functions

$$U_1 \triangleq \lambda_1 \|w\|^2, \qquad U_2 \triangleq \lambda_2 \|w\|^2 \tag{39}$$

where $\lambda_1 = \min\left\{\frac{1}{2}, \lambda_{\min}(M)\right\}$ and $\lambda_2 = \max\{1, \lambda_{\max}(M)\}$. After taking the time derivative of Equation (37) and utilizing Equation (20), Equation (21), Equation (30), and Equation (33), $\dot{V}(w,t)$ can be upper bounded as

$$\begin{aligned} \dot{V}(w,t) \leq & -\left(\alpha_1 - \frac{1}{2}\right)\|e_1\|^2 - \left(\alpha_2 - \frac{1}{2}\right)\|e_2\|^2 - \|r\|^2 \\ & -k_s\|r\|^2 + \rho_0\|z\|\|r\| \end{aligned} \tag{40}$$

where the bounds in Equation (26) were used, and the fact that $e_1^T e_2 \leq \frac{1}{2}\|e_1\|^2 + \frac{1}{2}\|e_2\|^2$ (i.e., Young's inequality) was utilized. After completing the squares in Equation (40), the upper bound on $\dot{V}(w,t)$ can be expressed as

$$\begin{aligned} \dot{V}(w,t) \leq & -\left(\alpha_1 - \frac{1}{2}\right)\|e_1\|^2 - \left(\alpha_2 - \frac{1}{2}\right)\|e_2\|^2 - \|r\|^2 \\ & -k_s\left(\|r\| - \frac{\rho_0}{2k_s}\|z\|\right)^2 + \frac{\rho_0^2}{4k_s}\|z\|^2 \end{aligned} \tag{41}$$

Since $k_s > 0$, the upper bound in Equation (41) can be expressed as

$$\dot{V}(w,t) \leq -\left(\lambda_0 - \frac{\rho_0^2}{4k_s}\right)\|z\|^2 \tag{42}$$

where $\lambda_0 \triangleq \min\left\{\alpha_1 - \frac{1}{2}, \alpha_2 - \frac{1}{2}, 1\right\}$.
The following expression can be obtained from Equation (42):

$$\dot{V}(w,t) \leq -U(w) \tag{43}$$

where $U(w) = c\|z\|^2$, for some positive constant $c \in \mathbb{R}$, is a continuous, positive semi-definite function.

It follows directly from the Lyapunov analysis that $e_1(t), e_2(t), r(t) \in \mathcal{L}_\infty$. This implies that $\dot{e}_1(t)$, $\dot{e}_2(t) \in \mathcal{L}_\infty$ from the definitions given in Equations (20) and (21). Given that $\dot{e}_1(t)$, $e_2(t)$, $r(t) \in \mathcal{L}_\infty$, it follows that $\ddot{e}_1(t) \in \mathcal{L}_\infty$ from Equation (21). Thus, Equation (19) can be used to prove that $p(t)$, $\dot{p}(t)$, $\ddot{p}(t) \in \mathcal{L}_\infty$. Since $p(t)$, $\dot{p}(t)$, $\ddot{p}(t) \in \mathcal{L}_\infty$, Equation (18) can be used to prove that $u(t) \in \mathcal{L}_\infty$. Since $r(t)$, $u(t) \in \mathcal{L}_\infty$, Equation (28) can be used to show that $\dot{u}(t) \in \mathcal{L}_\infty$. Given that $e_1(t)$, $e_2(t)$,

$r(t), \dot{u}(t) \in \mathcal{L}_\infty$, Equation (30) can be used along with Equation (26) to prove that $\dot{r}(t) \in \mathcal{L}_\infty$. Since $\dot{e}_1(t)$, $\dot{e}_2(t)$, $\dot{r}(t) \in \mathcal{L}_\infty$, $e_1(t), e_2(t), r(t)$, are uniformly continuous. Equation (27) can then be used to show that $z(t)$ is uniformly continuous. Given that $e_1(t), e_2(t), r(t) \in \mathcal{L}_\infty$, Equation (37) and Equation (42) can be used to prove that $z(t) \in \mathcal{L}_\infty \cap \mathcal{L}_2$. Barbalat's lemma [25] can now be invoked to prove that $\|z(t)\| \to 0$ as $t \to \infty$. Hence, $\|e_1(t)\| \to 0$ as $t \to \infty$ from Equation (27). Further, given that $V(w,t)$ in Equation (37) is radially unbounded, convergence of $e_1(t)$ is guaranteed, regardless of initial conditions—a global result.

4. Results and Discussion

A numerical simulation was created to demonstrate the performance of the control law developed in Equation (28). In order to develop a realistic stepping stone to high-fidelity numerical simulation results using detailed computational fluid dynamics models, the following simulation results are based on detailed dynamic parameters and specifications. The simulation is based on the dynamic model given in Equation (1) and Equation (11). The dynamic parameters utilized in the simulation are summarized in Table 1 and were obtained from [22].

Table 1. Constant simulation parameters.

$\rho = 1.225\frac{kg}{m^3}$	$a = -0.24$	$U = 18.25\frac{m}{s}, 19.5\frac{m}{s}, 20.5\frac{m}{s}$
$m = 2.55kg$	$b = 0.11m$	$v = 18\frac{m}{s}$
$S_\alpha = 10.4 \times 10^{-3} kg \cdot m$	$a_1 = 0.1650$	$a_2 = 0.0455$
$I_\alpha = 2.51 \times 10^{-3} kg \cdot m$	$b_1 = 0.3350$	$b_2 = 0.3000$
$k_h = 450\frac{N}{m}$	$k_\alpha = 9.3\frac{N}{m}$	$k_{\alpha^3} = 55\frac{N}{m}$
$\zeta_h = 5.5 \times 10^{-3}$	$\zeta_\alpha = 1.8 \times 10^{-2}$	

The following simulation results were achieved using control gains defined as

$$\beta = \begin{bmatrix} 0.001 & 0 \\ 0 & 25 \end{bmatrix}, \quad k_s = \begin{bmatrix} 0.00001 & 0 \\ 0 & 0.11 \end{bmatrix} \tag{44}$$

$$\alpha_1 = \begin{bmatrix} 1 & 0 \\ 0 & 35 \end{bmatrix}, \quad \alpha_2 = \begin{bmatrix} 1 & 0 \\ 0 & 35 \end{bmatrix} \tag{45}$$

The control gains given in Equations (44) and (45) were selected based on achieving a desirable response in terms of settling time and required control effort. To test the case where the input gain matrix B is uncertain, it is assumed in the simulation that the actual value of B is the 2×2 identity matrix, but the constant feedforward estimate $\hat{B}$ used in the control law is given by

$$\hat{B} = \begin{bmatrix} 0.9 & 0.1 \\ -0.1 & 1.1 \end{bmatrix}$$

Figures 2–3 and Figures 4–5 show the open-loop pitching and plunging response of the simulated system for the case where the free stream velocity U is 18.25 m/s and 19.5 m/s, respectively. These figures demonstrate that sustained LCO occur in the absence of feedback control. Figures 6 and 7 show the closed-loop pitching and plunging response for the case where $U = 18.25\ m/s$, and Figures 8 and 9 show the required control force and moment, respectively, in this case. These simulation results demonstrate the capability of the robust control law to asymptotically suppress LCO using a control force and moment that are within reasonable limits. This demonstrates the practical efficacy of the control design to suppress LCO using the limited control authority that is available from the SJA. It should be noted, however, that in actual application, arrays consisting of several SJAs can be utilized to realize a greater overall control force and moment.

Figure 2. Uncontrolled pitching response for free stream velocity $U = 18.25\ m/s$.

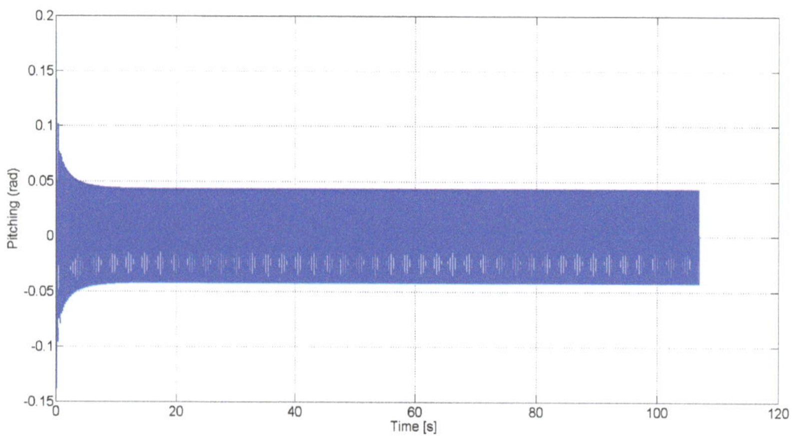

Figure 3. Uncontrolled plunging response for free stream velocity $U = 18.25\ m/s$.

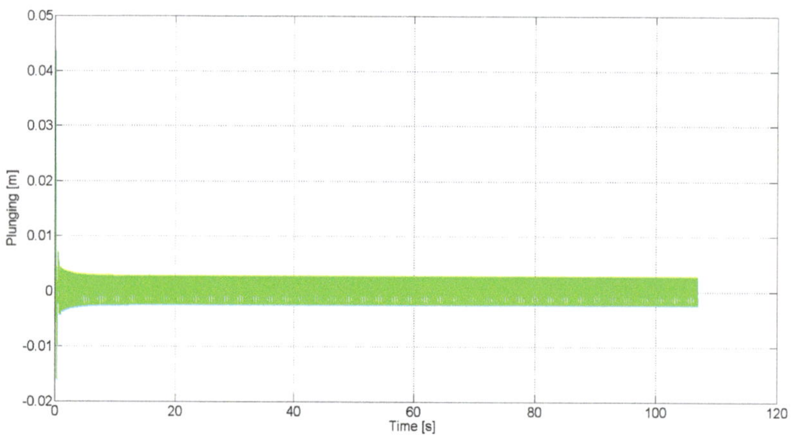

Figures 10–11 and Figures 12–13 show the closed-loop pitching and plunging response and the required control force and moment, respectively, for the case where the free stream velocity $U = 19.5\ m/s$. Figures 14–15 and Figures 16–17 show the closed-loop pitching and plunging response and the required control moment and force for the case where $U = 20.5\ m/s$. As shown in the figures, an increased control effort is required to suppress the LCO in the presence of the increased air velocity.

Figures 18 and 19 show the robustness of the proposed controller in the event of sudden unmodeled disturbances. In these plots, a sudden increase (*i.e.*, step function) in the pitching and plunging rates was programmed to occur at $10.9\ s$. Simultaneously, the free stream velocity U was increased from $19\ m/s$

to 25 m/s for approximately 0.2 seconds to simulate a disturbance due to a wind gust. The pitching and plunging responses in Figures 18 and 19 demonstrate the capability of the control law to compensate for the sudden and unexpected disturbances. Figures 20 and 21 show the commanded control force and moment, respectively, of the closed-loop system in response to the disturbance.

Figure 4. Uncontrolled pitching response for free stream velocity $U = 19.5\ m/s$.

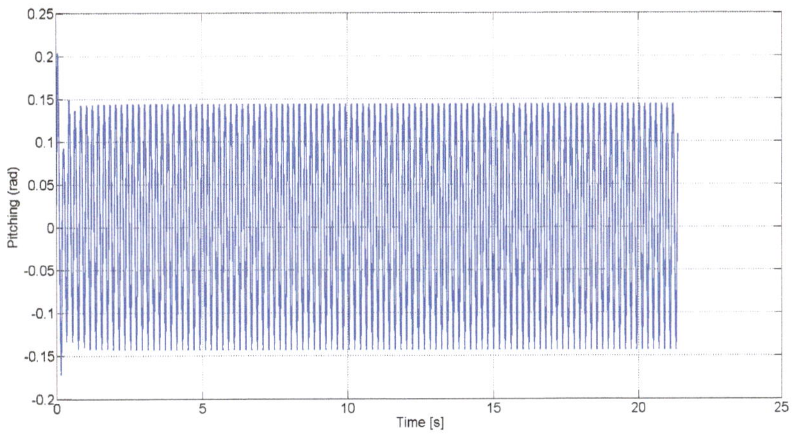

Figure 5. Uncontrolled plunging response for free stream velocity $U = 19.5\ m/s$.

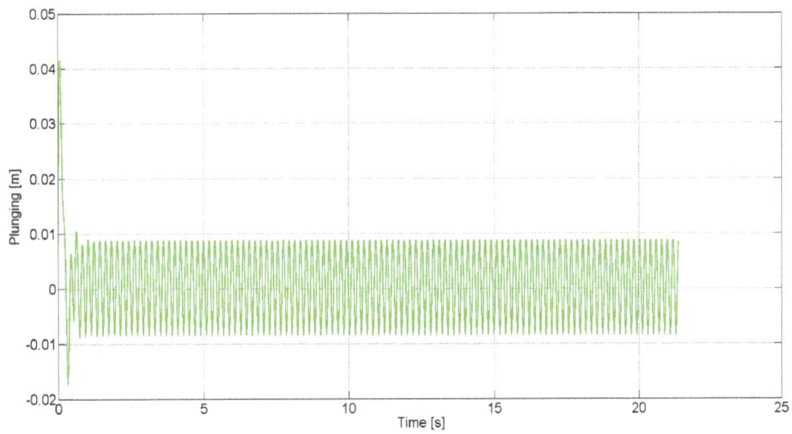

Figure 6. Closed-loop pitching response for free stream velocity $U = 18.25\ m/s$.

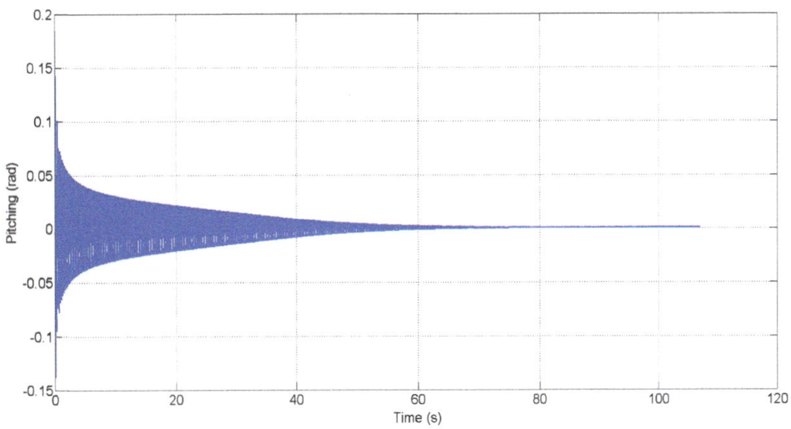

Figure 7. Closed-loop plunging response for free stream velocity $U = 18.25\ m/s$.

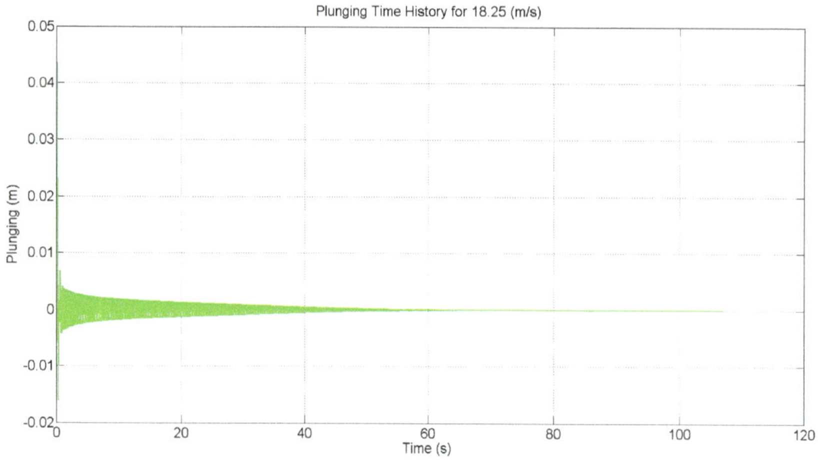

Figure 8. Actuator force required for free stream velocity $U = 18.25\ m/s$.

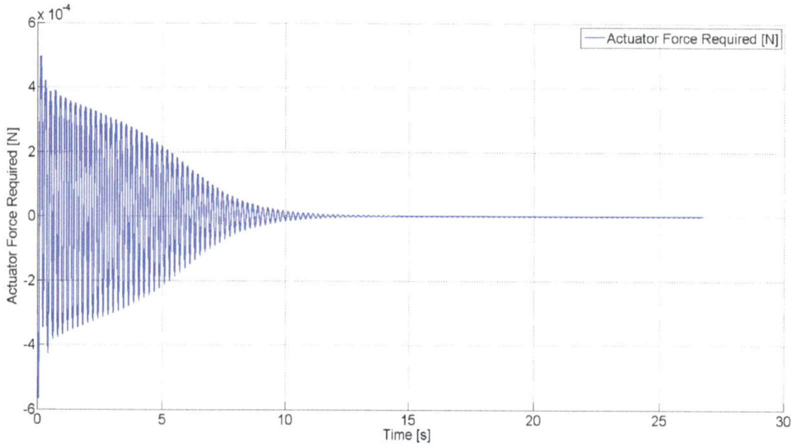

Figure 9. Actuator moment required for free stream velocity $U = 18.25\ m/s$.

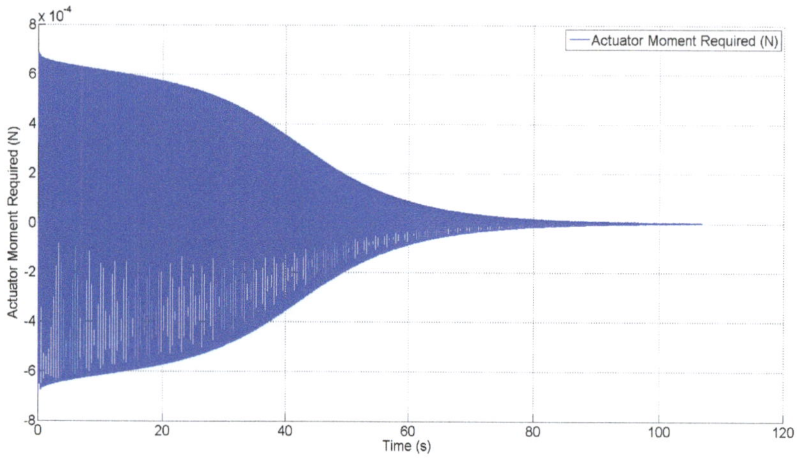

Figure 10. Closed-loop pitching response for free stream velocity $U = 19.5\ m/s$.

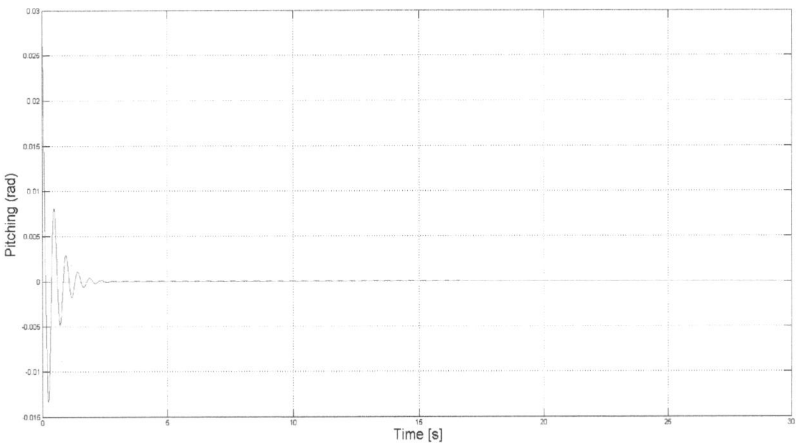

Figure 11. Closed-loop plunging response required for free stream velocity $U = 19.5\ m/s$.

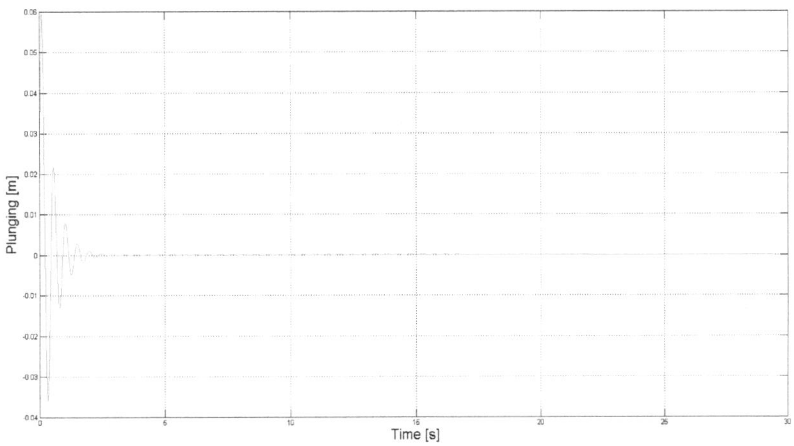

Figure 12. Actuator force required for free stream velocity $U = 19.5\ m/s$.

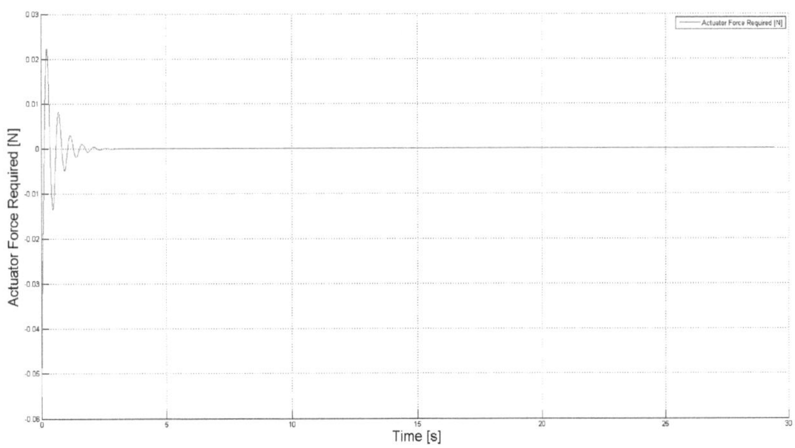

Figure 13. Actuator moment required for free stream velocity $U = 19.5\ m/s$.

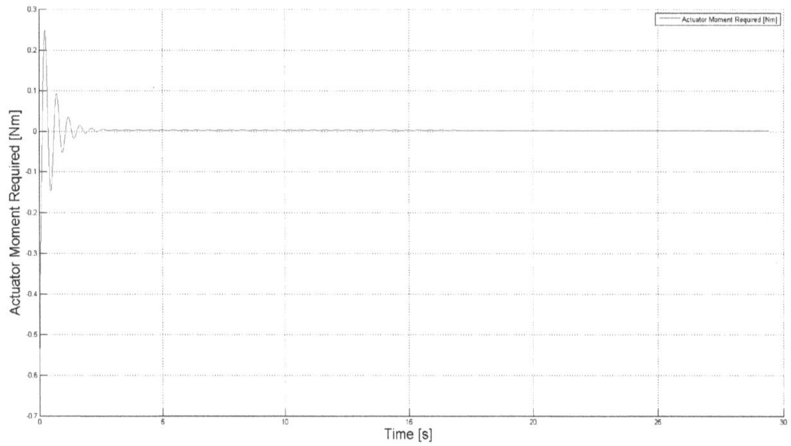

Figure 14. Closed-loop pitching response for free stream velocity $U = 20.5\ m/s$.

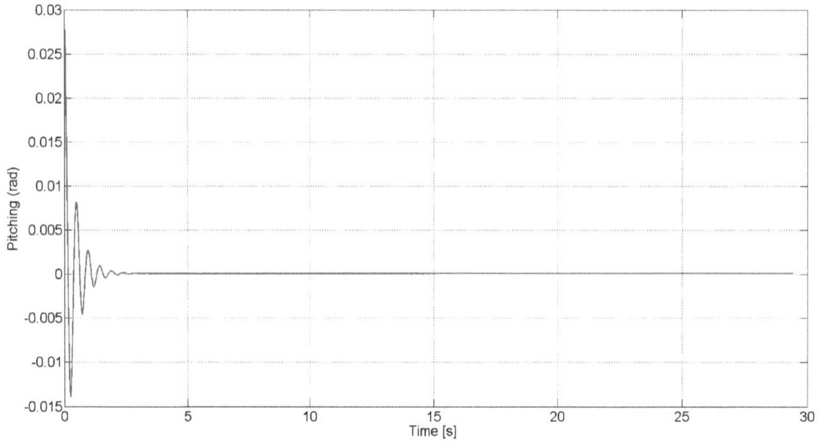

Figure 15. Closed-loop plunging response for free stream velocity $U = 20.5\ m/s$.

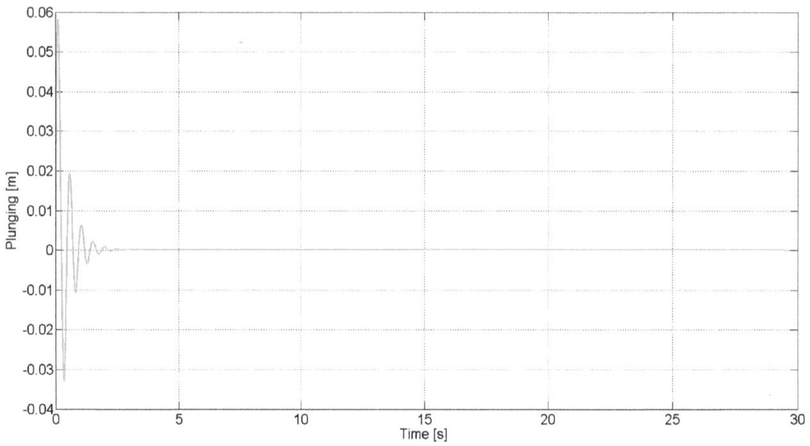

Figure 16. Actuator force required for free stream velocity $U = 20.5\ m/s$.

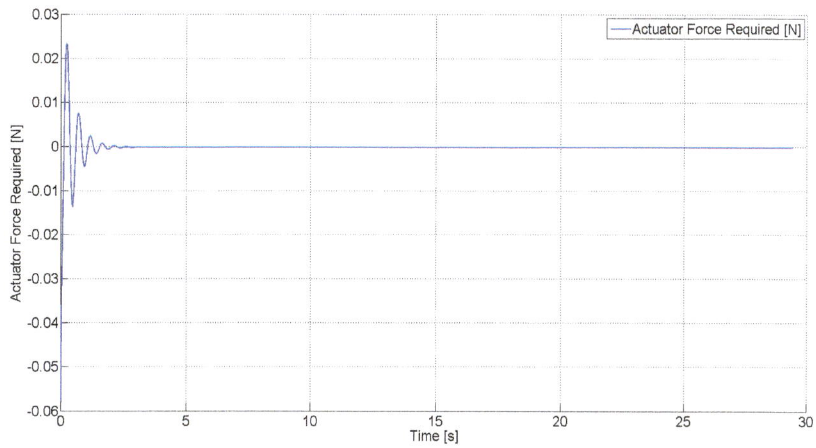

Figure 17. Actuator moment required for free stream velocity $U = 20.5\ m/s$.

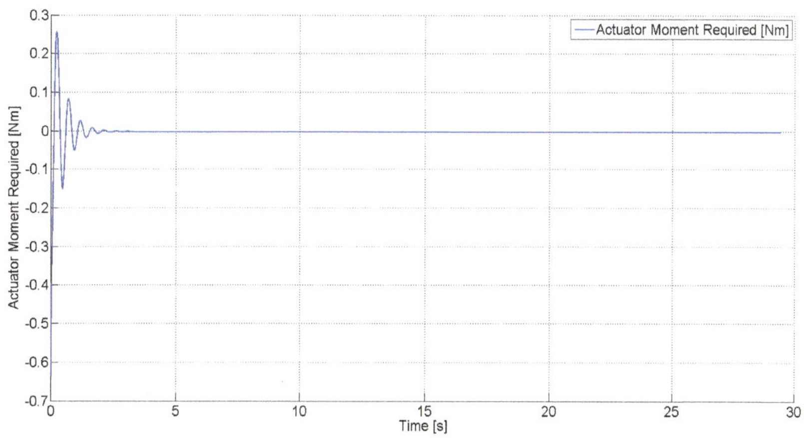

Figure 18. Closed-loop pitching response in the presence of a disturbance.

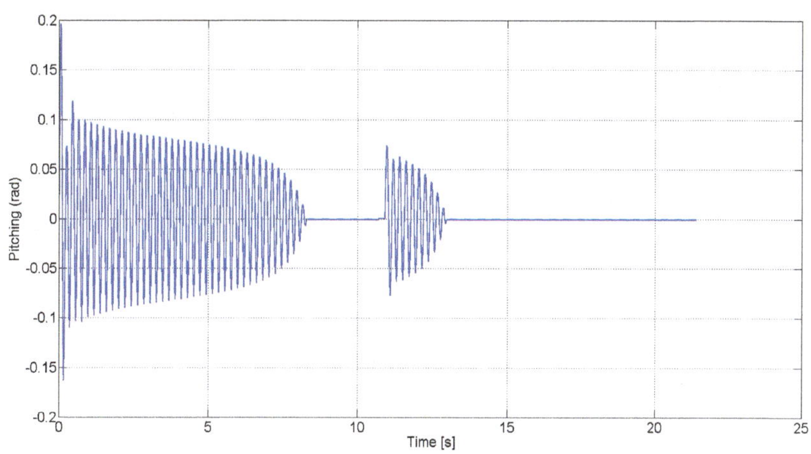

Figure 19. Closed-loop plunging response in the presence of a disturbance.

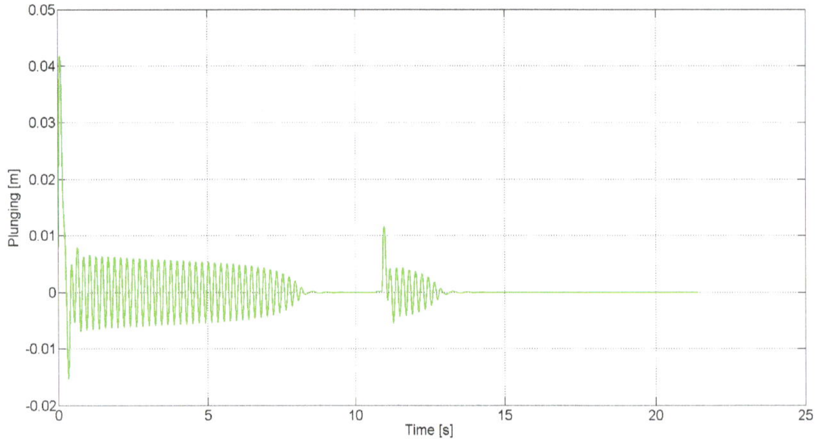

Figure 20. Actuator force use in the presence of a disturbance.

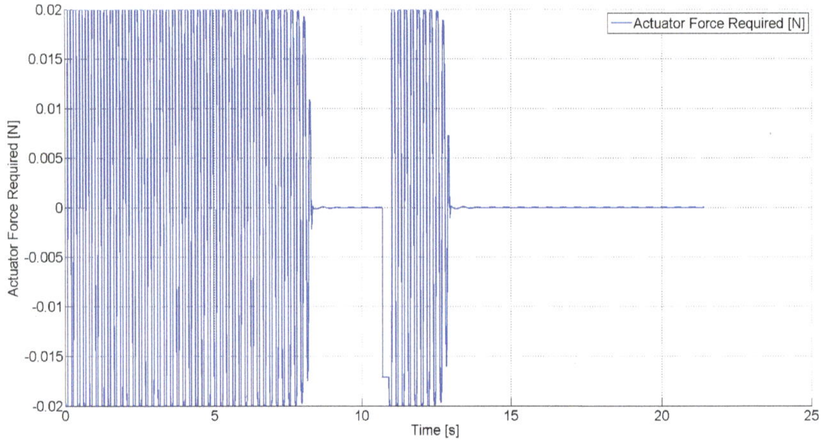

Figure 21. Actuator moment response in the presence of a disturbance.

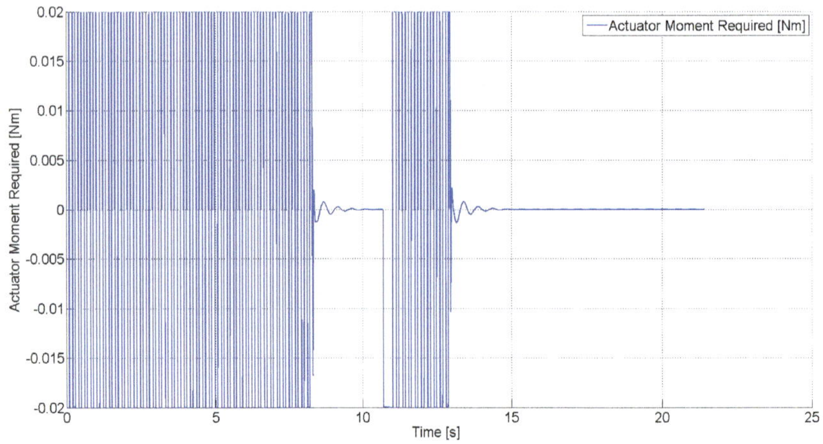

5. Conclusions

A nonlinear robust control law for SJA-based LCO suppression in UAV wings is presented. The proposed control law is rigorously proven to achieve global asymptotic regulation of the pitching and plunging displacements to zero in the presence of dynamic model uncertainty and parametric actuator uncertainty. Furthermore, the proposed control law is shown via numerical simulation to compensate for unmodeled external disturbances (*i.e.*, due to wind gusts and unmodeled effects). It is further shown that the robust control law can achieve suppression of LCO using minimal control effort. Future work will address LCO suppression control design, which can be proven to achieve simultaneous pitching and plunging regulation using only a scalar control input (*i.e.*, the underactuated control problem). In addition, future work will target controller validation via high-fidelity computational fluid dynamics simulations.

Acknowledgments

The authors would like to thank the Air Force Research Laboratory Munitions Directorate and the American Society for Engineering Education for their support in this research.

Author Contributions

Natalie Ramos Pedroza performed the literature review and prepared the preliminary version of the article; she also prepared the original controller derivation section and assisted with the simulation code-writing. William MacKunis thoroughly checked the mathematical development, and generated most of the numerical simulation results. Vladimir Golubev proofread the article, assisted with numerical simulation results, and performed preliminary testing to determine approximations of the force and moment magnitudes that can be delivered by synthetic jet actuators.

Conflicts of Interest

The authors declare no conflict of interest.

References

1. O'Donnell, K.S.; Marzocca, P.; Milanese, A. Design of a Wind Tunnel Apparatus to Assist Flow and Aeroelastic Control via Zero Net Mass Flow Actuators. In Proceedings of the Structures, Structural Dynamics, and Materials Conference, Fort Worth, TX, USA, 26 April 2007; pp. 1–12.
2. Satak, N.; Hernandez, E.A.P.; Hurtado, J.E. Rate-Free Control of Nonlinear Wing Section Using Pitch and Plunge Measurements Only. In Proceedings of the 53rd AIAA/ASME/ASCE/AHS/ASC Structures, Structural Dynamics and Materials Conference, Honolulu, HI, USA, 23–26 April 2012; pp. 1–7.
3. Rubillo, C.; Marzocca, P.; Bollt, E. Active aeroelastic control of lifting surfaces via jet reaction limiter control. *Int. J. Bifurc. Chaos* **2006**, *16*, 2559–2574.
4. Frampton, K.D.; Clark, R.L.; Carolina, N. Experiments on Control of Limit-Cycle Oscillations in a Typical Section Introduction. *J. Guid. Control Dyn.* **2000**, *23*, 956–960.

5. Strganac, T.W.; Kot, J.; Thompson, D.E.; Kurdila, A.J. Identifaction and control of limit cycle oscillations in aeroelastic systems. *J. Guid. Control Dyn.* **1999**, *23*, 1127–1133.
6. Platanitis, G.; Strganac, T.W. Control of a Nonlinear Wing Section Using Leading- and Trailing-Edge Surfaces. *J. Guid. Control Dyn.* **2004**, *27*, 52–58.
7. Mackunis, W.; Subramanian, S.; Mehta, S.; Ton, C.; Curtis, J.W.; Reyhanoglu, M. Robust Nonlinear Aircraft Tracking Control Using Synthetic Jet Actuators. In Proceedings of the IEEE American Control Conference, Firenze, Italy, 10–13 December 2013; pp. 1–8.
8. Jee, S.; Lopez Mejia, O.D.; Moser, R.D.; Muse, J.A.; Kutay, A.T.; Calise, A.J. Simulation of Rapidly Maneuvering Airfoils with Synthetic Jet Actuators. *AIAA J.* **2013**, *51*, 1883–1897.
9. Deb, D.; Burkholder, J.O.; Smith, D. An Adaptive Inverse Control Scheme for Synthetic Jet Actuator Arrays. In Proceedings of Infotech@Aerospace' 05, Arlington, VA, USA, 27 September 2005, pp. 1–11.
10. Milanese, A.; Breuker, R.D.; Marzocca, P.; Abdalla, M. Distributed Synthetic Jet Actuators for the Control of Nonlinear Aeroelastic Systems. In Proceedings of the 19th International Conference on Adaptive Structures and Technologies, Ascona, Switzerland, 6–9 October 2008; p. 12.
11. Amitay, M.; Smith, D.R.; Kilbens, V.; Parekh, D.E.; Glezer, A. Aerodynamic Flow Control over an Unconventional Airfoil Using Synthetic Jet Actuators. *AIAA J.* **2001**, *39*, 361–370.
12. Tchieu, A.A.; Kutay, A.T.; Muse, J.A.; Calise, A.J.; Leonard, A. Validation of a Low-Order Model for Closed-Loop Flow Control Enable Flight. In Proceedings of the 4th Flow Control Conference, Seattle, WA, USA, 23–26 June 2008; pp. 1–20.
13. Mondschein, S.T.; Tao, G.; Burkholder, J.O. Adaptive Actuator Nonlinearity Compensation and Disturbance Rejection Applied to Aircraft Models with Synthetic Jet Actuators. In Proceedings of the AIAA Guidance, Navigation, and Control Conference, Toronto, ON, Canada, 2–5 August 2010.
14. Deb, D.; Tao, G.; Burkholder, J.O.; Smith, D.R. An Adaptive Inverse Control Scheme for A Synthetic Jet Actuator Model. In Proceedings of the 2005 American Control Conference, Portland, OR, USA, 8–10 June 2005; Volume 82071, pp. 2646–2651.
15. Deb, D.; Burkholder, J.; Smith, D. Adaptive synthetic jet actuator compensation for a nonlinear tailless aircraft model at low angles of attack. In Proceedings of the 2006 American Control Conference, Minneapolis, MN, USA, 14–16 June 2006; p. 6.
16. Deb, D.; Tao, G.; Burkholder, J.O.; Smith, D.R. Adaptive Compensation Control of Synthetic Jet Actuator Arrays for Airfoil Virtual Shaping. *J. Aircr.* **2007**, *44*, 616–626.
17. Deb, D.; Tao, G.; Burkholder, J.O.; Smith, D.R. Adaptive Synthetic Jet Actuator Compensation for A Nonlinear Aircraft Model at Low Angles of Attack. *IEEE Trans. Control Syst. Technol.* **2008**, *16*, 983–995.
18. Liu, Y.; Ciuryla, M.; Amitay, M.; Chiman, K.; Myatt, J.H.; Zhang, X.; Ren, Z.; Casey, J.P. Integrated flight control and flow control using synthetic jet arrays. In Proceedings of AIAA Guidance, Navigation, and Control Conference and Exhibit, Keystone, CO, USA, 21 August 2006.
19. Singhal, C.; Tao, G.; Burkholder, J.O. Neural Network-Based Compensation of Synthetic Jet Actuator Nonlinearities for Aircraft Flight Control. In Proceedings of the AIAA Guidance, Navigation, and Control Conference, Chicago, IL, USA, 10–13 August 2009; pp. 1–19.

20. Tao, G.; Kokotovic, P.V. *Adaptive Control of Systems with Actuator and Sensor Nonlinearities*, 1st ed.; Wiley: New York, NY, USA, 1996; p. 294.
21. Jee, S.K.; Lopez, O.; Moser, R.D.; Kutay, A.T.; Muse, J.A.; Calise, A.J. Flow Simulation of a Controlled Airfoil with Synthetic Jet Actuators. In Proceedings of the 19th AIAA Computational Fluid Dynamics, San Antonio, TX, USA, 22–25 June 2009; pp. 1–13.
22. Golubev, V.V.; Dreyer, B.D.; Hollenshade, T.M.; Visbal, M.R. High-Accuracy Viscous Simulations of Gust-Airfoil Nonlinear Aeroelastic Interaction. In Proceedings of the 39th AIAA Fluid Dynamics Conference, San Antonio, TX, USA, 22–25 June 2009; pp. 1–15.
23. MacKunis, W.; Wilcox, Z.; Kaiser, M.K.; Dixon, W.E. Global Adaptive Output Feedback Tracking Control of an Unmanned Aerial Vehicle. *IEEE Trans. Control Syst. Technol.* **2010**, *18*, 1390–1397.
24. MacKunis, W.; Wilcox, Z.D.; Kaiser, M.K.; Dixon, W.E. Global Adaptive Output Feedback MRAC. In Proceedings of the IEEE Confernece on Decision and Control, Shanghai, China, 15–18 December 2009; pp. 3483–3488.
25. Khalil, H.K. *Nonlinear Systems*, 3rd ed.; Prentice Hall: Upper Saddle River, NJ, USA, 2002; p. 750.

6

Position Estimation of Small Robotic Fish Based on Camera Information and Gyro Sensors

Yogo Takada *, Keisuke Koyama and Takahiro Usami

Graduate School of Engineering, Osaka City University, 3-3-138, Sugimoto, Sumiyoshi-ku, Osaka 558-8585, Japan; E-Mails: yamadataro800@gmail.com (K.K.); us.a.3.tk.uoc@gmail.com (T.U.)

* Author to whom correspondence should be addressed; E-Mail: takada@mech.eng.osaka-cu.ac.jp

Abstract: Robotic fish are ideal for surveying fish resources and performing underwater structural inspections. If a robot is sufficiently fishlike in appearance and does not use a screw propeller, real fish will not be easily surprised by it. However, it is comparatively difficult for such a robot to determine its own position in water. Radio signals, such as those used by GPS, cannot be easily received. Moreover, sound ranging is impractical because of the presence of rocks and waterweed in places where fish spend a lot of time. For practical applications such as photographing fish, a robotic fish needs to follow the target fish without losing awareness of its own position, in order to be able to swim autonomously. We have developed a robotic fish named FOCUS (FPGA Offline Control Underwater Searcher) which is equipped with two CMOS cameras and a field-programmable gate array (FPGA) circuit board for data processing. The forward-facing camera is used to track red objects, since this is the color of the fish of interest. In addition, using visual information obtained with the bottom-facing camera, the robot can estimate its present position. This is achieved by performing real-time digital image correlation using the FPGA. However, until now, the position estimation accuracy has been poor due to the influence of yaw and roll. In the present study, the position estimation method has been greatly improved by taking into account the yaw and roll values measured using gyro sensors.

Keywords: robotic fish; swimming; position estimation; digital image correlation; gyro sensor

1. Introduction

The influence of global warming and acid rain is gradually changing the water quality in rivers and lakes. One method of checking water quality is by determining the amount of dissolved oxygen. Another approach is to investigate the number of creatures living in the water, and the species that are present. Creatures that can only thrive in good-quality water will not be found when the water quality is bad. Therefore, the water quality can be evaluated by classifying the species of living creatures such as fish.

Rather than have researchers enter the water to survey fish species, one interesting approach is to use a robotic fish that is capable of photographing real fish. Such a robot should be as small as possible in order to avoid frightening the target fish. It must also be able to swim autonomously and to track and photograph the target fish for the required period of time. After this, it can be retrieved and its data transferred to a computer. To achieve this level of autonomy, the robot must be constantly aware of its own position.

A number of underwater robots such as robotic fish have already been developed [1–6], including by our group [7–9]. In an event held in the Suma Aqualife Park in Japan, it was found that aquarium fish were not frightened by robotic fish when they were placed in the same water tank [10]. Another advantage of robotic fish is that they can swim even in constricted areas where conventional underwater robots used for oceanographic surveys cannot enter [11–16]. In addition, since their propulsion system is a fin just like in real fish, it does not become entangled in waterweed. However, the performance of a fin is still inferior to that of a screw propeller, and we are currently investigating ways of improving this. We are also comparing the propulsion mechanisms of real fish and robotic fish [17–21].

As stated earlier, the goal is to develop a robotic fish that is capable of identifying and photographing real fish. There are several methods by which such identification could be performed, such as using color information, shape information, and motion information. Of these, by far the simplest method is to identify fish based on their color, and this is the approach that we have been pursuing [22]. As we plan to evaluate the effectiveness of this method in the near future by tracking young carp, whose bodies are red, the present study focuses on identifying red objects.

The robotic fish used in this study is referred to as FOCUS (FPGA Offline Control Underwater Searcher), and is equipped with two small CMOS cameras and a field-programmable gate array (FPGA). It can track a red target autonomously using a forward-facing camera. A second CMOS camera points downwards and is used by FOCUS to identify its own position. Position estimation can be difficult underwater because radio signals such as those used by GPS cannot be easily received. Moreover, sound ranging is impractical because of the presence of rocks and waterweed in places where fish spend a lot of time, and the difficulty of accurately positioning sound sources. For these reasons, we have taken the approach of position estimation using visual images of the floor at the bottom of the water [23,24]. This is achieved using digital image correlation (DIC), which has proved useful in a variety of fields [25,26]. If the water is clear and shallow, this method is considered to be potentially simpler and more effective than approaches such as sound localization and particle filters [27–29]. In the present study, in order to improve the accuracy of position estimation using DIC, the effects of yaw and roll are taken into account using data from gyro sensors installed in the robotic fish.

2. Robotic Fish FOCUS

2.1. Structure of FOCUS

There has been a great deal of research on the propulsion method used by underwater creatures such as fish, which involves their caudal (tail) fin [30–33]. Fish generate a reverse Karman vortex street which allows them to swim fast and efficiently, and they move their entire body flexibly. Based on results previously obtained using test models, we believe that the flexibility of the caudal fin is an important factor for improving the propulsive performance of robotic fish. For this reason, FOCUS is equipped with a caudal fin made of a flexible material.

Figure 1 shows a photograph of FOCUS, and a schematic of its internal structure is shown in Figure 2. The robot has both forward- and downward-facing CMOS cameras. A 3D gyroscope with two bi-axial gyro sensors is installed to measure the pitch, roll and yaw motions of the robot. Image processing and motion control are carried out using an FPGA board containing a Spartan-6 XC6SLX9 (Xilinx Inc., San Jose, CA, USA).

Figure 1. Robotic fish FPGA Offline Control Underwater Searcher (FOCUS).

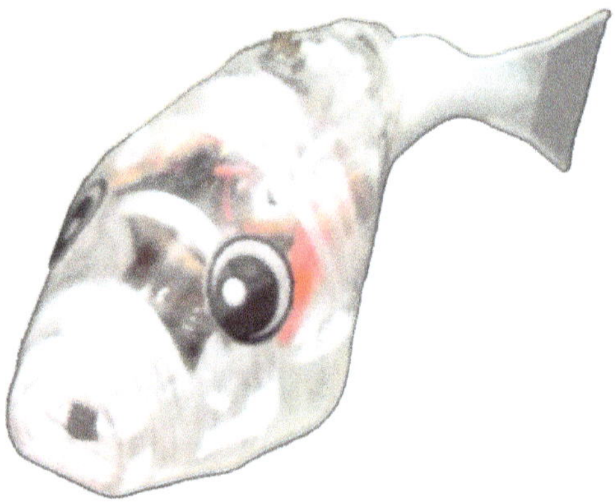

Figure 2. Structure of FOCUS.

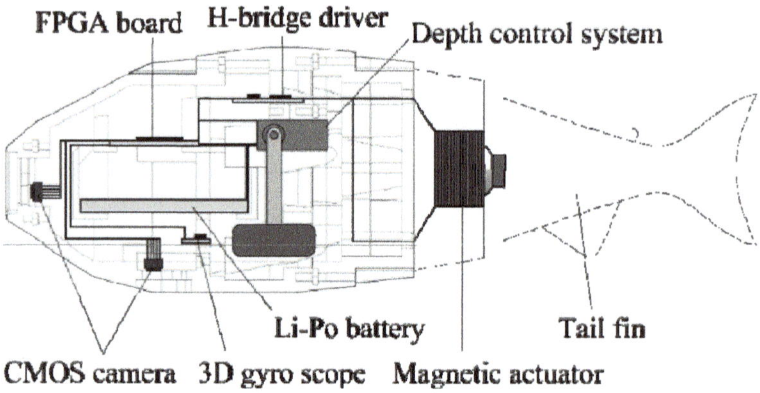

The outer shell is constructed from acrylic parts. The caudal fin is connected to the main body by a single joint, in which a magnetic actuator is installed, as illustrated in Figure 3. The coil in the actuator has 2000 turns of polyurethane wire with a diameter of 0.12 mm, and has a resistance of 175 Ω. A spherical neodymium magnet is installed inside the coil, and when a current passes through the coil, the tail fin attached to the magnet is driven by an electromagnetic force. A lithium polymer battery

(3.7 V, 1,000 mAh) is used as the electric power source. Power is supplied to the magnetic actuator for the fin drive through an H-bridge circuit containing a power MOSFET (SP8M4). The servomotor (Tahmazo TS-1002) in the depth control system is driven by the electric power from the same battery, and receives commands from the FPGA board. Figure 4 shows a schematic diagram of the depth control system. When the weight is moved forward, the robot sinks, and when it is moved backwards, the robot rises.

Figure 3. Magnetic actuator used for FOCUS.

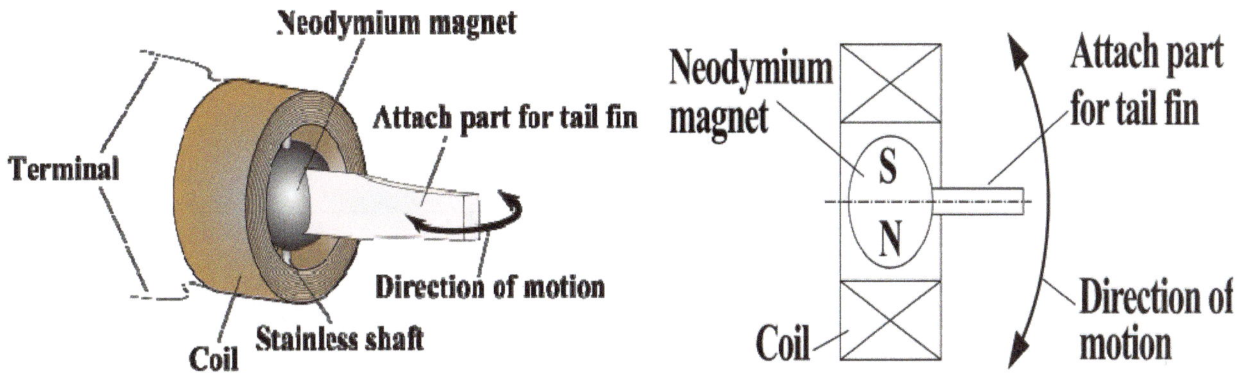

Figure 4. Depth control system. (**a**) sinking; (**b**) level; (**c**) rising.

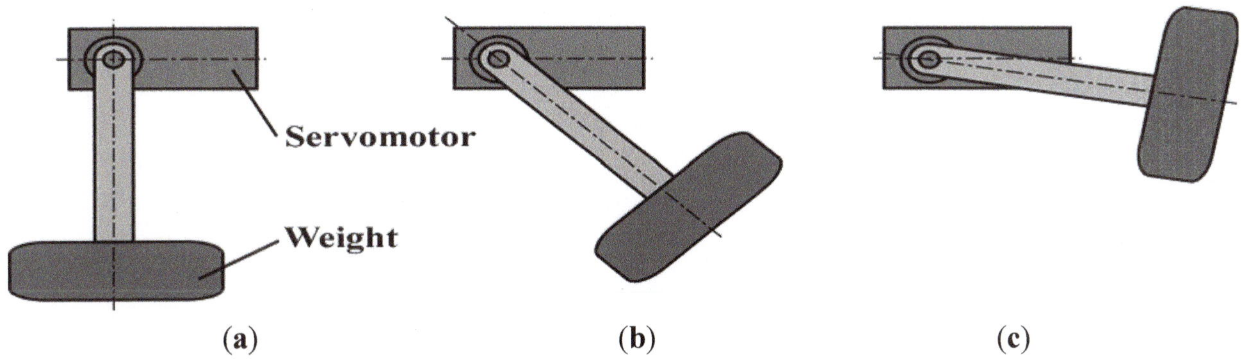

The magnetic actuator is controlled by signals from the FPGA. By outputting an alternating positive and negative voltage, the attached fin can be made to swing from side to side. The pulse-width-modulation duty ratio set by the VHDL program in the FPGA determines whether the robot swims straight ahead or turns right or left.

2.2. Target Tracking Method

The forward-facing camera on the head of the robot for target tracking has a resolution of 128 pixel × 96 pixel (Sub-QCIF), and the image data in RGB565 format are sent to the FPGA memory at 30 frames per second. The robotic fish can autonomously determine the correct traveling direction to track a red target based on the visual information obtained from the camera. It attempts to bring the center of the red area into the center of its own field of view, as shown in Figure 5.

Figure 5. Visual method for guiding FOCUS.

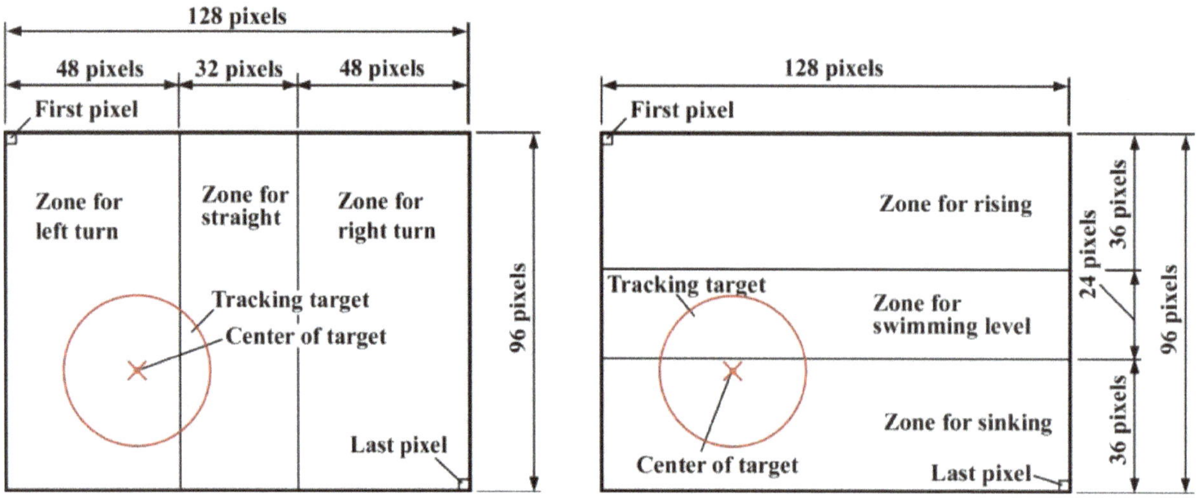

The procedure is as follows. For an image containing a red target, a summation (ΣX) is made of the X coordinates of pixels identified as red, for all pixels from the upper left to the lower right in Figure 5. The centroid X_g in the X direction for the red region is then determined by dividing ΣX by the total number of red pixels N. If X_g is larger than 80 pixels, the robot turns right, and if it is smaller than 48 pixels, it turns left. Otherwise, it swims straight. A similar analysis is carried out in the Y direction to determine whether the robot should rise, sink or swim level.

The ability of FOCUS to track a red target was experimentally investigated. The results are shown in Figure 6, and it was confirmed that tracking could be carried out successfully. The average swimming velocity of the robotic fish was 123 mm/s, which was obtained by measurements of 50 times target tracking swim experiment. The angular velocity of the robotic fish is 0.52 rad/s. When the target movement exceeds the kinematic performance of the robot, it becomes impossible to track the target.

Figure 6. FOCUS following a red target object.

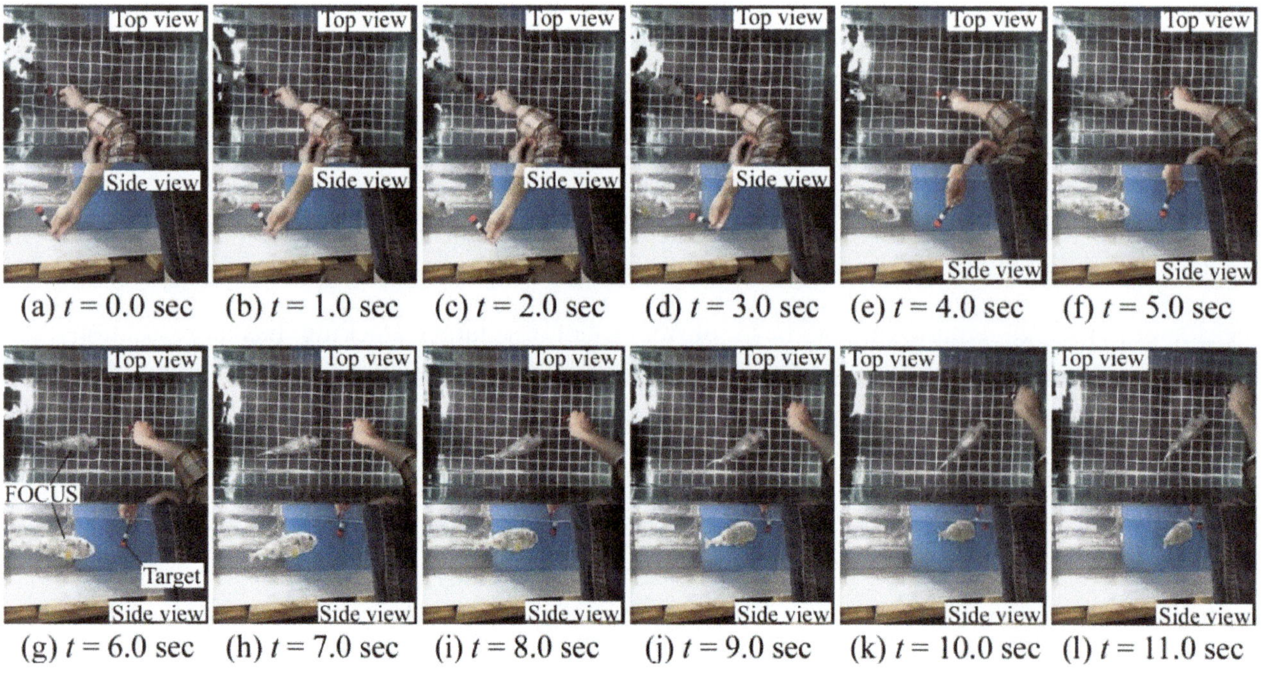

3. Position Estimation

To evaluate the position estimation capability of FOCUS, a launcher containing the robot was placed in a tank with a length of 900 mm, a width of 450 mm, and height of 450 mm, which contained tap water, as illustrated in Figure 7a. The robotic fish swam horizontally without a target, at a distance of 100 mm from the bottom of the tank, which was lined with eight A4 size photographs of small pebbles in a river. The composite image is shown in Figure 7b; its dimensions were 796 mm × 576 mm.

Figure 7. Position estimation experiment: (**a**) water tank and (**b**) image of small pebbles.

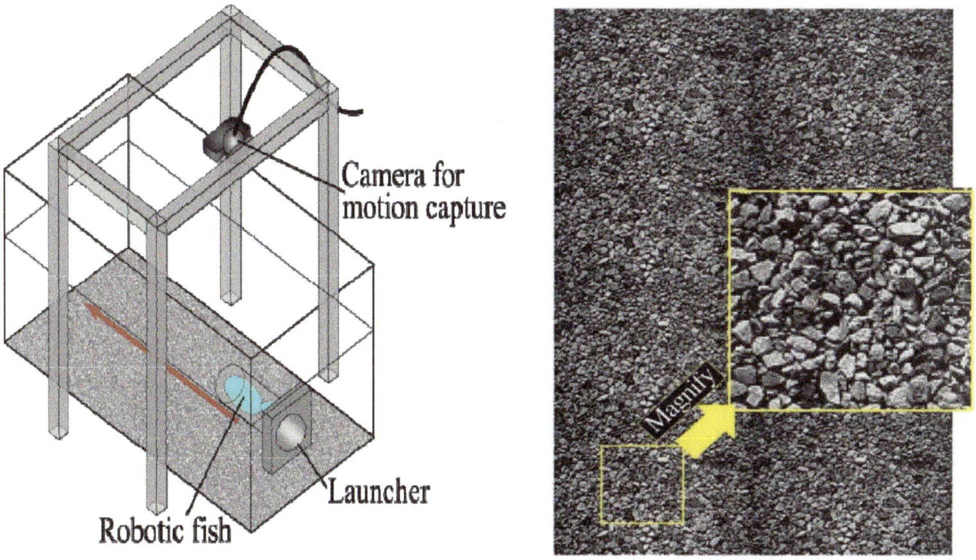

(**a**) (**b**)

Figure 8. Flow chart of motion capture process.

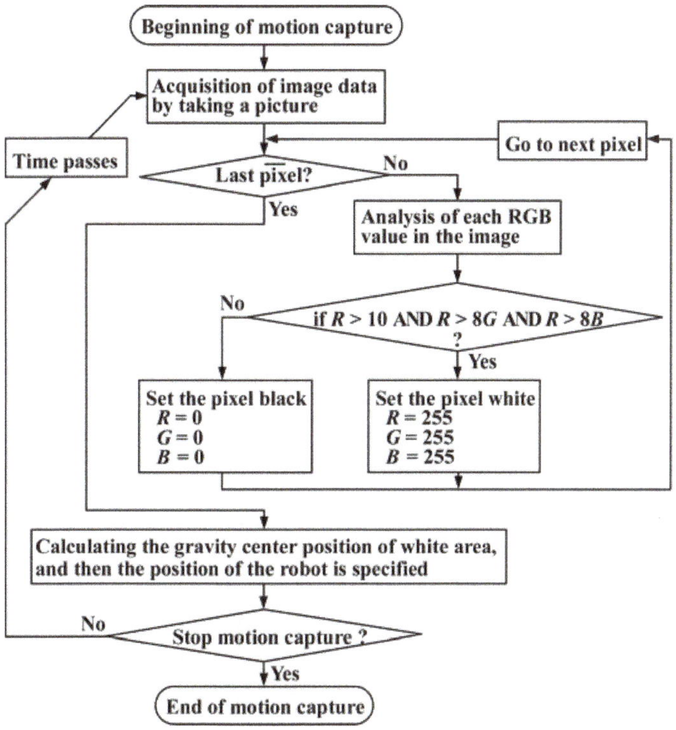

The accuracy of the position estimation technique was evaluated using a motion capture camera mounted above the water tank, which acquires 8-bit RGB images. Figure 8 shows a flow chart of the motion capture process. Its purpose was to track the center position of the red robotic fish. The color of each pixel was evaluated, and if $R > 10$, $R > 8G$, and $R > 8B$ then the pixel was judged to be red. Since the camera acquired images at 30 frames per second, the location of the robot could be determined at intervals of approximately 33.3 ms.

The method used by the robotic fish to estimate its position is as follows. The downward-facing camera captures images of the pebble pattern at a resolution of 64 pixel × 64 pixel. For the first image, the 32 pixel × 32 pixel central region is extracted and stored in the FPGA memory. After about 33 ms, a second 64 pixel × 64 pixel image is captured and stored in the FPGA. This 64 pixel × 64 pixel image can be thought of as containing a total of 1,089 (33 × 33) unique sub-images with dimensions of 32 pixel × 32 pixel, one of which correlates most closely with the first 32 pixel × 32 pixel image. Identifying this sub-image yields the distance (ΔX, ΔY) that the robotic fish swam during the 33-ms interval. To evaluate the correlation between images, the residual sum of squares was used:

$$R_{fg}(\Delta X, \Delta Y) = \sum_{i=1}^{N}\sum_{j=1}^{N}\{f(X_i, Y_j) - g(X_i + \Delta X, Y_j + \Delta Y)\}^2 \quad (1)$$

where R_{fg} is the correlation value, f is the luminance at pixel (X_i, Y_j) in the first 32 pixel × 32 pixel image, g is the luminance at pixel (X_i, Y_j) in the candidate region in the second image, and N is the number of pixels (32) in the X and Y directions. Figure 9 shows the flow chart to get the distance of ΔX, ΔY.

Figure 9. Flow chart of digital image correlation method.

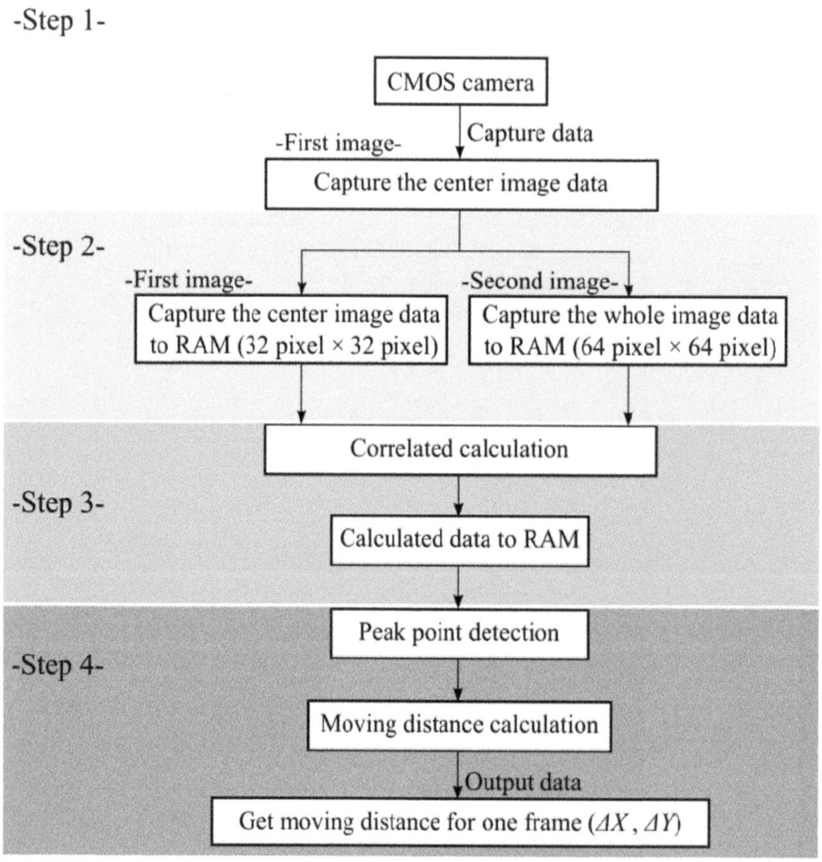

As the robotic fish swims, it is subject to yaw and roll motions. This can bring about a situation in which the most highly correlated region of the second image is far from the image center, or in extreme cases, is not contained within the 64 pixel × 64 pixel image at all. This can lead to a large increase in the number of correlation calculations that must be performed. This is particularly true for roll motion. However, increasing the number of candidate images considered is difficult due to the limited memory and number of multipliers in the FPGA. Therefore, to compensate for the effects of yaw and roll, the X and Y coordinates in the second image were transformed using the yaw and roll angles θ_{yaw} and θ_{roll}, respectively, obtained by integrating the angular velocities measured by the gyro sensors. We used LPR503AL and LPY503AL as gyro sensors. The moving average method was used to remove the sensor's drift. Figure 10 shows the method used for yaw compensation, and Figure 11 shows the case for both yaw and roll compensation.

Figure 10. Flow chart of yaw compensation method.

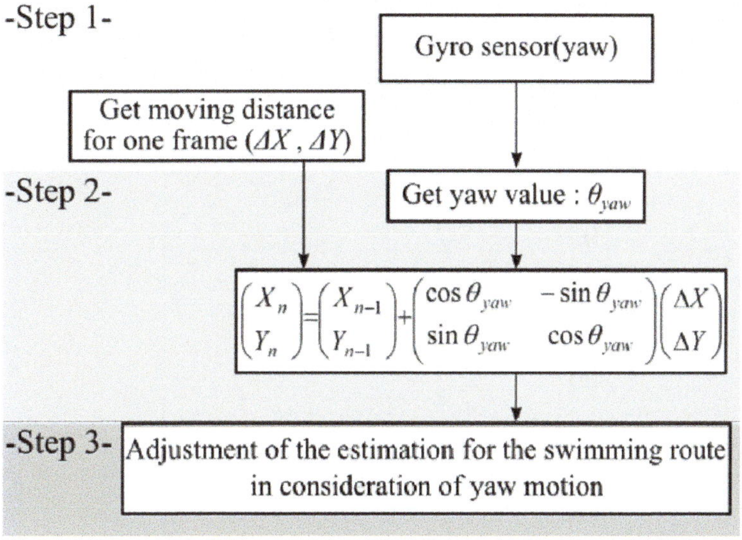

Figure 11. Flow chart of yaw and roll compensation method.

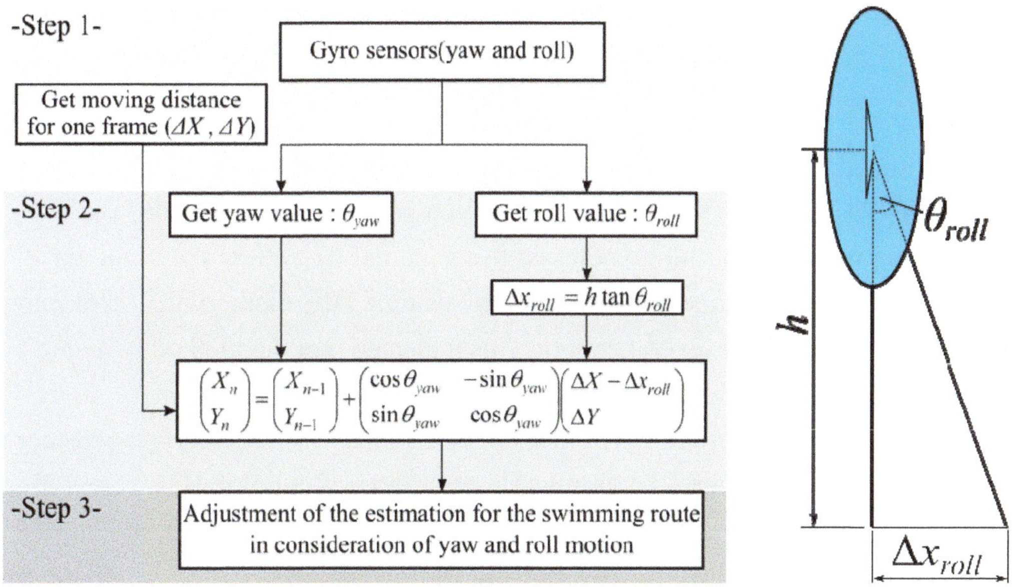

Although the yaw and roll angle calculations can be performed in the FPGA, and the position can be estimated, the robot cannot output the full set of results, including the raw DIC results, those following yaw compensation, and those following yaw and roll compensation. Since all of this information is needed in order to evaluate the relative effectiveness of the different position estimation methods, the robot outputs the X, Y, θ_{yaw} and θ_{roll} values to a notebook computer, which can also perform the position estimations. Thus, all of the data required to evaluate the effectiveness of the different methods can be obtained in a single swimming experiment.

Figure 12 shows the results of the accuracy evaluation experiment. In the absence of a tracking target, the robot tends to drift left or right due a slight imbalance in the horizontal forces produced by the magnetic actuator. In this experiment, it exhibited a small deviation to the left. The red curve represents the actual trajectory captured by the motion capture camera. The black curve indicates the trajectory determined using DIC only, and it is seen to deviate from the actual position. In contrast, the trajectory estimated using yaw and roll compensation (blue curve) is close to the true trajectory.

Figure 12. Accuracy of position estimation using various methods during free swimming.

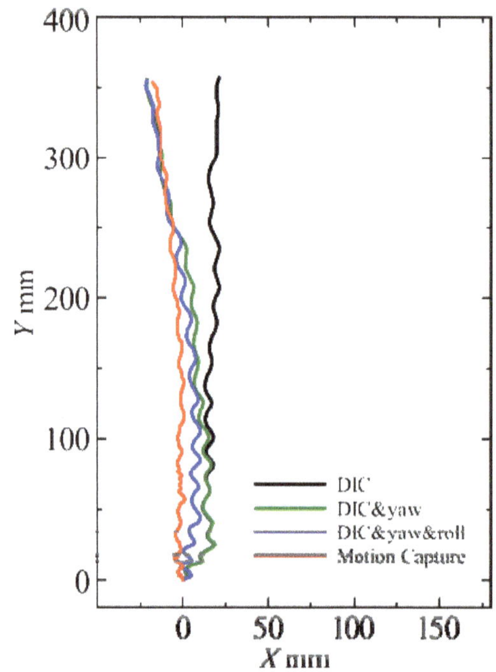

In the next experiment, the robotic fish was guided in a rightward arc using a red target object as shown in Figure 13a. The experimental results are shown in Figure 13b. Again, the blue trajectory obtained using yaw and roll compensation is closest to the actual trajectory shown in red. It should be noted that yaw compensation could also be performed by using a DIC method that takes image rotation into account. However, the results of several experiments indicated that the method based on gyro sensor measurements was superior.

For a robotic fish to be used under real environmental conditions, it is necessary to consider the effects of flowing water, particularly in rivers. To simulate these kinds of conditions, a screw propeller was placed in the water tank to produce a flow across the trajectory of the robot, as shown in Figure 14a. The robotic fish was again guided towards the opposite wall using a red target. When the robot enters this cross flow, it is knocked off balance because the hydraulic force acts on a point some distance from its center of gravity. However, as seen in Figure 14b, the trajectory obtained using yaw and roll compensation is quite close to the true trajectory.

Figure 13. Guided swimming experiment: (**a**) experimental setup and (**b**) experimental results.

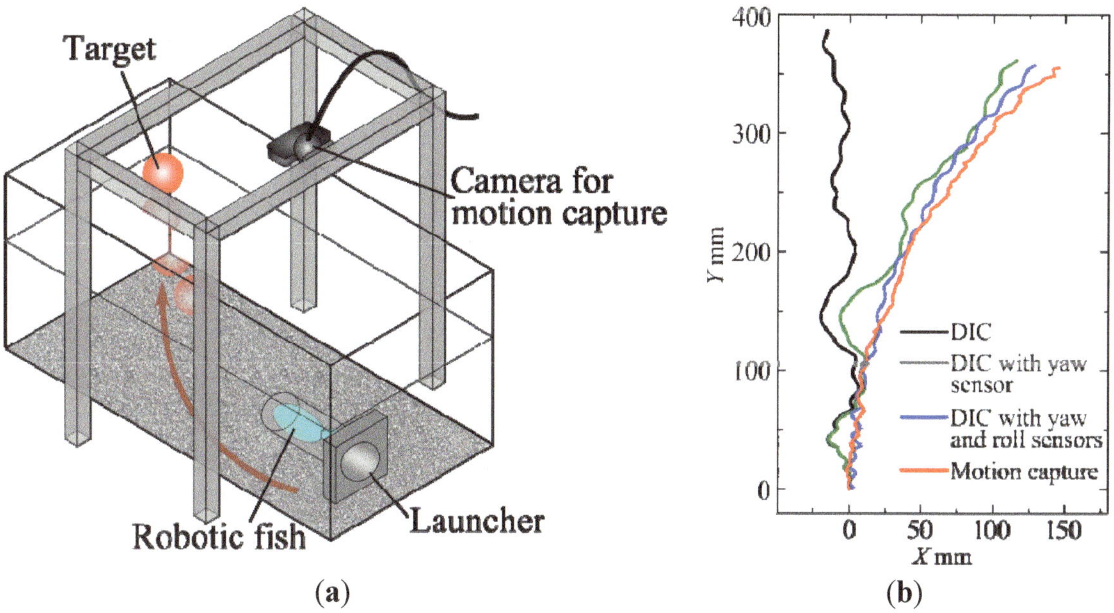

Figure 14. Swimming experiment with disturbed water flow: (**a**) experimental setup and (**b**) experimental results.

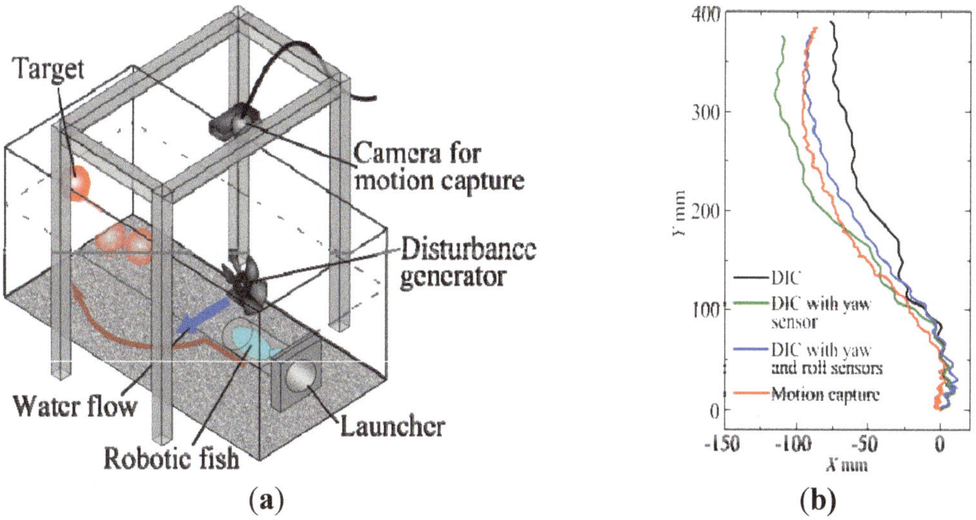

The position estimation with yaw and roll still had some errors in Figures 12 to 14. There are initial position errors and initial angle errors in every experiment because the person put the robotic fish in the launcher. Moreover, human errors might cause easily because a person pushes the button to start the measurement. Therefore, the experiments of Figures 12 to 14 were conducted 8 times respectively because there might be the man-made errors in each experiment. Figure 15 shows a comparison of the accuracy of the different position evaluation methods. Each bar represents the average error between the estimated positions during the experiments and the true positions. It can be seen that in each case, the smallest error is obtained when combined yaw and roll compensation is used.

Figure 15. Accuracy of different position evaluation techniques.

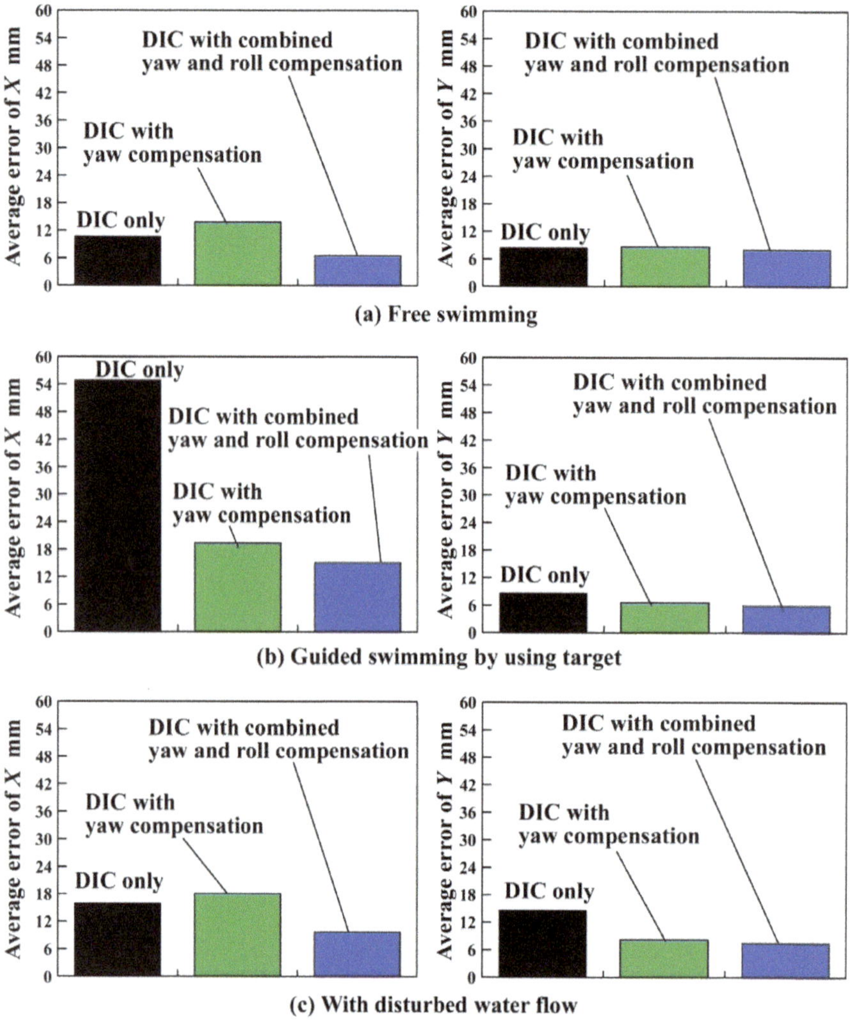

There are a lot of factors for the position errors; such as DIC, drift of gyro sensor, pitch motion and man-made errors. It is extremely important to reduce the error from the large error factor. Therefore, the magnitude relation among the errors should be clarified.

4. Conclusions

A small robotic fish (FOCUS) equipped with two CMOS cameras and an FPGA was fabricated in order to test its ability to swim autonomously in clear shallow water. The forward-facing camera was used to track and follow target objects. The downward-facing camera was used by the robot to estimate its position based on a DIC technique, and gyro sensors were installed to provide yaw and roll information. It was found that using DIC alone, with no consideration for image rotation due to yaw, the robot could not estimate its position accurately. However, the accuracy was greatly improved when both the yaw and roll values obtained by the gyro sensors were taken into account. Good experimental results were obtained for free swimming, guided swimming, and swimming through a cross flow.

The experiments in this study were performed at a constant distance from the bottom of the water tank. In future work, the effectiveness of the position estimation method will be evaluated in situations where the robotic fish is changing its vertical position.

Acknowledgments

The authors would like to express their gratitude to the Japan Society for the Promotion of Science for support in the form of Grant-in-Aid for Scientific Research (C), No. 24560301.

Author Contributions

Keisuke Koyama manufactured the robotic fish FOCUS. Keisuke Koyama and Takahiro Usami performed swimming experiments of FOCUS for the position estimation. Yogo Takada came up with the idea for the position estimation of robotic fish and the structure of FOCUS, and wrote the paper.

Conflicts of Interest

The authors declare no conflict of interest.

References

1. Hirata, K.; Takimoto, T.; Tamura, K. Study on turning performance of a fish robot. In Proceedings of the 1st International Symposium on Aqua Bio-Mechanisms, Honolulu, HI, USA, 28–30 August 2000; pp. 287–292.
2. Nakashima, M.; Kaminaga, K.; Ono, K. Experimental study of two-joint dolphin robot (propulsive characteristics of 2nd large model). In Proceedings of the International Symposium on Aqua Bio-Mechanisms, Honolulu, HI, USA, 28–30 August 2000; pp. 311–314.
3. Toda, Y.; Ikeda, H.; Sogihara, N. The motion of fish-like under-water vehicle two undulating side fins. In Proceedings of the 3rd International Symposium on Aqua Bio-Mechanisms, Okinawa, Japan, 3–7 July 2006.
4. Kobayashi, S.; Nakabayashi, M.; Morikawa, H. Bioinspired propulsion mechanism in fluid using fin with dynamic variable-effective-length spring. *J. Biomech. Sci. Eng.* **2006**, *1*, 280–289.
5. Yamamoto, I. Research and development of robotic fish. *J. JIME* **2008**, *43*, 99–102.
6. Low, K.H. Modelling and parametric study of modular undulating fin rays for fish robots. *Mech. Mach. Theory* **2009**, *44*, 615–632.
7. Takada, Y.; Araki, R.; Nonogaki, M.; Ebita, K.; Ishii, T.; Wakisaka, T. Development of small and ultra-light passive-type polymer electrolyte fuel cell and application to small fish robots. *Trans. Jpn. Soc. Mech. Eng. Ser. B* **2010**, *76*, 650–659.
8. Takada, Y.; Araki, R.; Nakanishi, Y.; Nonogaki, M.; Ebita, K.; Wakisaka, T. Development of small fish robots powered by small and ultra-light passive-type polymer electrolyte fuel cells. *J. Robot. Mechatron.* **2010**, *22*, 150–157.
9. Takada, Y.; Masuda, H.; Tajiri, T.; Wakisaka, T. Development of small fish robot with built-in fuel cells. In Proceedings of the Techno-Ocean, Kobe, Japan, 27 July 2010.
10. Techno-Ocean 2010 Pre Event, Ocean Science Seminar, Amazing World of Aqua Bio Mechanics. Available online: http://www.marine.osakafu-u.ac.jp/~lab06/pre-event/ (accessed on 28 January 2014).
11. Asakawa, K.; Kojima, J.; Ito, Y.; Takagi, S.; Shirasaki, Y.; Kato, N. Autonomous underwater vehicle AQUA EXPLORER 1000 for inspection of underwater cables. In Proceedings of the AUV'96, Monterey, CA, USA, 2–6 June 1996.

12. Ura, T. Construction of AUV R2D4 based on the success of full-autonomous exploration of Teisi Knoll by R-one robot. In Proceedings of the AUV Show Case, New Maiden, UK, 27–28 September 2002; pp.23–28.
13. Maeda, T.; Ishiguro, S.; Yokoyama, K.; Hirokawa, K.; Hashimoto, A.; Okuda, Y.; Tani, T. development of fuel cell AUV "URASHIMA". *Mitsubishi Heavy Ind. Tech. Rev.* **2004**, *41*, 344–347.
14. Nakatani, T.; Ura, T.; Ito, Y.; Kojima, J.; Tamura, K.; Sakamaki, T.; Nose, Y. AUV "TUNA-SAND" and its Exploration of hydrothermal vents at Kagoshima Bay. In Porceedings of the OCEANS 2008—MTS/IEEE Kobe Techno-Ocean, Kobe, Japan, 8–11 April 2008.
15. Hyakudome, T.; Furuyama, H.; Baba, T.; Kasaya, T.; Oomika, S.; Ishibashi, S.; Yoshida, H.; Tsukioka, S.; Matsuura, M.; Aoki, T. The Seafloors Which AUV "URASHIMA" Watched. Available online: http://jairo.nii.ac.jp/0153/00013762/en (accessed on 15 March 2014).
16. Sakagami, N.; Ibata, D.; Ikeda, T.; Shibata, M.; Ueda, T.; Ishimaru, K.; Onishi, H.; Murakami, S.; Kawamura, S. Development of a removable multi-DOF manipulator system for man-portable underwater robots. In Proceedings of the 21th International Offshore and Polar Engineering Conference, Maui, HI, USA, 19–24 June 2011; pp. 279–284.
17. Takada, Y.; Nakanishi, Y.; Araki, R.; Wakisaka, T. Investigation of propulsive force and water flow around a small fish robot by PIV Measurement and three-dimensional numerical analysis. *Trans. Jpn. Soc. Mech. Eng. Ser. C* **2010**, *76*, 665–672.
18. Takada, Y.; Nakanishi, Y.; Araki, R.; Nonogaki, M.; Wakisaka, T. Effect of material and thickness about tail fins on propulsive performance of a small fish robot. *J. Aero Aqua Bio-Mech.* **2010**, *1*, 51–56.
19. Takada, Y.; Araki, R.; Ochiai, T.; Tajiri, T.; Wakisaka, T. Effects of tail fin flexibility on propulsive performance in small fish robots (investigation by fluid-structure interaction analysis considering elastic deformation of tail fin). *Trans. Jpn. Soc. Mech. Eng. Ser. C* **2011**, *77*, 2351–2362.
20. Takada, Y.; Ochiai, T.; Fukuzaki, N.; Tajiri, T.; Wakisaka, T. Analysis of flow around robotic fish by three-dimensional fluid-structure interaction simulation and evaluation of propulsive performance. *J. Aero Aqua Bio-Mech.* **2013**, *3*, 57–64.
21. Takada, Y.; Fukuzaki, N.; Ochiai, T.; Tajiri, T.; Wakisaka, T. Evaluation of artificial caudal fin for fish robot with two joints by using three-dimensional fluid-structure simulation. *Adv. Mech. Eng.* **2013**, *2013*, ID:310432.
22. Takada, T.; Nakamura, T.; Koyama, K.; Tajiri, T. Target following control of small fish robot FOCUS based on color information. *Trans. Jpn. Soc. Mech. Eng.Ser. C* **2012**, *78*, 2924–2934.
23. Takada, Y.; Nakamura, T.; Koyama, K.; Wakisaka, T. Self-position estimation of small fish robot based on visual information from camera. *J. Jpn. Inst. Mar. Eng.* **2012**, *47*, 138–146.
24. Takada, Y.; Nakamura, T.; Koyama, K.; Fukuzaki, N.; Tajiri, T.; Wakisaka, T. Target following and self-position estimation of small fish robot FOCUS. *J. Jpn. Inst. Mar. Eng.* **2012**, 47, 108–113.
25. Kawasue, K.; Ishimatsu, T.; Shih, C. Fast correlation technique of particle image velocimetry. *Trans. Jpn. Soc. Mech. Eng. Ser. C* **1994**, *60*, 2084–2089.

26. Uneda, M.; Murata, S.; Narise, T.; Yamazaki, T.; Ohnishi, O.; Kurokawa, S.; Ishikawa, K.; Doi, T. Quantitative evaluation of slurry flow behavior using digital image processing. *Trans. Jpn. Soc. Mech. Eng. Ser. C* **2011**, *77*, 3891–3903.
27. Maki, T.; Kondo, H.; Ura, T.; Sakamaki, T. AUVnavigation with particle filter. *Month. J. Inst. Industrial Sci.* **2004**, *56*, 429–433.
28. Eriguchi, Y.; Ishii, K.; Watanabe, K. An underwater sound source localization system using probabilistic methods. *Dyn. Des. Conf.* **2009**, *11*, 244–247.
29. Tajiri, T.; Takada, Y.; Takada, M.; Kawai, T. Movement control including localization for a wheeled mobile robot. In Proceedings of the 16th Asia Pacific Symposium on Intelligent and Evolutionary Systems, Kyoto, Japan, 13–14 December 2012; pp. 3–6.
30. Gray, J. Studies in Animal Locomotion, VI. The Propulsive Power of the Dolphin. *J. Exp. Biol.* **1935**, *13*, 192–199.
31. Lighthill, M.J. Note on the swimming of slender fish. *J. Fluid Mech.* **1960**, *9*, 305–317.
32. Lighthill, M.J. Aquatic animal propulsion of high hydrodynamical efficiency. *J. Fluid Mech.* **1970**, *44*, 265–301.
33. Nagai, M.; Teruya, I.; Uechi, K.; Miyazato, T. Study on an oscillating wing propulsion mechanism. *Trans. JSME, Ser. B* **1996**, *62*, 200–206.

7

Design Issues and in Field Tests of the New Sustainable Tractor LOCOSTRA

Emanuela Elisa Cepolina [1], Michał Przybyłko [2], Giovanni Battista Polentes [3] and Matteo Zoppi [2,*]

[1] Snail Aid—Technology for Development, Via Montallegro5/3, Genoa 16145, Italy; E-Mail: patfordemining@gmail.com

[2] PMAR Robotics Group, Department of Mechanics, University of Genoa, Via All'Opera Pia 15/A, Genoa 16145, Italy; E-Mail: michal.p@dimec.unige.it

[3] Pierre Trattori snc, Via Novi 19, 15060 Silvano D'orba (AL), Italy; E-Mail: info@pierretra.com

* Author to whom correspondence should be addressed; E-Mail: zoppi@dimec.unige.it

---

Abstract: This paper presents a robotized small tractor designed to perform humanitarian demining operations and agricultural operations when demining is accomplished. This dual use makes this system unique. The focus is on the design of the machine, including the modeling for design and control, the characteristics of the tools operated by the machine, the motion control at tele-operated, and semi-autonomous levels. The mechatronic design process applied uses sustainable design strategies. Technical contributions are in the tractor architecture, designed to make automation easier, and in the control functions implemented on this architecture. Extensive field tests were performed in different sites; first, in Italy, focusing on the agricultural application of the machine, in natural scenarios with different ground and vegetation; then, in two real mine fields in Jordan focusing on the performance for technical survey in humanitarian demining. The tests have confirmed the performance for both task categories (agricultural and demining), confirming the correctness of the statement. For the demining application, the machine has been assessed by professional users confirming the acceptance to field use and the novelty of the concept.

Keywords: agricultural automation; ground processing; 4WD articulated steering mobile platform; mechatronic design; modular tractor; articulated steering dynamic model; field tests

---

1. Introduction

Agriculture involves different kinds of tasks, from accurate transplanting and meticulous individual plant curing [1] to heavy ground processing work. In the past, in more developed countries, and today in developing countries, agricultural mechanization through tractors contributed to improve the effectiveness and productivity of the field heavy works. Most existing agricultural tractors still need extensive human operation: a next step is transferring the results of tractor robotization to the fields ([2–6], together with many other works). As confirmed by current trends, today, in both developed and developing countries, the availability of human resources for farming is decreasing due to labor shortage both for lack of interest from young people and for weak or aging farming workforce: this means that each worker, sometimes weak, is responsible for larger extensions of land. These factors influence the development of local agriculture and open a market share for automation also of small machines (mass lower than three tons). Differently from heavy tractors, small machines with competitive cost cannot be effectively developed simply modifying existing manual driven models: their architecture should be rethink for automation. The paper presents the case of a small tractor called LOCOSTRA. A peculiarity of this machine is that it performs very well tasks related to agriculture and to humanitarian demining using same implements.

LOCOSTRA costs like an agricultural system, four times less than the cheapest machine for demining; furthermore, after demining, it can be used on the same land demined, operated by local people, in part already trained while participating to demining [7]. This contributes more than any other means or method to the development and mechanization of agriculture in areas that have been generally strongly hit by war events [8,9]. Moreover, these areas may benefit of foreign funding for demining that they would never have for agricultural development: today the funding for demining is used to pay huge machines, running tools not useful for agriculture, expensive to operate due to fuel and maintenance costs, that become useless at the end of the demining.

LOCOSTRA is designed to satisfy end-user requirements and sustainability paradigms [10]. Figure 1 shows a strategy wheel with the principles that qualitatively characterize the design of LOCOSTRA (stars) in comparison to a traditional ground processing machine (squares). The six characteristics considered are relevant for both agriculture and demining. They oriented the design effort and were used to improve step by step the solution. The plot of the strategy wheel of a new design was compared to the plot of the initial design, considering which strategies were effective, which were not so effective, and trying to assess where more improvements could be made in the following design iteration.

These six characteristics regard: the use of materials with respect to the functions, human friendly life cycle, ecological impact, extended services and functions, revalorization, health, and environment risk.

On the use of materials, the two peculiar characteristic of LOCOSTRA are the power to mass ratio, higher than the average of agricultural tractors, and the mass (between 2.5 and 3.5 metric tons, depending on the implement), much lower than the one of machinery for similar demining tasks (smart minimalist armoring and no one on board). For both characteristics, LOCOSTRA is an example of break innovation originating from a critical analysis of the requirements face to the task, in opposition to following a common practice.

Figure 1. Eco-design strategy wheel.

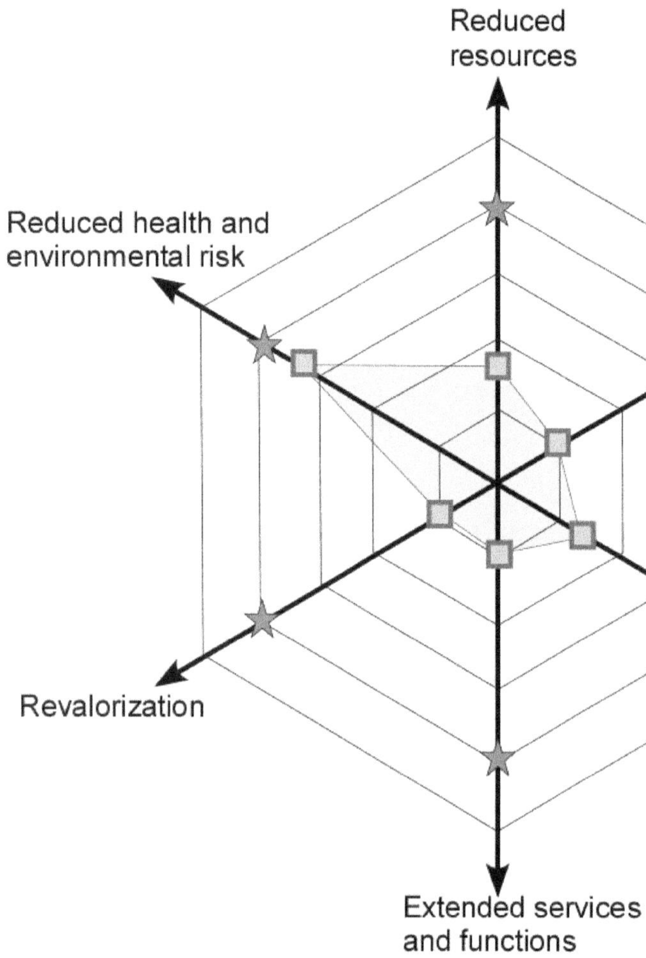

The machine design is user-and task-oriented: the user needs have been collected in questionnaires and normalized with respect to the local environment; the user interface is intuitive, and maintenance and repair operations can be done locally using available skills.

The ecological impact is low, either due to the use of Western standard powering, although the machine does not operate under strict regulation on emissions, and for the high efficiency of the engine and driveline. It is lighter than others with respect to its tasks, and stays in place for agriculture after demining.

The dual use (agriculture and demining) makes the machine immediately compliant to the point on extended services and functions [11]. Furthermore, the machine is small but can run implements with high power demand normally operated by tractors in the higher size class, so the same machine can perform over a wider range of activities. This is an advantage in a context of developing agricultural automation, where few machines are available for all tasks; the Western market of technologically advanced tractors does not satisfy this niche demand.

The design and realization of the machine can be pursued in a craft way, making easy the disassembly operations that allow, at the end of life, the recovery and re-use of components giving revalorization to the investment.

Certainly, the machine reduces health risk and hard work to humans in both agricultural and demining tasks. The reclaim, to agriculture, of mined land has a double social advantage [12].

The automation of an already existing agricultural tractor rather then designing a whole new robot for use in agriculture and demining has many advantages [13]. The battery of the tractor and the diesel engine supply all the electronic elements, included in the control system, thus, removing the problem of batteries and autonomy. The overall design can withstand harsh atmospheric conditions, the machine can traverse rough terrains, has working-day autonomy and is efficient in its tasks. Additionally the problem of heavy payloads, of significant limitation for small robots, has little or no significance. In addition, the cost of the robotization of an agricultural tractor is a relatively small percentage of the tractor cost.

The paper is organized as follows. Section 2 is dedicated to the LOCOSTRA functional design from the mechanical point of view. The foundation of the kineto-dynamic model used for the knowledge based definition of the LOCOSTRA parameters and control functions is presented. The design and implementation of the most innovative modules are the subject of the second sub-section. The motion control is illustrated into the third section in both the remote control level and the autonomy enhanced module. Section 4 reports on the experimental tests made in the field, in Italy, mainly aimed to check the agricultural operational performances of LOCOSTRA. Then the tests performed in two mine fields in Jordan are described.

2. Mechanical Architecture and Design

LOCOSTRA is a small, lightweight, four-wheel driving (4WD), electro hydraulic driven, articulated steering (AS) mini tractor. Different from agricultural automated tractors, that is based on human driven machines with sensors and actuators added to replace the driver, LOCOSTRA has been designed as a robotic vehicle with suitable architectures of the driveline and the control. A human can sit onboard and drive it, but with an additional interface that operates on the robotized layer.

LOCOSTRA can work on agricultural fields and AP mine-fields. These environments are similar, distinguished by the possible occurrence of AP mine blasts: the chassis design developed makes the vehicle insensible to mine fragments; blast resistant wheels protect the machine from blast shocks. Shaped mild steel deflection plates, 3 mm thick, cover the extremities of the axles and stop and deflect fragments projected by the agricultural implement or due to blasts.

The machine is designed to be easily transported over unimproved terrain without the need for a dedicated transporter. When medium to short distances have to be covered, the tractor can be equipped with traditional pneumatic wheels. The maximum travel speed is 30 km/h. A summary of possible tools and configurations of LOCOSTRA is shown in Figure 2.

2.1. Platform Driveline and Architecture

Almost all agricultural tractors use a mechanical driveline with the engine connected to a gear box and shafts transmitting motion to the axles; the drive involves the control of clutch and gear box; the torque at the wheels changes with the speed and can be controlled in a range with an engine load command.

Small construction machinery uses hydrostatic transmissions with the engine driving a pump and the pump feeding hydraulic motors; the motors, of which there are usually two or four, drive single wheels or an axle.

Figure 2. Modular view of LOCOSTRA.

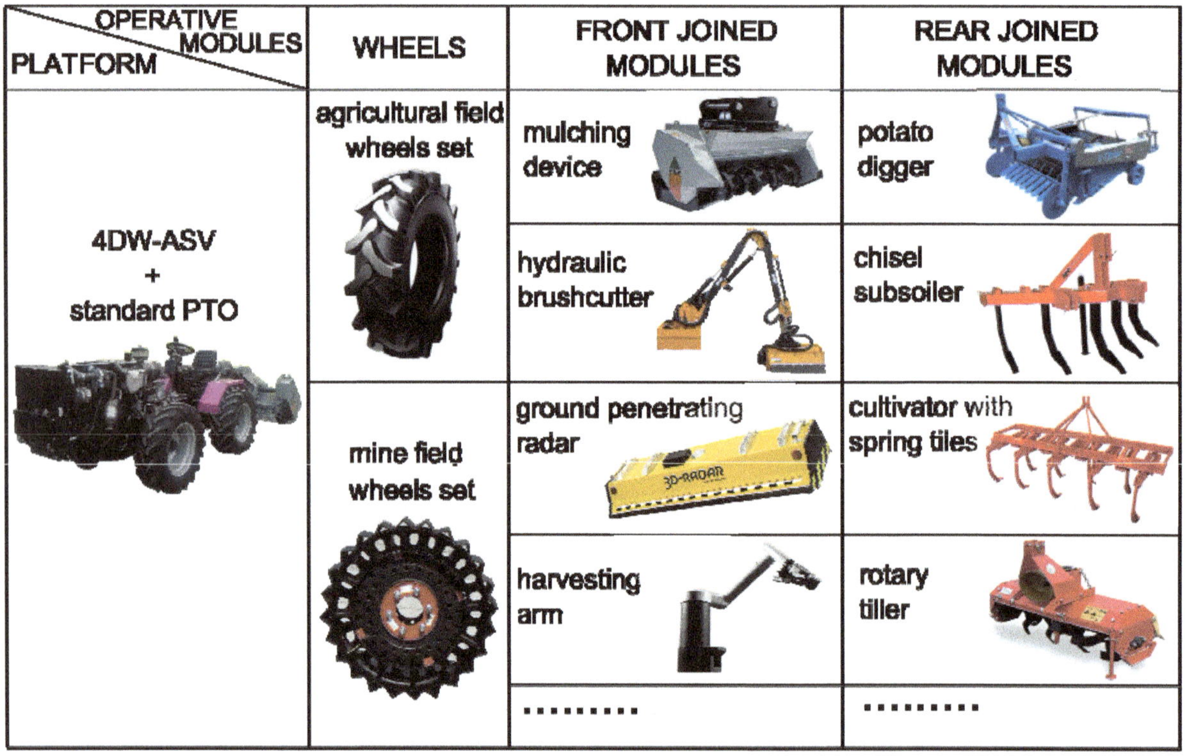

The LOCOSTRA driveline, shown in Figure 3, is a compromise between these two common architectures above: one pump driving one hydraulic motor, which drives mechanical transmissions to the axles; the power-take-off shafts (PTO; two, at the rear and front of the vehicle) are independently driven by the engine, with electro-hydraulic clutches. This simplifies, remarkably, the operation of the machine; the torque at the axles can be constant for any speed of the tractor and continuous variation of the speed is available from zero to maximum; on the other hand, the driveline efficiency remains high and close to the one of fully mechanical drivelines.

Figure 3. LOCOSTRA platform scheme.

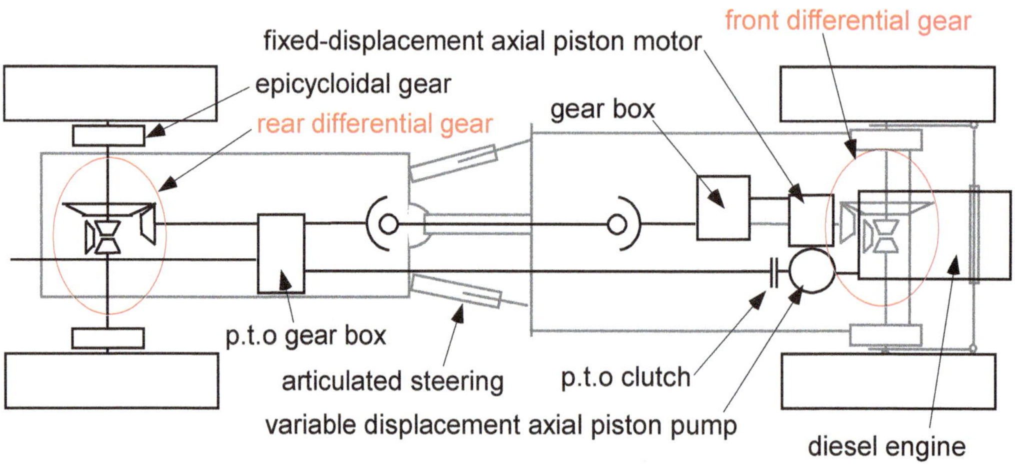

Another innovation is in the kinematic architecture of the articulation joint, which uses a spherical joint and a link with a contact pair to constrain pitch rotation of the front with respect to the rear of the tractor. This gives a higher resistance to impulsive loads generated by under-wheel blasts.

Two steering methods are available: front wheel steering and articulated steering. The two methods are combined by the control system to minimize the difference of average speed between the front and rear axles while the tractor is turning; thus, there is no need for a central differential gear and the driveline is more compact and simple: driving resistance forces are lower and tire wear is reduced.

A standard three point linkage attachment at the rear, interfacing implements in both agricultural class 1 and 2 standards, can lift and drive implements with mass up to half the mass of the tractor (and 1.5 m backward offset of the center of mass with respect to the rear axle) with the potential to operate dedicated robotic devices for agricultural or demining tasks (more complex than common agricultural implements) [14], see Figure 2.

Despite the above-mentioned advantages, the articulated frame steering layout introduces some problems to AS vehicles, including roll and lateral instabilities in different situations. As the tractor is used for both demining and agriculture and it is particularly meant for steep lands and very irregular terrain, the waving lateral motion is relevant. In addition, the control system could improve, in real time, the vehicle stability, a smart choice of the parameters aimed to improve lateral vehicle stability (especially while using the central steering) was considered at the vehicle design level. The lateral stability of the articulated vehicle was analyzed by simulating the dynamic behavior, based on the non-linear model, hereafter summarized.

A simple but realistic non-linear 3-DOF model was used to dimension the vehicle with intrinsic yaw stability. The model is illustrated in Figure 4 where the X–Y reference frame is the global coordinate system fixed to the ground.

Figure 4. LOCOSTRA kinematic scheme, yaw plane.

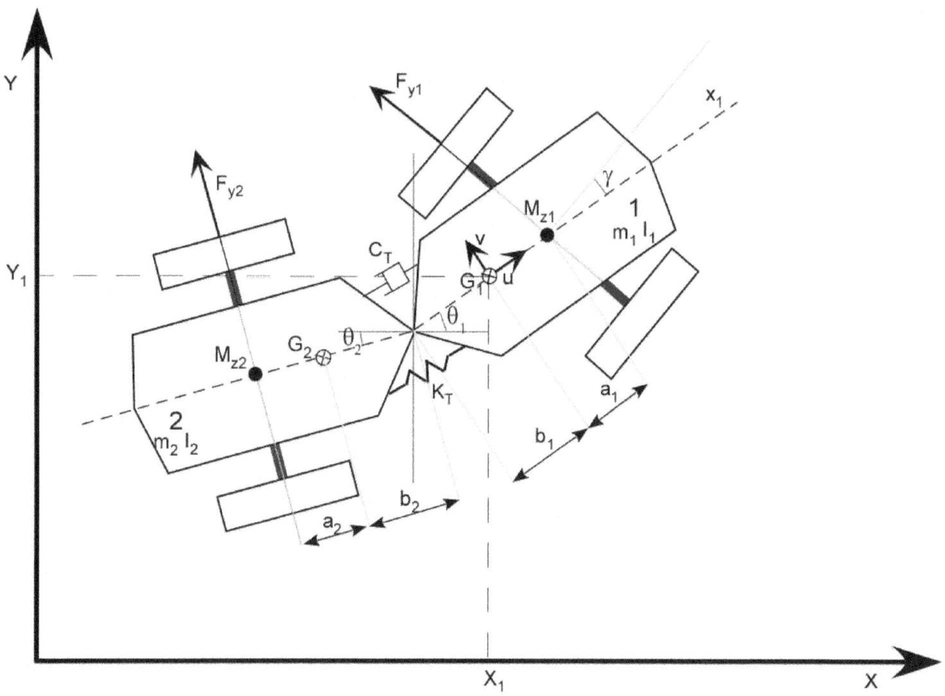

If it is assumed that the front part of the vehicle is not affected by the changes in the articulation angle, the only secondary motion is the yawing motion of the rear part. In addition, it is assumed that the tire rolling resistance is very small and, thus, it is neglected. Moreover, the aerodynamic resistance is neglected due to the low speed. In addition, it is assumed that the vehicle is travelling on a soil surface with constant tire-road contact parameters and the forward speed is assumed constant. The gyroscopic effects are neglected and the axles and wheels are assumed lumped in with their respective parts.

Following the work of Horton and Crolla [15], a torsional spring K_T and a torsional damper C_T at the articulation joint were considered to represent the hydraulic cylinders between the front and rear portions of the tractor.

For this model, θ_1 is the yaw angle between the front part centerline and the X axis. In addition, θ_2 is the angle between the rear part centerline and the X axis.

The model refers to the state vector $[X_1, Y_1, \vartheta_1, \vartheta_2]$ where the chosen state variables are the absolute coordinates of the front part center of mass X_1, Y_1 and the front and rear body yaw angles ϑ_1, ϑ_2. The velocity of the rear part center of mass is:

$$\begin{bmatrix} \dot{X}_2 \\ \dot{Y}_2 \end{bmatrix} = \begin{bmatrix} \dot{X}_1 \\ \dot{Y}_1 \end{bmatrix} + \begin{bmatrix} b_1 sen\vartheta_1 & b_2 sen\vartheta_2 \\ -b_1 \cos\vartheta_1 & -b_2 \cos\vartheta_2 \end{bmatrix} \begin{bmatrix} \dot{\vartheta}_1 \\ \dot{\vartheta}_2 \end{bmatrix} \tag{1}$$

where b_1 and b_2 are the distances from the articulated joint to respectively the centers of mass of the front and rear portions of the tractor.

The motion model is derived using the Lagrangian approach.

The kinetic energy T of the platform is obtained by summing the kinetic energy of the front and rear parts, respectively T_1 and T_2:

$$T = T_1 + T_2 = \frac{1}{2} m_1 \left(\dot{X}_1^2 + \dot{Y}_1^2 \right) + \frac{1}{2} I_1 \dot{\vartheta}_1^2 + \frac{1}{2} m_2 \left(\dot{X}_2^2 + \dot{Y}_2^2 \right) + \frac{1}{2} I_2 \dot{\vartheta}_2^2 \tag{2}$$

where m_i and I_i represent the front (i = 1) part mass and moment of inertia, respectively. The same parameters for the rear part are described by i = 2.

The potential energy U of the system is due to the torsional spring at the articulation joint:

$$U = \frac{1}{2} K_T \left(\vartheta_1 - \vartheta_2 \right)^2 \tag{3}$$

Where K_T is the equivalent torsional stiffness of the hydraulic cylinders, modeled as linear springs including joint mechanical compliance and flexibility of the hoses.

The dissipation energy R is modeled as in Equation (4) where the equivalent torsional damping C_T includes, beyond the joint torsional damping, the effects of friction, and leakage.

$$R = \frac{1}{2} C_T \left(\dot{\vartheta}_1 - \dot{\vartheta}_2 \right)^2 \tag{4}$$

The lateral motion of the articulated vehicle is due to the external forces F_{y1}, F_{y2} and to the alignment moments M_{z1} and M_{z2} due to the interaction of tire and soil. The total virtual work done by these external forces and moments is:

$$\begin{gathered} \delta W = F'_{y1} \delta(Y_1 + a_1 \sin\vartheta_1) + F'_{y2} \delta(Y_2 - a_2 \sin\vartheta_2) + M_{z1} \delta\vartheta_1 + M_{z2} \delta\vartheta_2 \\ \text{where} \quad F'_{y1} = F_{y1} \cos(\gamma + \vartheta_1); \quad F'_{y2} = F_{y2} \cos\vartheta_2 \end{gathered} \tag{5}$$

With γ the steering angle.

The generalized external forces are obtained by introducing the position Equation $Y_2 = Y_1 - b_1 \sin\vartheta_1 - b_2 \sin\vartheta_2$ into Equation (5), and collecting the terms in δY_1, $\delta\vartheta_1$ and $\delta\vartheta_2$; they result in:

$$\begin{aligned} Q_Y &= F_{y1}\cos(\gamma+\vartheta_1) + F_{y2}\cos\vartheta_2 \\ Q_{\vartheta_1} &= \left[F_{y1}a_1\cos(\gamma+\vartheta_1) - F_{y2}b_1\cos\vartheta_2\right]\cos\vartheta_1 + M_{z1} \\ Q_{\vartheta_2} &= -F_{y2}(a_2+b_2)\cos^2\vartheta_2 + M_{z2} \end{aligned} \tag{6}$$

The off road tire model based on the mobility number was used [16]. The mobility number was first defined by using dimensional analysis to model the tire forces; it models the effects of the tire size, load, and soil conditions [17]. Due to the modularity of the approach and of the model implementation in Simulink, other tire models can be adopted. The one used considers a modified definition of slip *s'* and computes the tire longitudinal force F_X in function of the vertical load on the axle F_Z as:

$$s' = \frac{\omega R - v_x}{\omega R} \qquad F_x = F_z C_s\left(1 - e^{-k_1 s'}\right); \quad \text{for small s'} \;\; F_x = F_z C_s k_1 s' \tag{7}$$

where ω is the wheel angular velocity, R is the tire effective rolling radius; C_S and k_1 are constants depending on the tyre and on soil hardness [18]. The values used in the design of the vehicle are for compact hard soil (0.67 and 0.16, respectively); this is also correct for blast resistant wheels with the design we used (inner rubber core and outer steel envelope).

Called α the slip angle [19], the lateral side force can be computed analogously:

$$F_y = F_z C_\alpha\left(1 - e^{-k_2\alpha}\right); \qquad \text{for small } \alpha \;\; F_y = F_z C_\alpha k_2 \alpha \tag{8}$$

with C_α and k_2, respectively, 0.67 and 0.21.

The tire self alignment moments are computed with analogous equations.

The model of the lateral motion is derived as three second order non-linear equations in the independent variables $q_1 = Y_1$, $q_2 = \theta_1$ and $q_3 = \theta_2$, applying the Lagrange approach:

$$\frac{d}{dt}\left(\frac{\partial T}{\partial \dot{q}_i}\right) - \frac{\partial T}{\partial q_i} + \frac{\partial U}{\partial q_i} + \frac{\partial R}{\partial \dot{q}_i} = Q_i \tag{9}$$

where all the energy components are computed using Equations (1–8).

The model was symbolically solved using Maple and then written in Simulink modules. The simulation results of the adopted non-linear model in the case of small angles were compared with the results obtained by Lashgarian Azad [19] using a linearized model for validation purposes.

From the simulation tests, for a given set of realistic parameters, it could be observed an undamped oscillatory yaw motion or unstable snaking mode over a speed value. As the forward speed *u* increases, the amplitude of the unstable snaking mode grows at a higher rate. If K_T increases, the stability of the vehicle oscillatory yaw motion improves; if the torsional damping C_T at the articulation joint increases, the vehicle will be stable up to a given speed. These results are in accordance with the ones for similar AS vehicles, discussed and reported in the literature. They have been used to select the design parameters of the hydraulic rams actuating the articulation of the tractor. In the final design, we use two plunger type hydraulic cylinders with half-inch hoses and flow control valves. The flow control valves have been tuned experimentally to damp yaw oscillations in the speed range of the machine. The design parameters of the whole machine are summarized in Table 1.

Table 1. LOCOSTRA main data.

Length	3.5 m	**Max mass of implement**	900 kg
Width	1.45 m	**Armor thickness (steel plates)**	5 mm
Normal clearing/working width	1.50 m	**Max engine power at flywheel**	55 kW
Front to rear axle distance	1.50 m	**Max engine power at working tool**	48 kW
Inner distance between wheels on same axle	0.95 m	**Fuel capacity**	50 lt
Height (overall)	2.3 m	**Oil capacity**	12 lt
Mass (basic)	3,100 kg	**Cooling system**	Water

2.2. Implements

The work has focused on a set of typical, almost purely mechanical, agricultural implements and on two complex implements driven by PTO shaft. All implements have been tested in the agricultural design and in a modified version dedicated to technical survey in humanitarian demining. The application in humanitarian demining is in the phase called technical survey, part of the land release process: the machine explores land where landmines might be present; the scope of the exploration is finding evidence of mines on portions only of this larger initial area. Typical operations involved are vegetation cutting, land breaking, mine sweeping. The tractor may accidentally run over a mine that may explode under the wheel, so special wheels are required resisting the blast and damping the energy transmitted to the axle and chassis.

2.2.1. Standard Wheels

Wheel characteristics affect both the performance of the tractor in terms of maximum drawbar pull and tractor manoeuvrability.

The wheel diameter of LOCOSTRA was decided as a compromise between manoeuvrability (the smaller wheel radius, the smaller minimum radius of curvature) and drawbar pull (the higher wheel diameter, the higher adherence but higher torque required).

For the radius of curvature, a simplified model of the machine was used with all degrees of freedom modeled with their spans.

The wheel diameter with respect to drawbar pull D_p has been computed with the following equations in terms of the wheel diameter D [m] and width b [m], engine output torque T_e [Nm], overall transmission efficiency η reduction ratio τ, and vehicle body mass m:

$$D_p = \frac{2T_e\tau\eta}{D} - \sum_{i=1}^{4} F_i \qquad F_i = f_i(n, b, m, D) \tag{10}$$

The equation to calculate the rolling resistance F_i of the wheel for rigid wheels over unprepared terrain is derived empirically by Bekker [20], who introduced the pressure sinkage parameter n typical of the interaction ground-wheel. The n parameter value was chosen from the table of data collected by Bekker for different types of soil. In particular in our case $n = 0.5$, $T_e = 230$ Nm, $\tau = 49$, $\eta = 0.95$, $b = 0.244$, $m = 2{,}450$ kg.

The compromise wheel diameter selected is 950 mm. The steering radius is 2,370 mm with articulation, and 1,800 mm with articulation and front wheels steering.

2.2.2. AP Mine Resistant Wheels

The wheels mounted on the tractor during demining related operations are designed to withstand the detonation of 240 g of TNT and to resist at least five explosions before some maintenance may be required. The wheels must keep the tractor safe by reducing the shockwave transmitted to the hub to harmless levels; they have to be inexpensive and easy to repair locally, to allow efficient traction and to be lightweight. The same wheels are mounted on the front and rear axles.

Figure 5. Wheel tests with ballistic pendulum (**a**,**b**) and example accelerometer acquisition for a 240 g TNT blast (**c**).

(**a**)

(**b**)

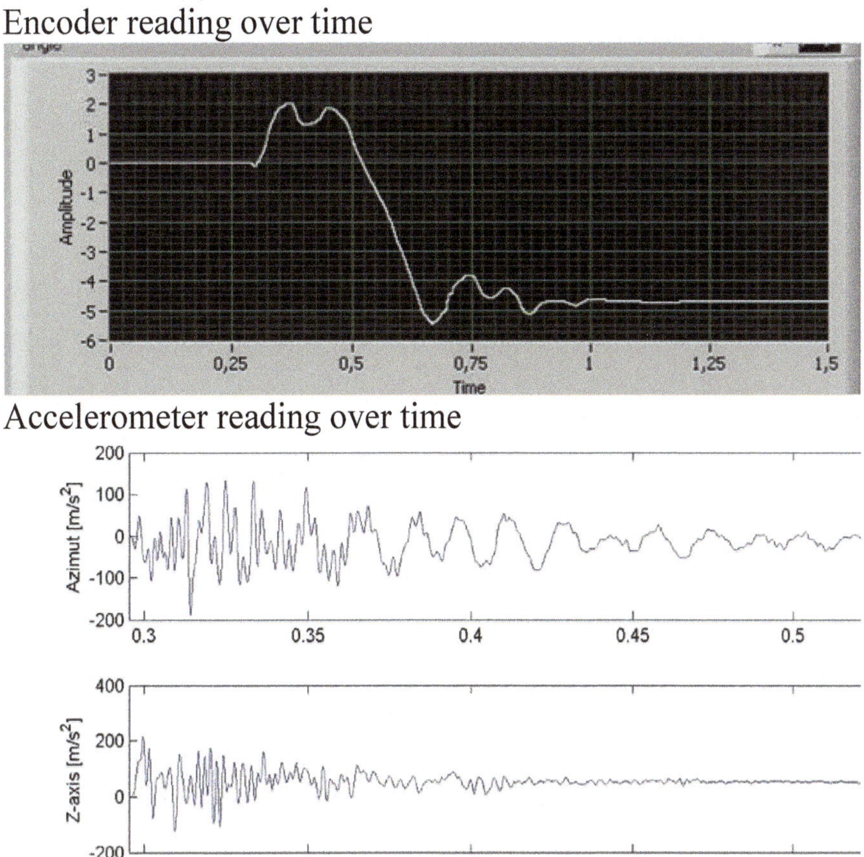

(**c**)

The wheel architecture adopted comprises a solid rubber tire embedded in an outer steel envelope providing ventilation of the blast gas and protection of the inner tire. The inner tire dampens the blast shock-waves. Four different wheel designs with the same outer diameter were manufactured and tested with explosions on a ballistic pendulum and on the tractor in real conditions [21,22], Figure 5. The wheel selected was further modified to simplify the design of the steel envelope.

2.2.3. Vegetation Cutting Implements

The mulcher is the main tool used to clean vegetation over an area before technical survey and any other demining operation can start. The same tool is commonly used in agriculture, as an alternative to cutter bars, to cut the vegetation over land before the ground is processed. Different cutting principles are used. Two implements to cut the vegetation have been developed, one using the principle of the mulcher, the second one using cutter bars.

The mulcher implement is based on a modified commercial mulcher. The modifications comprise the adaptation of the enclosure and frame to withstand possible explosions and the implementation of two depth sensors at the sides, with oscillating bars, to control the height of the tool with respect to the average ground level. When the mulcher is operated, the control system of the tractor switches automatically to speed-control mode. The speed of the machine is controlled to maintain constant power of the engine: the power required by the mulcher to run at constant speed while processing is very sensible to the velocity of advancement in the vegetation as the quantity of vegetation processed in the unit of time depends on this velocity.

The implement using cutter bars uses three bars, one horizontal and two vertical at the sides: the horizontal bar cuts the stems while the vertical bars cut branches to open a corridor. Compared to the mulcher, this tool leaves the material cut on the ground.

2.2.4. Side-Arm Brushcutter

The side-arm brushcutter developed is based on a commercial agricultural arm boom actuated by hydraulic cylinders operated by a hydraulic flow divider and simple on-off electro-hydraulic valves.

Angular sensors have been added to measure the current arm configuration; a pressure sensor enables the control of the arm in floating mode during cutting. The original hydraulic was replaced with solenoid operated proportional control valves that allows smooth and accurate control of the arm motion and make trajectory generation of the bush-cutting tool easier. The valves are directly operated by PWM signals generated by the electronics.

2.2.5. Ground Processing

The ground processing tool developed is based on a modified potato digger. This tool can be used as a higher performance potato digger for agricultural tasks, as well as to expose mines in the soil. Compared to in use sifter implements, carried by 20-ton class dozers, the potato digger can be used over lands with any local surface geometry guaranteeing the desired work depth.

The agricultural potato digger was reinforced in a way to lower the level of damages resulted by the mine explosions and to prevent fragmentation of the damaged parts. This tool is equipped in with a sensor measuring the working depth. The control system having the information of the working depth of the tool is adjusting by commanding the tractor three-point linkage lift to ensure it is proper for the ongoing activity (digging potatoes or digging mines). In demining activities, it is very important to assure the proper digging depth, as it is required for further operations taking place in the processed areas. An additional feature specifically intended for the demining is a metal detector bar mounted on the rear part of the potato digger. The metal detector passes signals of detected metal objects (possibly mines) to the control system. In the future, having already installed DGPS positioning system, the control system will pass information about the detected objects and its position to the higher level management system (implemented in the ROS environment) in order to build a map with the suspected mine positions.

3. Control System

3.1. Overview

A master-slave architecture is used with the tractor as slave; master can be the implement, in the case of complex implements, which have a dedicated control and that can command the operation of the tractor based on their task requirements; master can also be a supervising layer, for example, planning agricultural or demining operations over an area.

The control system is distributed: the control hardware resident in the tractor and the control algorithms there implemented manage all tractor onboard functions and make available the higher level functions of: trajectory following, speed control, steering control, power control, PTO shaft control, three point linkage lift control, and control of the oil distributors. The tractor can operate at constant speed, constant power, constant pull-bar force.

The control variables are: engine velocity, engine torque, engine working modes, engine fan air draught direction, pull-bar force, three point linkage lift level, hydraulic motor, and pump displacement.

Figure 6. Control system elements (core components).

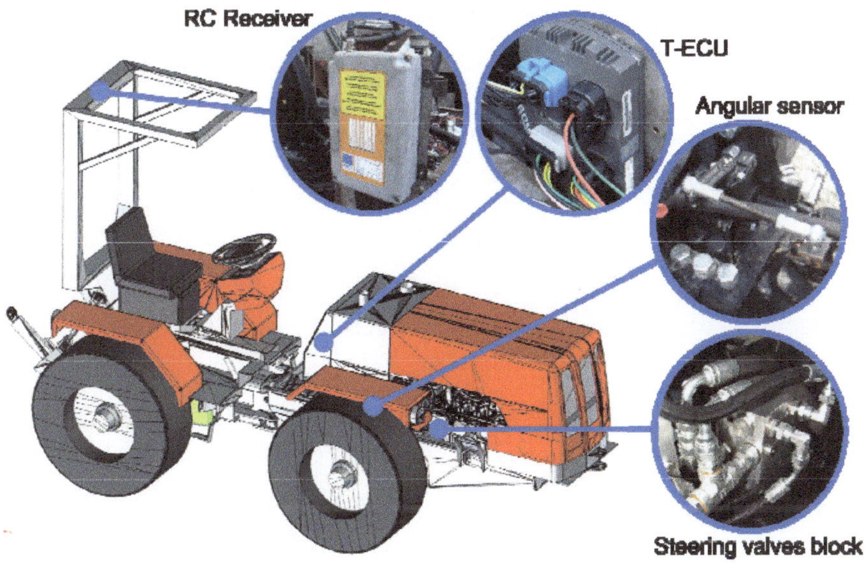

The hardware components of the control are shown in Figure 6; a schematic is provided in Figure 7 [23]. An EPEC 2024 module is used as main control board as it is widely applied in mining, agricultural, forest, and construction machines. It comprises 52 I/O pins (Digital, Analog, PWM) and two CAN ports. The main board is connected to a set of devices: (1) powertrain control devices (via CAN-BUS, J1939 protocol); (2) actuators (via I/O pins); (3) sensors (via I/O pins); (4) human machine interface (HMI) devices (via CAN-BUS, J1939/CANopen/ISObus-ISO11783 protocols); and (5) auxiliary devices (via I/O pins and CAN-BUS using J1939/CANopen/ ISObus-ISO11783 protocols). The basic operation of the tractor (ignition, steering, adjusting velocity, switch between steering axis—front wheels and central articulation) is carried out by controlling electro-hydraulic solenoid valves (PWM or ON/OFF signals for proportional and open/closed valves, respectively), which operate the onboard actuators. The control of the engine is provided by a powertrain control module (PCM) interfaced to the T-ECU over the CAN-BUS (J1939 protocol). The PCM provides to the T-ECU a vast set of information used to manage the operation of the tractor (in particular: engine oil pressure, malfunction alarms, engine speed, engine load at current speed); the T-ECU can set parameters in the PCM to adapt the engine operation to the current tractor operation (desired speed/speed limit, desired torque/torque limit, and others).

Figure 7. Diagram of the tractor electronic control system (TECS).

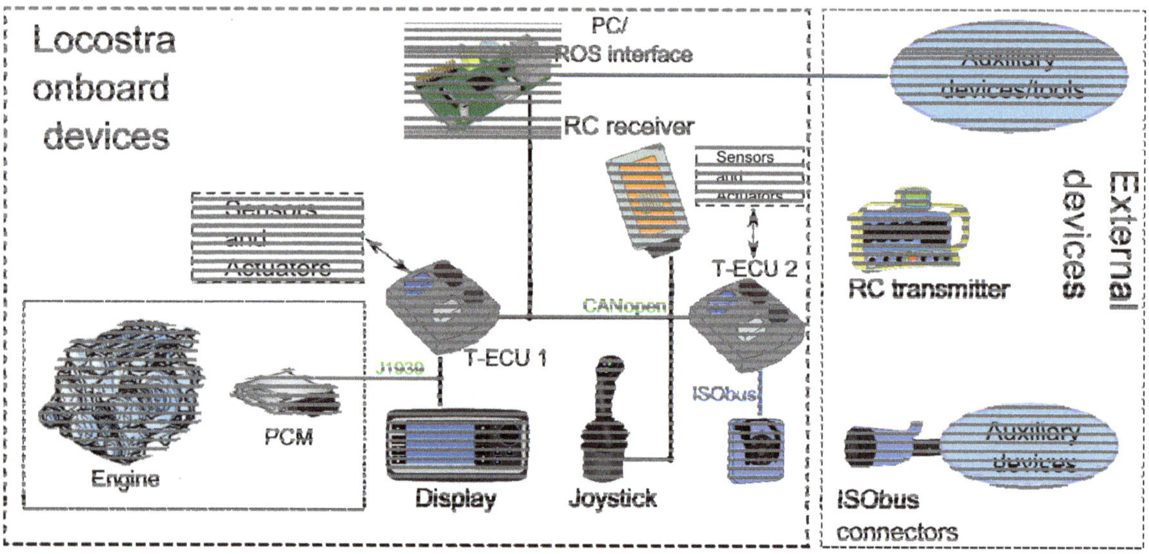

Having implemented the ISObus protocol in the control system it is possible to profit from the use plug and use type implements (ISObus ready). The agricultural implements with ISObus interface can directly command some of the tractors' parameters, such as: requested velocity, use of PTO shaft, use of oil distributors, three point linkage lift level, engine working mode; and present its own working parameters on the tractor's display. In a second direction, with ISObus interface, some of the implement's functions can be commanded using tractor's onboard controllers, e.g., no additional controller is necessary to work with the side-arm brushcutter, it can be commanded using the joystick onboard of the tractor.

At the logical level, ROS middleware is used because it supports well the modular and scalable control architecture adopted. In order to profit from the use of ROS environment we have introduced a module (PC) running ROS middleware. This module provides a tractor higher-level communication interface in a form of services for all the robotized tools and other units cooperating with the tractor (other ROS nodes in the network). It will make easier having one or more tractors part of a multirobot system, and implement and apply algorithms developing its autonomy.

This ROS module providing all the tractor control functions (steering, power control, implement control, *etc.*) and operational data (tractor's position, velocity, *etc.*) makes the tractor fully controllable by the master units – robotized tools which need to control the tractor behavior in order to proceed with their work. An example of such a master unit is a 5DOF manipulator with a metal detector or other kind of sensor or actuator controlling the tractor's movements and actuators in order to perform planned actions (e.g., autonomously checks a selected area in a field for mine presence or dig holes and plant trees in specified positions).

3.2. Control Modes and Human Machine Interface

The tractor is equipped with three human machine interface devices that operate on the slave control level. Three control modes are available: human onboard, human remote, and ISObus.

In the human onboard mode, the driver uses an onboard joystick, a conventional steering wheel, a 4.3″ industrial display and pedals with electric signal transduction. The display (Wachendorff OPUS A3S) provides information about the state of the tractor (e.g., velocity, engine speed, coolant temperature), it visualizes alarms and warnings (e.g., low fuel level, low oil pressure, sensors malfunctions) and allows to input data and changes T-ECU settings (e.g., engine work parameters, tractor operational parameters). With the onboard joystick, the operator can control the velocity of the tractor, the actuators controlling the position of the implement, the blockage of the differential gears of the axles, the rear power take-off (PTO) shaft, and operate singularly the steering axes.

In the human remote mode, the driver uses an industrial remote control (RC) system connected and interfaced to the TECS. The RC system is composed of two units: receiver and transmitter. The receiver is mounted on-board and is connected to the internal tractor CAN bus (CANopen protocol) providing a bi-directional communication between T-ECU and the RC transmitting unit. It also ensures a high level of safety during the work introducing to the system a safety circuit having a Pl:e (Performance level) category 4 and/or SIL 3 (Safety Integrity Level) according to the standards ISO 13489-1 and ISO 62061, respectively. The transmitter, Figure 8, gives the operator control over a majority of tractor functions and visualizes a feedback information from the T-ECU. It allows control of the tractor velocity, the range of speed, the steering, the differential gear blockage, the front and rear oil distributors, the three-point linkage lift, the PTO shaft. In normal operation, front wheel, and articulated steering are automatically combined: the operator can choose to take the control of each of the two independently. Furthermore, the transmitter allows to control the tools (implements) attached to the tractor (side-arm brushcutter as default) using a set of four unidirectional proportional joysticks. The LED indicators and the display mounted on the transmitter provide feedback information from the T-ECU (e.g., active steering switching mode, engine parameters, working parameters, warnings, and alarms).

The most advanced control mode is an autonomous control mode. In this mode the tractor performs specific actions under a control of higher level implemented using the ROS framework, or it works as a slave machine, controlled by the intelligent implements attached to the tractor. The units controlling the tractor can command velocity, driving direction, engine working mode and speed, oil distributors, three-point linkage lift, and PTO shaft.

In the next step of the development the LOCOSTRA tractor will be equipped with a RTK DGPS positioning system in order to implement the localization and navigation systems and make it able to perform fully autonomous tasks in the field [24].

Figure 8. Remote control transmitting unit.

3.3. Control System Features

The tractor control system developed provides innovative and original functionalities not available today all together in small size tractors. The most relevant ones are: (1) the control of the speed on the base of the engine load (to maintain the engine automatically at a desired, set, power rate); (2) steering assistance to combine front wheels and articulated steering for minimum wheel skid; (3) remote control up to 200 m from the tractor; and (4) fully automatic operation.

The velocity of the tractor depends on the hydraulic oil flow provided by the main pump run by the tractor engine. The operational power, with any implement, is a function of the tractor speed and it tends to zero when the speed goes to zero. In cases when the attached implement, under a heavy load (e.g., mulcher processing compact soil or cutting very thick vegetation), absorbs an amount of power reaching the 93% of the engine maximum capacity, the engine speed lowers to the level at which the hydraulic pump is not able to produce sufficient hydraulic oil flow enough to propel the hydraulic motors continuously. When this situation occurs the tractor starts moving in a flutter way. In severe cases, the engine can stop. The speed control implemented reads the current engine load and adjusts the velocity of the tractor to guarantee smooth and uninterrupted movement. The hydraulic pump for the traction has variable displacement: the control system maintains the speed of the engine constant while the displacements of hydraulic pump and motor are adjusted to work at constant maximum power.

The steering assistance (SA) uses a model of the tractor kinematics and the current tractor configuration known from sensors positioned in the front wheels and in the central articulation. Two steering options are available: automatic steering (automatic combination of front and articulated steering) and manual steering.

3.4. Tractor Applications

Figure 9 shows two applications of the tractor in agriculture in tasks that could not be carried out in the same way without the robotic functions developed. In the case of tractor equipped with orchard platform, Figure 9a, the remote control allows the operator to control the tractor directly from the platform attached to the tractor. Platforms designed to be operated (lifted or lowered) from the cabin of the tractor using hydraulic oil distributors can be safely operated via the remote control using the transmitter from the board of the platform (the remote control system conforms to ISO 13849-1 Pl:e and ISO 62061SIL 3 for the E-STOP safety function and Pl:d category 3 and/or SIL 2 for the steering functions).

Figure 9b shows the use of a side-arm brushcutter operated from remote with no need of a separate control device, using the joysticks already available in the remote control board. With an additional control module in the tractor, working as a gateway between the TECS internal bus and the auxiliary device interface bus, it is possible to command implements equipped with electronic modules using the CAN-BUS ISO 11783 (ISOBUS) agricultural machine protocol (based on the J1939 protocol).

A planner operating up to four tractors, through the autonomous navigation layer, is under implementation. It comprises, for each tractor, also the control of the side-arm brushcutter: (a) navigation of the tractor along specified trajectory; (b) positioning of the brushcutter head along specified trajectory adapting to the geometry of the terrain.

Figure 9. Example applications of the tractor with TECS: (**a**) with orchard platform; (**b**) with side-arm brushcutter.

(**a**) (**b**)

In the first task the software developed in the ROS framework is responsible for the localization of the tractor and its navigation along the specified trajectory. The localization part of the software will use signals coming from the RTK DGPS system and odometry sensors to localize the tractor in the work coordinate system (a coordinate system tied to the base of the stationary receiver of the RTK DGPS system).

The second task controls the actuators of the arm, Figure 10. The arm is in floating mode (laid on the surface of the terrain) and the center of the wrist of the brush-cutter is moved in a plane orthogonal to the direction of motion of the tractor to follow the desired trajectory.

Figure 10. Autonomous operation mode example.

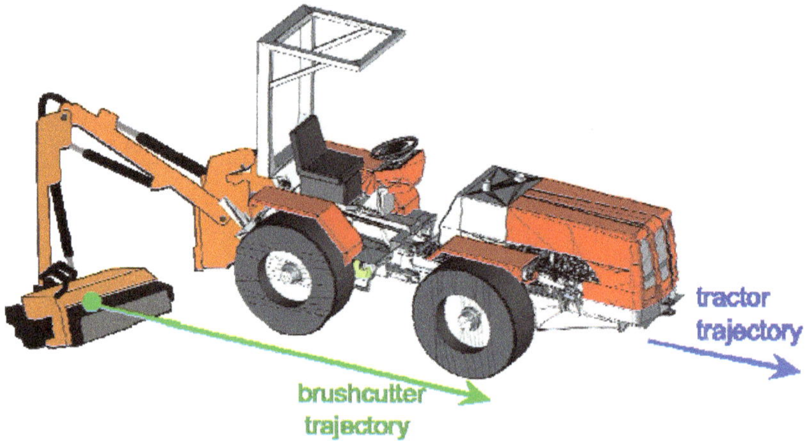

4. Field Tests

The machine was tested, first, in Italy, to confirm functions and performance. At the end of this domestic campaign, the machine was shipped to Jordan and tested in minefields in real operative conditions and by professional demining personnel. All the tests done in the two campaigns, and reported in the following, focus on specific performance figures of the machine; campaigns aimed at collecting data for statistical assessment of the success rate of the machine, for instance with implements for mine exposure, or the rate of blasted mines during, for instance, vegetation cutting, will be organized in the next future as a part of the program of development of the machine.

4.1. Italy in-Field Pre-Test

The tractor ground pressure measured varies depending on the terrain conditions: from 1,813 kPa on class III soil to 453.25 kPa on class I soil (soil classification according to CEN CWA 15044:2009).

The maximum drawbar pull was measured in different soil conditions. The comparison of the experiment data with the computed figures was satisfactory: average 13,700 N in class III soil with moisture content 93% using first gear, average 6,400 N in class III soil with moisture content 93% using second gear; average 19,900 N in class I soil with moisture content 78% using first gear, average 6,400 N in class I soil with moisture content 78% using second gear.

The tractor demonstrated able to climb until 30° of natural hill ground.

Yaw motion stability showed high in the whole working range 0 to 15 km/h for different classes of terrain with best performance in grounds whose characteristic sinkage number was used to design the tractor.

4.2. Jordan in-Field Test

The tests took place at two different locations: in the safe area in front of the minefield Sabha12 along the border with Syria, near Jabir village (later referred to as Jabir test site) and in the suspected hazardous area (SHA) near to the border with Israel, close to Karama village, in the Jordan Valley (later on referred to as Karama test site). The Jabir test site had no vegetation and class I soil; the Karama test site had vegetation and class I soil.

The test protocol followed the CWA15044:2009 and took place in three subsequent phases: performance test, survivability test and acceptance test. Depending on the vegetation and on the landmines used, LOCOSTRA was equipped with ground processing tool (GPT) (potato digger, Figure 11a) or vegetation cutting tool (mulcher, Figure 11b) or with no tools.

Figure 11. LOCOSTRA equipped with GPT (**a**) and with mulcher (**b**).

(**a**)

(**b**)

The overall results were very satisfactory: as reported by the demining personnel supervising the tests, the tractor accomplished correctly all operations of site preparation for technical survey and demining; the tractor performed well, also, the agricultural operations typical of the recovery of land to agriculture once released as safe from mines. Market agricultural implements available locally were also tested and operated by the tractor using its remote controlled and automatic functions with very good results.

4.3. Performance Tests

The aim of these tests was to evaluate the capability of the GPT on digging buried mines out and relocate them on the surface. 33 dummy AP mines (11 of type M14, 11 of type M35, and 11 of type A10) were laid at different depths between 0 and 15 cm in soft ground, at different distances from each other. Three mm rectangular fibreboards of size 180 cm × 35 cm were placed in the soil to register the penetration profile. The results are summarized in Table 2.

Table 2. Performance test results.

Land Mines	Results
11 of A10 mines were laid direct under the surface	1. All mines removed to the surface. 2. One min was broken by the tines of the GPT.
11 of M35 mines were laid at the depth of 5 cm	1. 8 mines removed to the surface. 2. 2 mines touched and moved few centimeters of their location but stayed 5 cm under the surface.
11 of M14 mines were laid at depth of 10 cm	1. No mines were removed to the surface. 2. 6 mines were only touched by tines and remained almost at the same depth. 3. 5 mines were not touched at all.
Fibreboards	**Results**
Three 3 mm fiberboards with dimensions of 180 cm × 35 cm were put in the soil to give the penetration profile	All fiberboards were affected by the tines of the GPT at depth varying from 13–16 cm with average of 15 cm.

4.4. Survivability Tests

Different types of live mines were laid to evaluate the survivability of the blast resistant wheels, the tractor and the GPT against AP mine blasts. LOCOSTRA resisted well, without needing any repair, to four explosions, as described in Figure 12a.

4.5. Acceptance Tests

The planned tests are the following; the main results are collected in Table 3.

- Five live mines (M14) were laid at different depths between 0 and 10 cm in soft ground. Two mines were passed over by the wheels and three ones were processed by the GPT.
- Two live mines (M35) were laid in a safe area and passed over by the wheels. The test was repeated in an active mine field containing mines of type PMN (containing 240 g of TNT explosive).
- Two days of tests using the mulcher on suspected hazardous areas covered by medium and tall vegetation in the Jordan Valley.

Table 3. Survivability tests results.

Activities	Results
One live M14 (28 g) was laid at depth of 5cm to be activated by the rear wheel	(1). The mine was activated. (2). No damage reported.
One live M14 (28 g) was laid at depth of 5cm to be activated by the front wheel	(1). The mine was not activated.
Two live M14 were laid to be processed by the GPT. One of them was lad directly under the surface and one was laid at depth of 10 cm	(1). One mine was removed to the surface. (2). One mine exploded under the wheel of the GPT.
Two live M35 (100 g) were laid, the first one was laid on the surface one meter from the front wheel, the second mine was laid one meter from the rear wheel at 5cm depth. Wheels past over the mines	(1). The first mine was activated, no damage to the wheel was reported. (2). The second mine was not activated. The mine was removed out by the GPT and it was found pressed down and cracked.
Two live PMN (240 g) were laid, the first one was laid on the surface one meter from the front wheel, the second mine was laid one meter from the rear wheel at 5 cm depth. Wheels past over the mines	(1). The first mine was activated, only light damage was caused to the wheel. (2). The second mine was not activated, it was disposed by another mine.

The tests with the GPT opened up space for further study and a proposal on the possible application of agricultural tools in demining. Even if the GPT would benefit from further studying, it proved to be a very efficient tool for ground processing and mine removal. In fact, the GPT successfully removed mines down to a depth of 5 cm without exploding them, Figure 12b. When tested with target wooden mines, after adjusting the length of the third-point linkage, it reached the working depth of 15 cm. Damages recorded after the survivability tests were very localized and easy to repair, Figure 12c.

Figure 12. LOCOSTRA test campaign in Jordan: reference test mine location (**a**); (**b**) Explosions occurred under LOCOSTRA, an M14 mine removed and processed by the GPT without explosion (**c**); the worst damage caused to blast resistant wheels (very light) after a PMN explosion (**d**); medium vegetation before (**e**); and after clearance (**f**).

The combination of LOCOSTRA with the mulcher proved to be ready to be exploited in mine action activities. LOCOSTRA with the mulcher successfully cut low, Figure 12e, medium and tall vegetation up to small trees 15 cm in diameter. Especially during the pre-tests in Italy, where LOCOSTRA equipped with the mulcher was tested in forest-like areas, the high agility and great efficiency of the machine was appreciated. Thanks to the double steering system, the tractor could turn around the bigger trees (that generally do not have to be cut) and cut all the others. Moreover, the mulcher capacity to slightly process the soil was also appreciated, as it can be seen from Figure 12f.

5. Conclusions

This paper presents the first phase of the development of the tractor robotic platform called LOCOSTRA. This machine was designed to accomplish agricultural and demining tasks, a unique case ever. It was built and tested using implements typical of agriculture, with modifications to make them operable by a semi-autonomous robotic tractor, without missing the original agricultural application. The work done to develop the LOCOSTRA machine results in providing solutions valid both for demining and agriculture activities. The features of the machine: light weight, small size, great manoeuvrability, remote and semi-autonomous control, and automated functions; give new quality tools for use as well in agriculture as in demining.

The modular approach adopted in the development process results in a flexible framework, divided into smaller subsystems (Tractor Electronic Control System, Remote Control System, intelligent auxiliary tools, higher level control modules), which can be easily transferred and applied to other agricultural vehicles in order to make them controlled remotely and add them some autonomy.

The use of LOCOSTRA for humanitarian demining is confirmed by the results of the tests performed which also confirm all the initial statements and the expectations of the end-users involved.

The work on the system continues with the development of the navigation and autonomous operation, with the aim to achieve in field autonomy. A peculiarity of the LOCOSTRA system is the high readiness level, already 9, remarkably high for the field of application.

The LOCOSTRA machine has been included within the technologies for technical survey on which the European project TIRAMISU (Toolbox Implementation for Removal of Antipersonnel Mines, Submunitions and UXO) research is based.

Acknowledgments

The Italian Ministry of Economic Development and the Italian Institute for Foreign Trade that co-funded the LOCOSTRA project, coordinated by the University of Genoa, are gratefully acknowledged.

Authors Contributions

Emanuela Elisa Cepolina, prime researcher at Snailaid, developer of the demining application of LOCOSTRA and expert on machine architecture and implements for humanitarian demining, responsible for in-field tests planning. Michał Przybyłko, developer of the automation and control part of LOCOSTRA, developer of the control hardware infrastructure. Giovanni Battista Polentes, manufacturer of the tractor, designer of the machine with focus on the mechanical components, responsible for the preparation and development of the in-field tests. Matteo Zoppi, development coordinator, responsible for the integration, responsible for part of the design of the machine.

Conflicts of Interest

The authors declare no conflict of interest.

References

1. Acaccia, G.M.; Michelini, R.C.; Molfino, R.M.; Razzoli, R.P. Mobile Robots in Greenhouse Cultivation: Inspection and Treatment of Plants. In Proceedings of the 1st International Workshop Advanced in Service Robot, Bardolino, Italy, 13–15 March 2003.
2. Kondo, N.; Morita, M.; Noguchi, N. *Agricultural Robots: Mechanisms and Practice*; Trans Pacific Press: Melbourne, Australia, 2011; pp.1–360.

3. Yukumoto, O.; Matsuo, Y.; Noguchi, N. Robotization of agricultural vehicles (Part 1). *Jpn. Agri. Res. Q.* **2000**, *34*, 99–105.
4. Yukumoto, O.; Matsuo, Y.; Noguchi, N. Robotization of agricultural vehicles (Part 2). *Jpn. Agri. Res. Q.* **2000**, *34*, 107–114.
5. Matsuo, Y.; Yukumoto, O.; Yamamoto, S.; Noguchi, N.; Hara, Y. Review enhanced adaptability of tilling robot (2nd report)—Execution of various operations by tilling robot. *Jpn. Agri. Res. Q.* **2013**, *47*, 153–164.
6. Ismail, W.I.W.; Hudzari, R.M.; Saufi, M.K.M.; Fung, L.T. Computer-controlled system for autonomous tractor in agricultural application. *J. Food Agri. Environ.* **2012**, *10*, 350–356.
7. Cepolina, E.E. Power tillers and snails for demining in Sri Lanka. *J. Mine Action* **2006**, *10*, 76–79.
8. Dixon, R. A Concept of Implementing Technology to Encourage Economic Growth in Demining. In Proceedings of the International Workshop on Robotics and Mechanical Assistance in Humanitarian Demining and Similarly Risky Interventions, Brussels, Belgium, 16–18 June 2004.
9. Bottomley, R. Returning Life to Field and Forest: Demining by Villagers in Cambodia. Available online: http://www.jmu.edu/cisr/journal/5.1/Focus/Ruth_Bottom/bottom.html (accessed on 10 March 2014).
10. Knight, P.; Jenkins, J.O. Adopting and applying eco-design techniques: A practitioners perspective. *J. Clean. Prod.* **2009**, *17*, 549–558.
11. Cepolina E.E.; Zoppi M. Could local agricultural machines make a country "impact free" by 2010? *Int. J. ERW Mine Action* **2009**, *13*, 47–52.
12. Rovira-Más, F. Sensor Architecture and task classification for agricultural vehicles and environments. *Sensors* **2010**, *10*, 11226–11247.
13. Delhay, S.; Lacroix, V.; Idrissa, M. PARADIS: Focusing on GIS Field Tools for Humanitarian Demining. In Proceedings of the Fifth IARP WS HUDEM, Brussels, Belgium, 16–18 June 2004.
14. Horton, D.N.L.; Crolla, D.A. Theoretical analysis of the steering behaviour of articulated frame steer vehicles. *Veh. Syst. Dyn.* **1986**, *15*, 211–234.
15. Crolla, D.A.; El-razaz, A.S.A. A review of the combined lateral force generation of tires on deformable surfaces. *J. Terramechanics* **1987**, *24*, 199–225.
16. Gee-Clough, D.; Sommer, M.S. Steering forces on undriven, angled wheels. *J. Terramechanics* **1981**, *18*, 25–49.
17. Metz, L.D. Dynamics of four-wheel-steer off-highway vehicles. *SAE Paper* **1993**, doi:10.4271/930765.
18. Iida, M.; Fukuta, M.; Tomiyama, H. Measurement and analysis of side slip angle of articulated vehicle. *Eng. Agric. Environ. Food* **2010**, *3*, 1–6.
19. Azad, N.L.; Khajepour, A.; Mcphee, J. Robust state feedback stabilization of articulated steer vehicles. *Veh. Syst. Dyn.* **2007**, *45*, 249–275.
20. Bekker, M.G. *Off-the-Road Locomotion, Research and Development in Terramechanics*; University of Michigan Press: Ann Arbor, MI, USA, 1960.
21. Cepolina, E.E.; Zoppi, M.; Belotti, V. Locostra: Blast-resistant wheels test. *Int. J. ERW Mine Action* **2011**, *15*, 75–81.

22. Naselli, G.A.; Polentes, G.; Cepolina, E.E.; Zoppi, M. A simple procedure for designing blast resistant wheels. *Procedia Eng.* **2013**, *64*, 1543–1551.

23. Przybyłko, M.; Cepolina, E.E.; Zoppi, M.; Polentes, G. A Robust, Simple, Low-Cost Autonomy Enhancement Module for Locostra: A Remotely Controlled Demining Machine. In Proceedings of the 9th International Symposium HUDEM 2012, Sibenik, Croatia, 24–26 April 2012.

24. Livatino, S.; Muscato, G.; Sessa, S.; Koffel, C.; Arena, C.; Pennisi, A.; Di Mauro D.; Malkondu, E. Depth-enhanced mobile robot teleguide based on video images. *IEEE Robot. Automa. Mag.* **2008**, *15*, 58–67.

8

A Miniature Robot for Retraction Tasks under Vision Assistance in Minimally Invasive Surgery

Giuseppe Tortora *, Tommaso Ranzani, Iris De Falco, Paolo Dario and Arianna Menciassi

The BioRobotics Institute, Scuola Superiore Sant'Anna, viale R. Piaggio 34, 56025, Pontedera (Pisa) 56025, Italy; E-Mails: t.ranzani@sssup.it (T.R.); i.defalco@sssup.it (I.D.F.); p.dario@sssup.it (P.D.); a.menciassi@sssup.it (A.M.)

* Author to whom correspondence should be addressed; E-Mail: g.tortora@sssup.it

---

Abstract: Minimally Invasive Surgery (MIS) is one of the main aims of modern medicine. It enables surgery to be performed with a lower number and severity of incisions. Medical robots have been developed worldwide to offer a robotic alternative to traditional medical procedures. New approaches aimed at a substantial decrease of visible scars have been explored, such as Natural Orifice Transluminal Endoscopic Surgery (NOTES). Simple surgical tasks such as the retraction of an organ can be a challenge when performed from narrow access ports. For this reason, there is a continuous need to develop new robotic tools for performing dedicated tasks. This article illustrates the design and testing of a new robotic tool for retraction tasks under vision assistance for NOTES. The retraction robots integrate brushless motors to enable additional degrees of freedom to that provided by magnetic anchoring, thus improving the dexterity of the overall platform. The retraction robot can be easily controlled to reach the target organ and apply a retraction force of up to 1.53 N. Additional degrees of freedom can be used for smooth manipulation and grasping of the organ.

Keywords: miniature robot; minimally invasive surgery; retraction robot; internal motors actuation

---

1. Introduction

Surgical procedures have advanced over the past years and many solutions have been proposed with the aim of reducing the invasiveness of current medical techniques [1]. Minimally Invasive Surgery (MIS) provides the same quality of surgery with smaller incisions compared to an open approach, thus reducing trauma for the patient and consequently improving acceptability and cosmetics as well as reducing recovery times. Laparoscopy is currently considered the MIS technique par excellence, and is performed using long instruments that are introduced in the abdomen through dedicated accesses. From a medical viewpoint, laparoscopy requires high skills, due to the intrinsic technical limitations that result in the loss of hand-eye coordination and reduced dexterity.

Robotic surgery has been introduced in the clinical practice to make difficult surgical tasks (*i.e.*, suturing tasks) easier for the surgeon, recovering hand-eye coordination and dexterity. The Da Vinci surgical robot was the first example of medical robot that was successfully applied to the medical practice [2]. Through a dedicated console, the surgeon can intuitively and precisely control the laparoscopic tools, while a stereoscopic 3D camera provides images of the surgical scenario. Despite its unquestionable advantages, the Da Vinci system is bulky, difficult to set up in the operating room and expensive. Above all, the same number of incisions (*i.e.*, the same level of invasiveness) is required for a robotic laparoscopic surgery and a surgery performed in traditional laparoscopy. Other more compact robots have been developed for approaching traditional laparoscopy with a robot, but they still have the major limitation of requiring the same number of incisions as in traditional laparoscopy [3–5].

In order to reduce the invasiveness of minimally invasive procedures, Natural Orifice Transluminal Endoscopic Surgery (NOTES) was introduced in 2004 [6]. NOTES has attracted growing interest because it eliminates abdominal incisions in the diagnosis and treatment of abdominal pathologies, by combining endoscopic and surgical techniques. Nowadays, NOTES procedures can be performedclinically using conventional or modified endoscopes [7,8], hence exploiting the flexibility of such instruments for reaching the surgical target from different natural orifices. On the other hand, the main drawbacks are related to their low dexterity, reduced force generation, and poor mechanical stability, thus making the surgical procedure extremely challenging [9]. In order to overcome stability limitations, miniaturized robotic systems have been proposed worldwide, that can be introduced in the body and driven to the surgical target, exploiting magnetic coupling between internal and external magnets [10–12]. These robotic platforms have undergone experimental trials on live animals and have shown promising potential to become clinical realities.

Although these solutions allow the successful anchoring of the robotic system to the abdominal wall, the fine control of the Degrees of Freedom (DOFs) has been only partially demonstrated during research studies [13]. For this reason, other robotic systems have been proposed, such as [14,15], in which dedicated miniaturized surgical tools have been proposed to perform basic surgical tasks in NOTES. Other solutions exploit advanced endoscopic robots, such as [16], or dexterous miniature robots with integrated motors [17,18]. As demonstrated in experimental trials on animals, these robotic systems allow tissue manipulation and the performance of simple surgical tasks, but they cannot be used to retract an organ or tissue while manipulating it. Therefore, the retraction task is typically performed through an additional incision to allow triangulation [19]. Several solutions have been proposed and tested on real working scenarios in order to perform retraction tasks by exploiting magnets as in [20–24]. Although the use of magnets improves the stability of the robotic device and allows the retraction of organs, it is generally difficult to accurately steer the internal robot.

In this work, we present the design of a novel miniature modular robot for retraction tasks in MIS. The robot combines the advantages of magnetic anchoring for positioning and anchoring purposes; furthermore it is integrated with motors to enable two active DOFs, thus improving the dexterity of the platform. The design of the robot took inspiration from preliminary works and was optimized to allow its use on bench applications in cooperation with a camera robot already developed [25]. The retraction robot was equipped with a magnetic anchoring system in order to add basic movement functionalities, as described further below.

2. Materials and Method

2.1. Medical Rationale

The role of modern medical practice is moving increasingly closer to the concept of early diagnosis, making it possible to fight diseases at their early stage and to enable minimally invasive surgical approaches when needed. Early diagnosis allows the therapeutic and surgical treatment of diseases, with lower risks and greater chances of successful outcomes. The concept we propose is based on modular robotic units aiming to provide current NOTES with smaller and less complicated devices capable of performing basic surgical tasks on relatively small operating areas (*i.e.*, low forces, small workspace required).

The aim of this work is to identify specific basic tasks that can be usefully performed during NOTES, and to adapt the robotic modules to them. Each robot is provided with specific functionalities that participate in the complete surgery all together. In previous work [25], we proposed the basic design of the modules by describing the development and manufacturing of a camera robot, paving the way to the development of other modules to be used during surgical NOTES procedures.

2.2. Retraction Robot

In this work, we describe another essential module for surgical procedures, which was tested in terms of generated force and possible motion in the surgical scenario.

The retraction robot, having a diameter of 12 mm and a total length of 52 mm including the magnetic base, was designed with a serial kinematic chain, since it can be more easily introduced through a surgical trocar compared to a parallel robot. The diameter and the length of the robot allow introduction through an esophageal access port, as demonstrated in [17]. Since the main goal of the devised task is to retract a tissue to allow its manipulation by another tool (*i.e.*, an endoscope or another robotic unit), we made the technical choice to internally integrate the robot with two motors for one single pitch DOF in addition to the opening/closing mechanisms of the gripper. The retraction robot relies on three additional external DOFs for the correct positioning of the robot in the abdomen, as described further below. Camera and retraction robot 3D sketches are shown in Figure 1.

Figure 1. (**Top-Left**) 3D sketches of the camera robot and their Degrees of Freedom (DOFs). (**Top-Right**) 3D sketches of the retraction robot. (**Bottom**) Detail of the retraction robot and embedded motors.

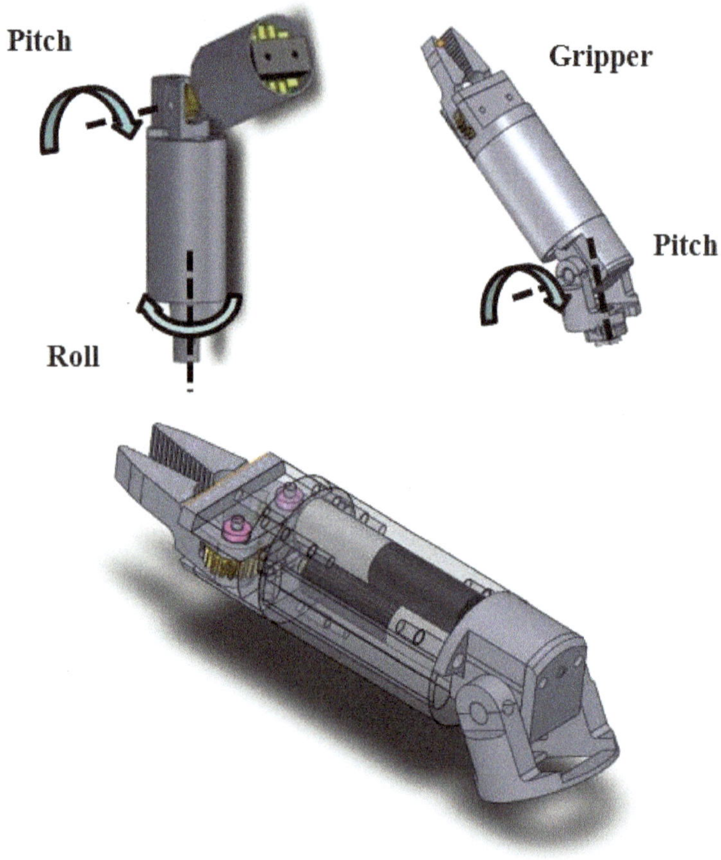

Brushless micromotors SBL04 (Namiki, Akita, Japan) with 1:337 reduction ratio were selected as actuators for the rotational DOFs, due to the large generated torque (up to 5.7 mNm) compared to the small size (4 mm in diameter and 17.4 mm in length including the gearbox). These motors can be easily controlled in two directions by using commercial boards mounted in a circuit architecture with buttons (for activating one motor) and switches (for choosing the motion direction).

The actuators inside the robot determine the design of the robot itself, since the proximal link motors must withstand a high torque; consequently, a high reduction ratio mechanism must be considered. The design of the mechanisms for the pitch joint consists of a helical gear mounted on a worm gear. This is used to transmit motion between the axis of the motor and the orthogonal pitch joint axis with a high reduction ratio that makes the system non-backdrivable.

The retraction robot has two DOFs: a pitch actuated joint and an end-effector consisting of an opening/closing mechanism for the gripper. Its workspace is thus one-dimensional as its task is to pull the retracted tissue towards the abdominal wall in order to expose the target site. The motion range of the pitch DOF is −180°; 180°. Although the retraction unit only provides 2 internal DOFs, a full tridimensional workspace can be swept by moving the robot magnetically attached to the abdominal wall; the abdominal wall can be also pushed, in order to gain additional mobility due to the magnetic coupling. The assembled retraction robot is shown in Figure 2.

Figure 2. Assembled prototype of the retraction robot.

2.3. Camera Robot

The camera robot consists of a 2 DOFs robotic unit, conceived to enable vision during NOTES. The camera robot design is based on a roll/pitch module linked to a passive support that holds the vision system.

Vision System

The vision system was integrated in the distal part of the camera robot. It includes a camera, necessary for each MIS technique, and a source of illumination to optimize the lighting conditions in a dark environment, such as the abdominal cavity.

The camera (Misumi Electronics corp., New Taipei City, Taiwan) has a frame rate of 30 fps (frame per second) and resolution of 320 × 240 pixels. It is 10 mm in length and 8.6 mm in diameter and has a power consumption of 100 mA at 3.3 V. These features are compatible with the devised application in terms of frame rate and resolution (they are sufficient for the proposed tasks), dimensions (the camera can be easily integrated in the robotic module) and power supply and consumption (there are no critical energetic constraints because the camera is wire-supplied from an external power source at 3.3 V). Appropriate lighting conditions are guaranteed by a PCB integrating four white LEDs (light emission diodes) by Nichia Corporation. It was properly designed and positioned close to the camera on the same plane in order to avoid artifacts of the image, such as shadows and highlights. Each LED has dimensions of $2 \times 1.2 \times 1.3$ mm^3 and a power consumption of 10 mA at 3.3 V. LEDs were dimensioned and integrated on a circular PCB having an external diameter of 12 mm, corresponding to the robot diameter, and an internal diameter of 8.6 mm, corresponding to the camera diameter, thus ensuring stability of the vision system inside the robot which is very important for the quality of the images during video acquisition. The camera sends the video to an external frame grabber that connects the camera with the USB of the PC and sends the images. The video streaming is displayed on the user interface allowing real-time control of the robotic modules.

2.4. Anchoring and Positioning of the Modules

In this work, we describe another essential module for a surgical procedure which was tested in terms of generated force and possible motion in the surgical scenario. The proposed design is inspired from previous work [25], and it was optimized to obtain a basic unit capable of performing retraction tasks.

2.4.1. Anchoring Frame

In order to provide stable adhesion of the camera module to the abdominal wall and guarantee accuracy and precision in surgical tasks, a dedicated anchoring frame was used. The magnetic attraction force between the internal frame and an external hand-held magnetic module was exploited. The anchoring frame was conceived to host, in the future, up to three miniature robots for vision and tissue manipulation. The possibility of docking the camera robot to the anchoring frame has been demonstrated in [26]. While the docking of vision and manipulators modules is essential for enabling a reliable position control of multi-DOFs robots, the two internal DOFs retraction robots do not need to be attached to the same frame. Although the attachment of the retraction robot to the anchoring frame would not represent an issue in terms of technical and medical efforts, this would limit its tissue retraction effectiveness. For this reason, we preferred to directly link this robot to the abdominal wall by using an additional magnet in order to provide additional DOFs. This allows the robot to move in the whole abdominal cavity, thus improving the ability to reach all interested areas by moving the external magnets and pushing/stretching the insufflated abdomen as needed, without any problems for the patient. A possible application of the retraction robot is highlighted by the red circle in Figure 3, where the retraction robot is used for exposing the target point in the gallbladder (black circle), lifting part of the liver and thus allowing operation of the other modules.

Figure 3. Possible working scenario of the retraction robots for exposing the gallbladder during Natural Orifice Transluminal Endoscopic Surgery (NOTES) cholecystectomy.

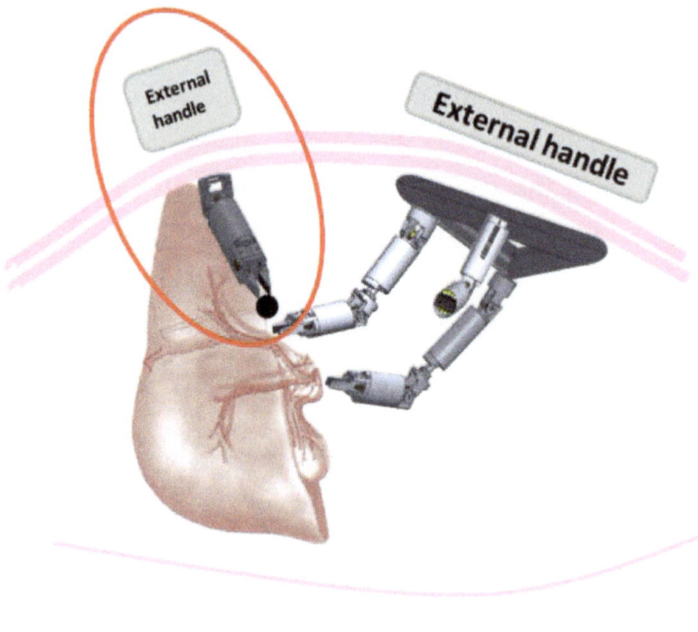

2.4.2. Anchoring and Positioning of the Modules

In order to enable the anchoring and positioning of the retraction robot in the working scenario, a magnetic system was used to attach the robot to the phantom abdominal wall. As the thickness increases, it is sufficient to increase the dimension of the external magnets to have a proper anchoring force. Neodymium-Ferrum-Boron (NdFeB) N52 magnets with diametric magnetization were used as internal and external magnets. The retraction robot was equipped with a cylindrical magnet placed on its base, having dimensions of 9.5 mm in diameter and 6 mm height. This module is anchored to the abdominal wall through a coupling with a larger cylindrical external magnet (51 mm in diameter and 25 mm height), which can be easily handled by the user to position and orient the retraction robot's three external DOFs. The external magnets were integrated in plastic handles in order to allow movement of the robotic modules and avoid magnetic attraction between them.

In order to assess whether the magnetic anchoring could effectively hold the retraction robot during basic retraction tasks, magnetic simulation was performed to quantify the attraction force between the internal and external magnets. The magnetic link in the current configuration was simulated with Finite Element Method (FEM) using COMSOL Multiphysics, as shown in Figure 4. In particular, a mesh consisting of 200,000 elements was considered adequate for the devised scenario.

The magnetic attraction force was calculated to support up to 160 g, considering an abdominal wall thickness of 25 mm (*i.e.*, the typical thickness considered for a normal patient). This value is compatible with basic retraction tasks, e.g., in cholecystectomy, the retractor robot could be used for exhibiting the gallbladder or the appendix by retracting part of the liver, while the removal tasks could be performed by using other miniaturized robots. Even if the total weight of the liver can be up to 1.5 kg, the retraction robot is not expected to retract the entire organ, but will be used for exposing the gallbladder.

Figure 4. Finite Element Method (FEM) simulation showing the magnetic flux density (T) for the magnetic anchoring of the retraction robot.

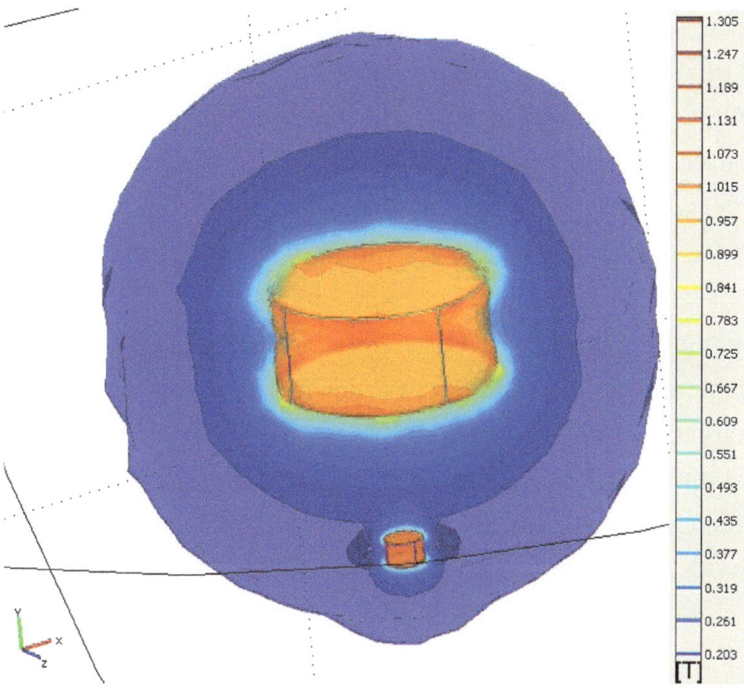

3. Experiment

The aim of this experiment was to demonstrate possible use of the robot for basic grasping and motion tasks by directly observing the real-time video streaming from the camera robot. Quantitative tests were also carried out to characterize the performance of the retraction robot.

Force experiments were performed to evaluate maximum lifting force and gripping force. A commercial six-axis load cell (Nano17, ATI, Industrial Automation, Apex, NC, USA) having a resolution of 3.18 mN was used. The maximum measured lift force was 1.53 N whereas the maximum force of the gripper was 5.3 N. Some snapshots from the experimental session are reported in Figure 5. The weight of the robot (12 g) slightly affects the maximum lift force.

Figure 5. (**Left**) Retraction robot during lifting experiments; (**Right**) Detail of the gripper.

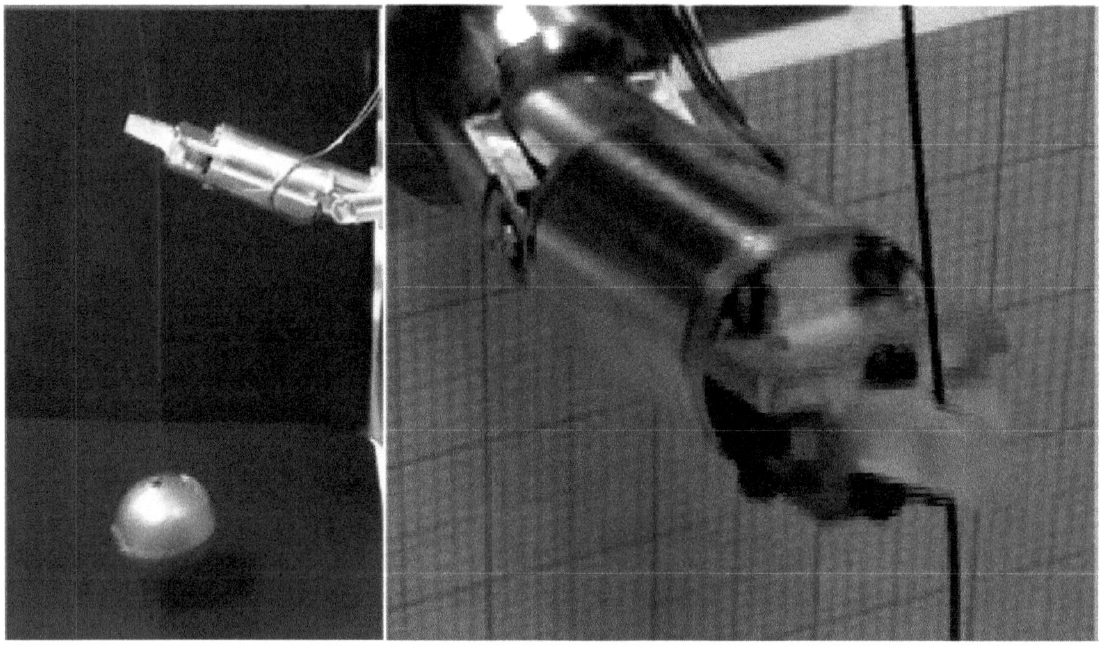

Afterwards, a simulated scenario was set to demonstrate the feasibility of a simple procedure. The procedure consisted of introducing all the robotic components for performing NOTES through an esophageal access port into the abdominal cavity, as in Figure 6. Once all the components were set-up, the retraction robot was used to grasp a tissue under vision assistance provided by the camera robot. The procedure we describe was performed by a user driving the retraction robot and an assistant user following the scene with the camera robot. The activation of both robots was performed by activating the corresponding buttons and switches, in an open loop control closed by the user looking at the real-time video streaming on a PC screen. Once all the components were placed in the abdomen in a starting position, the camera robot was moved in order to frame the retraction robot moving in the direction of a circular target, where a tissue simulator was placed. The retraction robot was then activated in order to grasp the object and release it in a second circular target placed on the other side of the abdomen. The video streaming on the screen allowed the used to steer the retraction robot toward the target in a smooth and intuitive way.

The quantitative tests and the experimental procedure on bench are shown in the companion video.

Figure 6. Simulated abdominal scenario with all the robotic units.

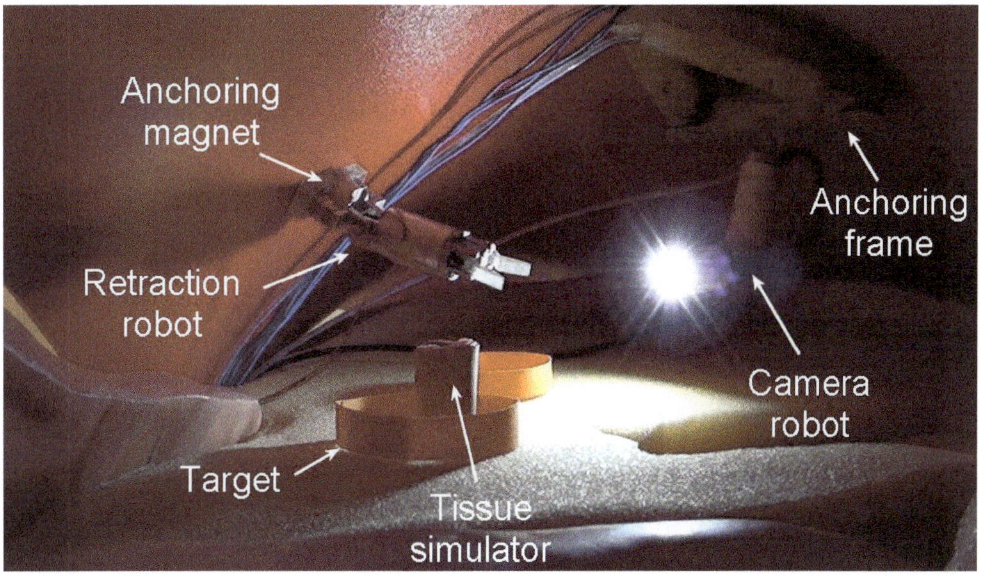

4. Discussion

The design of the retraction robot is a further step in the development of a complete intracavitary robotic platform for dedicated tasks in NOTES. This work has demonstrated the cooperation between two robotic units performing a simple retraction and grasping task.

The camera robot was docked to an anchoring frame following a well-established procedure, while the retraction robot was tested for the first time in a simulated scenario consisting of a human phantom torso. The camera robot was activated to view the entire procedure, while the user was able to move the retraction robot in correspondence of the target to grasp the tissue simulator. After driving the retraction robot toward the final target, the tissue simulator was successfully released. Despite the retraction robot has a low number of internal DOFs, additional mobility can be obtained by actively positioning the robot through the magnetic link.

The camera robot workspace was sufficient to visualize the working area and follow the entire procedure. The retraction robot's external DOFs were sufficient for positioning the robot close to the target, while the internal DOF and the gripper were able to pitch the object successfully. Force tests were performed to assess the maximum force generated by the retraction robot, which was 1.53 N. The gripping force was also evaluated, showing a maximum grip force of 5.3 N. These force values are compatible with the lifting and grasping of tissues performed to expose organs, as for example the gallbladder during cholecystectomy. The magnetic link between the robot and an external magnet was evaluated by means of FEM simulations, showing the reliability of anchoring for basic retraction tasks. Moreover, in case of practical needs (*i.e.*, for obese patients), the handle can be pushed against the insufflated abdomen so as to enhance the magnetic coupling with the retraction robot. On the other hand, in order to augment the magnetic coupling force, the dimensions of the external magnets can be increased as well. These force values are compatible with other systems commonly used in current NOTES procedures, as reported in Table 1. In addition, the retraction robot introduces additional flexibility thanks to the embedded active DOFs. Moreover, the external handle can be used without problems on the abdomen since it is embedded in a soft case, thus improving safety during the contact with the patient.

Table 1. Maximum force sustained by different retraction systems [20].

Retraction System	Main Features	Sustained Force
Single-channel gastroscope (linear)	*clinical tool*	3236 mN
Single-channel gastroscope (retroflexed)	*clinical tool*	1392 mN
Double-channel colonoscope (linear)	*clinical tool*	2373 mN
Double-channel colonoscope (retroflexed)	*clinical tool*	1245 mN
Pure magnetic system [20]	*research tool, pure magnetic*	882 mN
Retraction robot	*research tool, magnets for anchoring and embedded motors for internal DOFs*	1530 mN

It is worth noting that, whenever needed, the retraction robot, in cooperation with an array of robots, can be used as an independent robotic tool for dedicated tasks in NOTES, and is able to perform essential surgical tasks such as manipulation, vision and cutting. The main goal of retraction tasks is to retract an organ in order to expose the underlying tissues and allow another robot to perform dedicated manipulation, as shown in Figure 3. For this reason, we made the choice of integrating two motors to provide basic manipulation and smooth grasping and to provide the robot with a compact size in order to be used in a real scenario.

The workspace is one-dimensional since the only task consists of pulling the retracted tissue towards the abdominal wall in order to expose the target site. The motion range of the pitch DOF is ±180°, while the maximum pulling and grasping forces are 1.53 N and 5.3 N, respectively. The retraction robot reachable workspace can be enhanced thanks to the motion of the external magnetic handle. In fact, thanks to the magnetic link, the retraction robot relies on three additional external DOFs for the correct positioning of the robot in the abdomen. In addition, the handle can be pushed against the insufflated abdomen for the approximation of the target tissue with the robot. Retraction robot and anchoring frame external handles can be used without problems on the abdomen since they are embedded in soft cases that limit magnetic interferences. In a real scenario, both handles will supposedly be held in position by medical assistants, who will move the handles according to the surgeon's indications.

5. Conclusion

In this paper, the suitability of the retraction robot for NOTES has been successfully demonstrated. With the retraction robot, a specific task of lifting a target object in a simulated scenario, under visual assistance of a camera robot has been performed. The users were able to control the camera robot and the internal and external DOFs of the retraction robot in order to localize an object and move toward the target position. The retraction robot can be indeed considered as a further essential device within the scenario of an array of robots performing dedicated surgical tasks in NOTES. In addition, the retraction robot could be also used as is in the clinical scenario. Future evaluation of the robot *in vivo* will be performed, once the effectiveness on bench has been successfully demonstrated.

Acknowledgements

This work was supported by the ARAKNES FP7 European Project no. 224565.

Authors Contributions

Arianna Menciassi and Paolo Dario inspired the idea of array of robots for augmenting the kinematic of endocavitary surgery. Giuseppe Tortora defined the scenario including a dedicated robot for retractions tasks for minimally invasive surgery and a camera robot for visualizing images. Giuseppe Tortora and Tommaso Ranzani designed the retraction robot and performed force experiments and simulations. Iris De Falco was involved in the integration of the camera and in the experimental testing of the overall platform. Giuseppe Tortora wrote the paper.

Conflicts of Interest

The authors declare no conflict of interest.

References

1. Forgione, A. *In vivo* microrobots for natural orifice transluminal surgery. Current status and future perspectives. *Surg. Oncol.* **2009**, *18*, 121–129.
2. Alterovitz, R.; Desai, J.P. Surgical robotics. *IEEE Robot. Autom. Mag.* **2009**, *16*, 16–17.
3. Lum, M.J.H.; Friedman, D.C.W.; Rosen, J.; Sankaranarayanan, G.; King, H.; Fodero,K., II; Leuschke, R.; Sinanan, M.N.; Hannaford, B. The RAVEN: Design and validation of a telesurgery system. *Int. J. Robot. Res.* **2009**, *28*, 1183–1197.
4. Yamashita, H.; Iimura, A.; Aoki, E.; Suzuki, T.; Nakazawa, T.; Kobayashi, E.; Hashizume, M.; Sakuma, I.; Dohi, T. Development of endoscopic forceps manipulator using multi-slider linkage mechanisms. In Proceedings of the 1st Asian Symposium on Computer Aided Surgery, Ibaraki, Japan, 28 April 2005.
5. Hagn, U.; Konietschke, R.; Tobergte, A.; Nickl, M.; Jörg, S.; Kübler, B.; Passig, G.; Gröger, M.; Fröhlich, F.; Seibold, U.; *et al.* DLR MiroSurge: A versatile system for research in endoscopic telesurgery. *Int. J. Comput. Assist. Radiol. Surg.* **2010**, *5*, 183–193.
6. Scott, D.J.; Tang, S.J.; Fernandez, R.; Bergs, R.; Goova, M.T.; Zeltser, I.; Kehdy, F.J.; Cadeddu, J.A. Completely transvaginal NOTES cholecystectomy using magnetically anchored instruments. *Surg. Endosc.* **2007**, *21*, 2308–2316.
7. Yeung, B.P.M.; Gourlay, T. A technical review of flexible endoscopicmultitasking platforms. *Int. J. Surg.* **2012**, *10*, 345–354.
8. Reavis, K.M.; Melvin, W.S. Advanced endoscopic technologies. *Surg. Endosc.* **2008**, *22*, 1533–1546.
9. Loeve, A.; Breedveld, P.; Dankelman, J. Scopes too flexible...and too stiff. *IEEE Pulse* **2010**, *1*, 26–41.
10. Lehman, A.C.; Dumpert, J.; Wood, N.A.; Redden, L.; Visty, A.Q.; Farritor, S.; Varnell, B.; Oleynikov, D. Natural orifice cholecystectomy using a miniature robot. *Surg. Endosc.* **2009**, *23*, 260–266.

11. Joseph, R.A.; Salas, N.A.; Donovan, M.A.; Reardon, P.R.; Bass, B.L.; Dunkin, B.J. Single-site laparoscopic (SSL) cholecystectomy in human cadavers using a novel percutaneous instrument platform and a magnetic anchoring and guidance system (MAGS): Reestablishing the "critical view". *Surg. Endosc.* **2012**, *26*, 149–153.
12. Kobayashi, T.; Gotohda, T.; Tamakawa, K.; Ueda, H.; Kakizoe, T. Magnetic anchor for more effective endoscopic mucosal resection. *Jpn. J. Clin. Oncol.* **2004**, *24*, 118–123.
13. Di Natali, C.; Ranzani, T.; Simi, M.; Menciassi, A.; Valdastri, P. Transabdominal active magnetic linkage for robotic surgery: Concept definition and model assessment. In Proceedings of the IEEE ICRA, St. Paul, MN, USA, 14–18 May 2012.
14. Tortora, G.; Salerno, M.; Ranzani, T.; Tognarelli, S.; Dario, P.; Menciassi, A. A modular magnetic platform for natural orifice transluminal endoscopic surgery. In Proceedings of Engineering in Medicine and Biology Society conference (EMBC), Osaka, Japan, 3–7 July 2013.
15. Tortora, G.; Dario, P.; Menciassi, A. Array of robots augmenting the kinematics of endocavitary surgery. *IEEE/ASME Trans. Mechatron.* **2014**, doi:10.1109/TMECH.2013.2296531.
16. Abbott, D.J.; Becke, C.; Rothstein, R.I.; Peine, W.J. Design of an endoluminal NOTES robotic system. In Proceedings of the IROS, San Diego, CA, USA, 29 October–2 November2007.
17. Lehman, A.C.; Wood, N.A.; Farritor, S.; Goede, M.R.; Oleynikov, D. Dexterous miniature robot for advanced minimally invasive surgery. *Surg. Endosc.* **2011**, *25*, 119–123.
18. Phee, S.J.; Kencana, A.P.; Huynh, V.A.; Sun, Z.L.; Low, S.C.; Yang, K.; Lomanto, D.; Ho, K.Y. Design of a master and slave transluminal endoscopic robot for natural orifice transluminal endoscopic surgery. *J. Mech. Eng. Sci.* **2010**, *224*, 1495–1503.
19. Dominguez, G.; Durand, L.; de Rosa, J.; Danguise, E.; Arozamena, C.; Ferraina, P.A. Retraction and triangulation with neodymium magnetic forceps for single-port laparoscopic cholecystectomy. *Surg. Endosc.* **2009**, *23*, 1660–1666.
20. Ryou, M.; Thompson, C.C. Magnetic retraction in natural-orifice transluminal endoscopic surgery (NOTES): Addressing the problem of traction and countertraction. *Endoscopy* **2009**, *41*, 143–148.
21. Park, P.O.; Bergstrom, M.; Ikeda, K. Experimental studies oftransgastric gallbladder surgery cholecystectomyand cholecystogastric anastomosis (videos). *Gastrointest. Endosc.* **2005**, *61*, 601–606.
22. Kume, M.; Miyazawa, H.; Abe, F.; Iwasaki, W.; Uchinami, H.; Shibata, S.; Sato, T.; Yamamoto, Y. A newly designed magnet-retracting forceps for laparoscopic cholecystectomy in a swine model. *Minim. Invasive Ther. Allied Technol.* **2008**, *17*, 251–254.
23. Steinemann, D.C.; Limani, P.;Clavien, P.A.; Breitenstein, S. Internal retraction in single-port laparoscopic cholecystectomy: Initial experience and learning curve. *Minim. Invasive Ther. Allied Technol.* **2013**, *22*, 171–176.
24. Cho, Y.B.; Park, C.M.; Chun, H.K.; Yi, L.J.; Park, J.H.; Yun, S.H.; Kim, H.C.; Lee, W.Y. Transvaginal endoscopic cholecystectomy using a simple magnetic traction system. *Minim. Invasive Ther. Allied Technol.* **2011**, *20*, 174–178.
25. Tortora, G.; Dimitracopoulos, A.; Valdastri, P.; Menciassi, A.; Dario, P. Design of miniature modular *in vivo* robots for dedicated tasks in minimally invasive surgery. In Proceedings of the AIM, Budapest, Hungary, 3–7 July 2011.

26. Tognarelli, S.; Salerno, M.; Tortora, G.; Quaglia, C.; Dario, P.; Menciassi, A. An endoluminal robotic platform for Minimally Invasive Surgery. In Proceedings of the IEEE RAS/EMBS BioRob Conference, Rome, Italy, 24–27 June 2012.

Image-Based Navigation for the *SnowEater* Robot Using a Low-Resolution USB Camera

Ernesto Rivas [1,*], Koutarou Komagome [1], Kazuhisa Mitobe [1] and Genci Capi [2]

[1] Department of Mechanical Systems Engineering, Yamagata University, Yamagata Prefecture, Yonezawa 992-8510, Jo-nan 4-3-16, Japan; E-Mails: tat71119@st.yamagata-u.ac.jp (K.K.); mitobe@yz.yamagata-u.ac.jp (K.M.)

[2] Graduate School of Science and Engineering, University of Toyama, Toyama 930-8555, Gofuku 3190, Japan; E-Mail: capi@eng.u-toyama.ac.jp

* Author to whom correspondence should be addressed; E-Mail: tnn21238@st.yamagata-u.ac.jp

Academic Editor: Huosheng Hu

Abstract: This paper reports on a navigation method for the snow-removal robot called *SnowEater*. The robot is designed to work autonomously within small areas (around 30 m^2 or less) following line segment paths. The line segment paths are laid out so as much snow as possible can be cleared from an area. Navigation is accomplished by using an onboard low-resolution USB camera and a small marker located in the area to be cleared. Low-resolution cameras allow only limited localization and present significant errors. However, these errors can be overcome by using an efficient navigation algorithm to exploit the merits of these cameras. For stable robust autonomous snow removal using this limited information, the most reliable data are selected and the travel paths are controlled. The navigation paths are a set of radially arranged line segments emanating from a marker placed in the environment area to be cleared, in a place where it is not covered by snow. With this method, by using a low-resolution camera (640 × 480 pixels) and a small marker (100 × 100 mm), the robot covered the testing area following line segments. For a reference angle of 4.5° between line paths, the average results are: 4° for motion on hard floor and 4.8° for motion on compacted snow. The main contribution of this study is the design of a path-following control algorithm capable of absorbing the errors generated by a low-cost camera.

Keywords: snow-removal robot; autonomous mobile robot; field robotics; image-based feedback; visual localization

1. Introduction

Robot technology is focused on increasing the quality of life by the creation of new machines and methods. This paper reports on the development of a snow-removal robot called *SnowEater*. The concept for this robot is the creation of a safe, slow, small, light, low-powered and inexpensive autonomous snow-removal machine for home use. In essence, *SnowEater* can be compared with autonomous vacuum cleaner robots [1], but instead of working inside homes, it is designed to operate on house walkways, and around front doors or garages. A number of attempts to automate commercial snow-blower machines have being made [2,3], but such attempts have been delayed out of concerns over safety.

The basic *SnowEater* model is derived from the heavy snow-removal robot named Yukitaro [4], which was equipped with a high-resolution laser rangefinder for the navigation system. In contrast to the Yukitaro robot, which weights 400 kg, the *SnowEater* robot is planned to weigh 50 kg or less. In 2010, a snow-intake system was fitted to *SnowEater* [5] that enables snow removal using low auger and traveling speeds while following a line path. In 2012, a navigation system based on a low-cost camera was introduced [6]. The control system was designed using linear feedback to ensure paths were followed accurately. However, due to the reduced localization accuracy, the system did not prove to be reliable. Among the many challenges in the realization of autonomous snow-removal robots, is the navigation and localization issue, which currently remains unsolved.

An early work on autonomous motion on snow is presented in [7]; in which four cameras and a scan laser were used for simultaneous localization and mapping (SLAM); and long routes in polar environments were successfully traversed. The mobility of several small tracked vehicles moving in natural and deep-snow environments is discussed in [8,9]. The terramechanics theory for motion on snow presented in [10] can be applied to improve the performance of robots moving on snow.

The basic motion models presented in [11–18] can be used for motion control. In addition, Gonzalez *et al.* [19] presented a model for off-road conditions. Also, the motion of a tracked mobile robot (TMR) on snow is notably affected by slip disturbance. In order to ensure the performance of path tracking or path following control against disturbances, robust controllers [20,21] and controllers based on advanced sensors [22] exist. However, advanced feedback compensation based on precise motion modeling is not necessary for this application because strict tracking is not required in our snow-removal robot.

One of the goals of this study is to develop a simple controller without the need for precise motion modeling. Another goal is to use a low-cost vision-based navigation system that does not require a large budget or an elaborate setup in the working environment. In contrast to other existing navigation strategies that use advanced sensors, we use only one low-cost USB camera and one marker for motion controlling.

This paper presents an effective method of utilizing a simple directional controller based on a camera-marker combination. In addition, a path-following method to enhance the reliability of navigation is proposed. Although the directional controller itself does not provide asymptotic stability of the path to follow, the simplicity of the system is a significant merit.

The rest of this paper is organized as follows. The task overview and prototype robot are presented in Section 2. The control law, motion and mathematical model are shown in Section 3. The experimental results are shown in Section 4 and our conclusions are given in Section 5.

2. Task Overview and Prototype Robots

The *SnowEater* robot prototype was developed in our laboratory. The prototype is a tracked mobile robot (TMR). Its weight currently is 26 kg, and the size is 540 mm × 740 mm × 420 mm. The main body is made of aluminum and has a flat shape for stability on irregular terrain. The intake system consists of steel and aluminum conveying screws that collect and compact snow at a low rotation speed. Two 10 W DC motors are used for the tracks, and two 15 W DC motors are used for the screws.

The front screw collects snow when the robot is moving with slow linear motion. The two internal screws compact the snow into blocks for easy transportation. Once the snow is compacted, the blocks are dropped out from the rear of the robot. In addition, the intake system compacts the snow in front of the robot reducing the influence of the track sinkage. Since the robot requires a linear motion to collect snow [5], the path to follow consists of line segments that cover the working area. In our plan, another robot carries the snow blocks to a storage location. Figure 1 shows the prototype *SnowEater* robot, and Figure 2 shows the line arrangement of the path-following method.

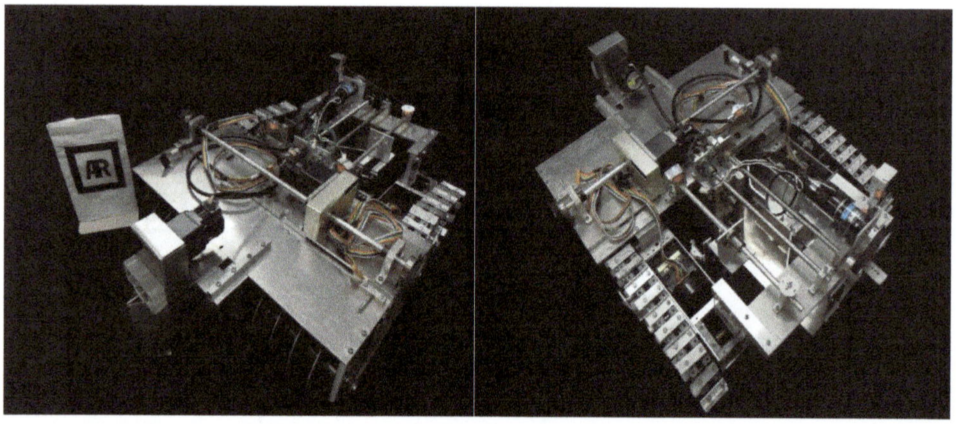

Figure 1. *SnowEater* robot prototype.

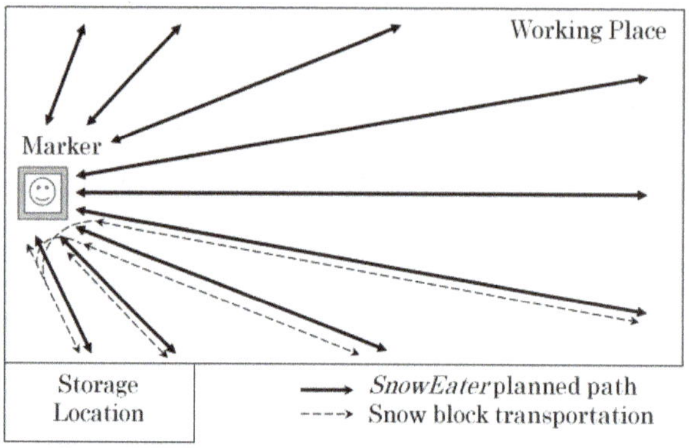

Figure 2. Line paths that cover the snow-removal area.

3. Image-Based Navigation System

One of the objectives of the *SnowEater* project is to use a low-resolution camera as the sole navigation sensor. In our strategy, a square marker is placed in the center of the working area, so as to be visible from any direction, and a camera is mounted on the robot. Robot camera-marker position/orientation is obtained by using the ARToolKit library [23]. This library uses the marker four corners position in the camera image and the information given *a priori* (camera calibration file and marker size) to calculate the transformation matrix between the camera and marker coordinate systems. Later the transformation matrix is used to estimate the translation components [24]. However, with low-resolution cameras, the accuracy and reliability of the measurement vary significantly, depending on the camera-marker distance. This is one of the problems associated with localization when using only vision. Gonzales *et al.* [25] discusses this issue and provides an innovative solution using two cameras. A solution for the localization problem using stereoscopic vision is given in [26,27], while [28] presents a solution using rotational stereo cameras. In [29], a trajectory-tracking method using an inexpensive camera without direct position measurement is presented.

Some studies have utilized different sensors such as a laser scanner to map outdoor environments [30]. Our study places a high priority on keeping the system as a whole inexpensive and simple; hence the challenge is to exploit robot performance using a low-resolution camera.

In this section, the term "recognized marker position" refers to the marker blob in the captured image, and the term "localization" means identification of the robot location with respect to the marker coordinate system.

3.1. Localization Performance Evaluation

To evaluate localization performance, outdoor and indoor experiments were carried out using a low-cost USB camera (Sony, PlayStation Eye 640 × 480 pixels) mounted on a small version of the *SnowEater* robot. Results with different cameras can be seen in Appendix A. The dimensions and weight of the small version are 420 mm × 310 mm × 190 mm and 2.28 kg, respectively. Because only a path-following strategy with visual feedback is considered, both robots (*SnowEater* and the small version) behave similarly. Figure 3 shows the small version of the *SnowEater* robot.

Figure 3. Monochromatic square marker and small version of the *SnowEater* robot.

The ARToolKit library uses monochromatic square markers to calculate the camera-marker distance and orientation, and the system can be used under poor lighting conditions [31]. Our objective is to exploit the library localization merits and performance with low-resolution cameras. Localization control calculations and track commands are completed every 300 ms. Following the results presented in [6] and Appendix A, a monochromatic square marker measuring 100 × 100 mm provides enough accuracy to be used in our application; hence the following experiments are done with this marker.

The camera is mounted on the robot facing toward the marker. The robot was placed above the X-axis of the marker-based coordinate system at different x_n positions for 1 min. The outdoor experiment was carried out in snow environments; with air and snow temperatures of −5 °C and −1 °C, respectively. Figure 4 shows the experimental setup.

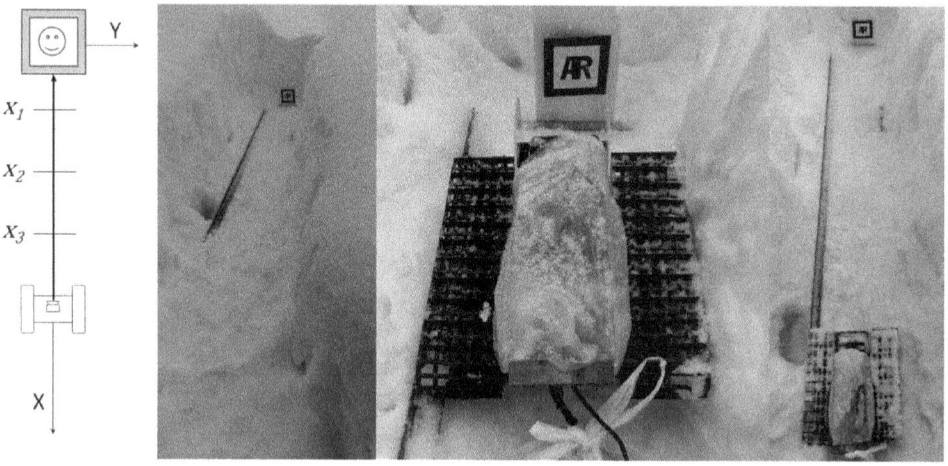

Figure 4. Experimental setup and different camera-marker position during the outdoor experiments.

Figure 5 shows the results for the robot position.

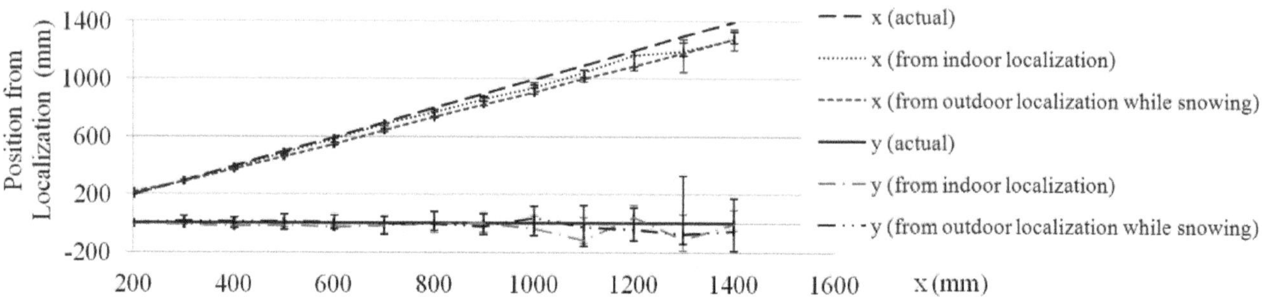

Figure 5. Localization position *vs.* actual robot-marker distance.

The results show that the robot x-coordinate is more reliable compared to the y-coordinate. The average values are stable within 800 mm. However, to use the average value the robot must remain in a static state. For motion-state applications, the x-value (camera-marker distance) is more reliable because its variance is smaller than the y-value.

Figure 6 shows the orientation angle experimental results. Since the robot was oriented toward the marker, the angle is 0°.

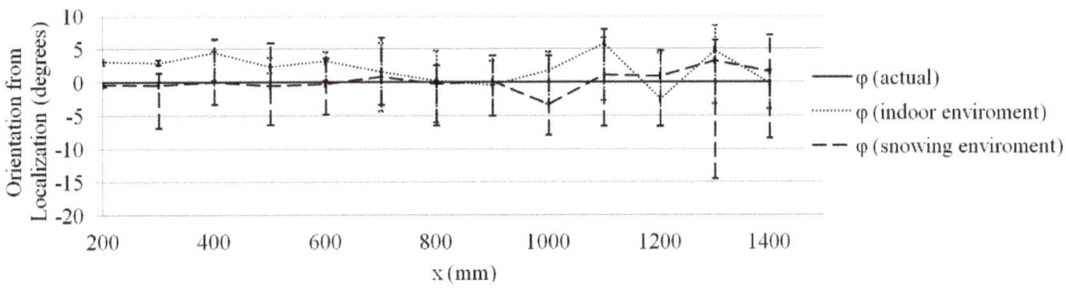

Figure 6. Orientation angle *vs.* robot-marker distance.

The orientation angle variance increases with the camera-marker distance, and its application is limited to the marker proximity (x < 300 mm) when using a 100 × 100 mm marker.

A simple way to obtain the camera-marker direction is to count the number of pixels between the center of the marker blob and the image center in the camera image. These pixel counts are converted into degrees by a simple pixel-degrees relation. This relation is found by finding the number of pixel counts related to a fixed angle. Figures 7 and 8 show the results.

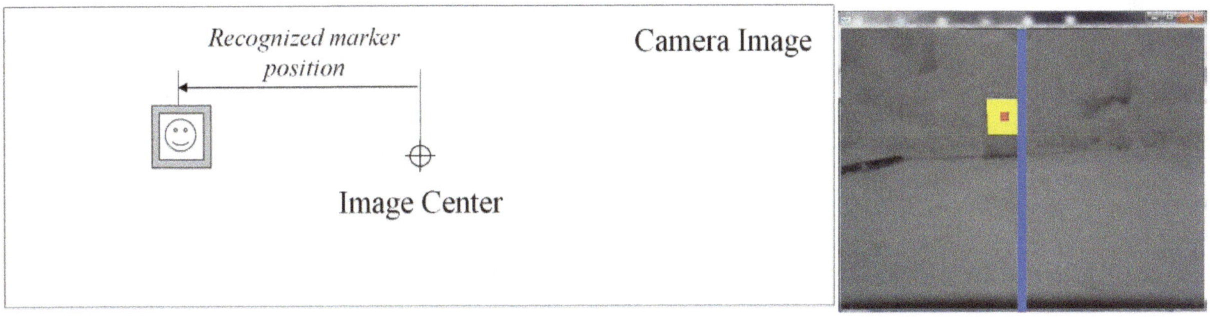

Figure 7. Recognized marker position in the camera image.

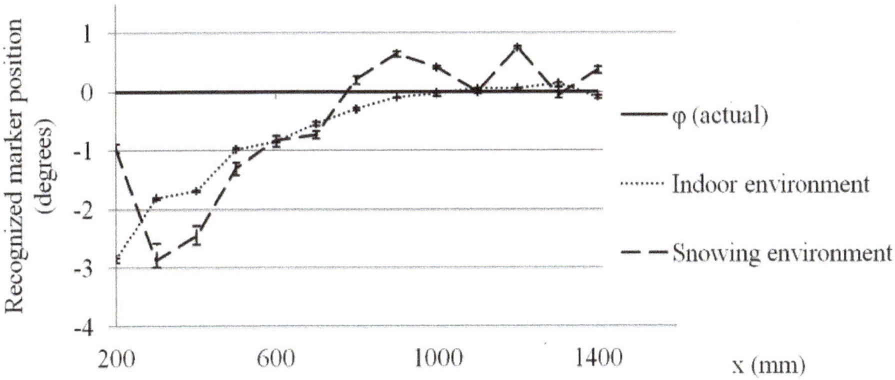

Figure 8. Recognized marker position *vs.* robot-marker distance.

Figure 8 shows the results for the recognized marker position and, as can be seen, there is not a large variance, even when x = 1400 mm. Therefore, the data are reliable and can be used during motion.

In contrast to the orientation results shown in Figure 6, as the camera-marker distance becomes larger, the recognized marker position error becomes smaller. This characteristic is very useful for our research because the relative angle between the marker direction and robot forward direction can be obtained directly from the image.

In summary, from the controlling perspective, the recognized marker position is reliable when the camera-marker distance is large. When the camera-marker distance is small, localization data can be used. The x-coordinate of the camera-marker distance can be obtained more accurate than the y-coordinate. Based on these properties, a navigation method for distant and vicinity regions can be created. Section 3.2 describes a novel navigation algorithm based on these properties.

3.2. Motion Model

Although the *SnowEater* is a TMR, a differential-drive robot model [11–18] is used in the motion algorithm.

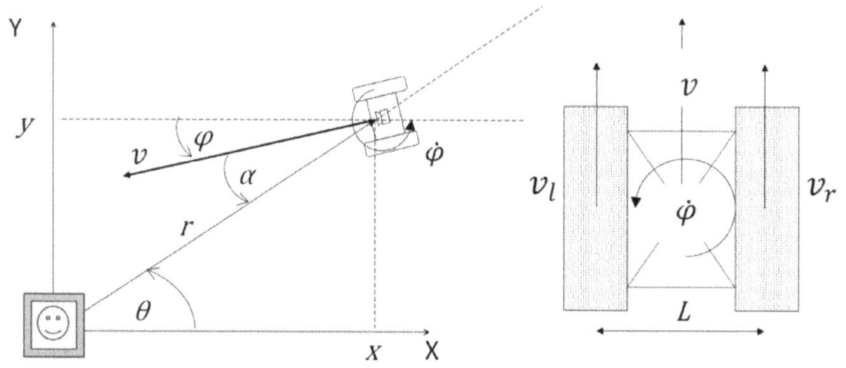

Figure 9. Motion model of the tracked robot.

Using the marker coordinate system, the robot coordinates are defined as shown in Figure 9. The longitudinal and angular velocities of the robot, v and $\dot{\varphi}$, are related to the right and left track velocities v_r and v_l, respectively, by:

$$v = \frac{v_r + v_l}{2} \tag{1}$$

$$\dot{\varphi} = \frac{v_r - v_l}{L} \tag{2}$$

where L represents the distance between the left and right tracks. Among these values, $\dot{\varphi}$ is used as the control input signal for path following. Velocity v is selected in relation to the snow-processing mechanism of the *SnowEater* robot [5].

Using the robot orientation angle φ, the robot velocity in the Cartesian coordinates is expressed as

$$\begin{aligned} \dot{y} &= -v \sin\varphi \\ \dot{x} &= -v \cos\varphi \end{aligned} \tag{3}$$

By differentiating $y = r\sin\theta$, $x = r\cos\theta$ and using Equation (3), the velocity in polar coordinates is expressed as:

$$\begin{aligned} \dot{r} &= -v\cos(\theta - \varphi) \\ \dot{\theta} &= \frac{v}{r}\sin(\theta - \varphi) \end{aligned} \tag{4}$$

where r is the distance from the marker. The relative angle $\theta - \varphi$ $(= \alpha)$ corresponds to the recognized marker position, and it is used in the motion controller described below.

3.3. Recursive Back-Forward Motion

A recursive back-forward motion was created to cover the working area. Figure 10 shows the travel route of the robot. The robot trajectory consists of two motions: *Motion 1* follows a linear path, while *Motion 2* is used to connect two consecutive linear paths through a curved motion in the vicinity region.

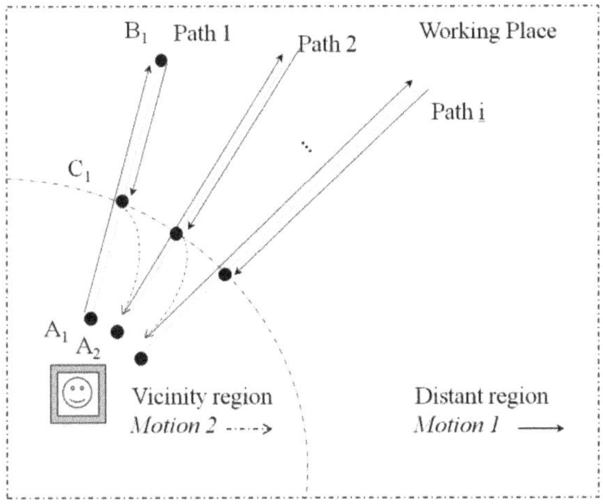

Figure 10. Recursive back-forward algorithm.

Point A_1 is the initial point of the task. The robot moves in the line segment from A_1 to B_1 using a straight motion (*Motion 1*). Point B_1 is in the limit of the working area in the distant region. After reaching B_1, the robot approaches point C_1 in the vicinity region using a straight motion (*Motion 1*). Later, the robot travels to point A_2 using a curve motion (*Motion 2*). Then, the robot embarks upon a new linear path.

Motion 1 control is executed using only the recognized marker position, while *Motion 2* uses the distance from the marker in addition to the recognized marker position. With this combined motion, the low-resolution limitation is overcome.

The camera-marker direction is obtained from the information in the image using a geometric relation. Figure 11 show this relation, where α_{image}, θ_{image}, φ_{image} are the distance in the camera image corresponding to α, θ and φ.

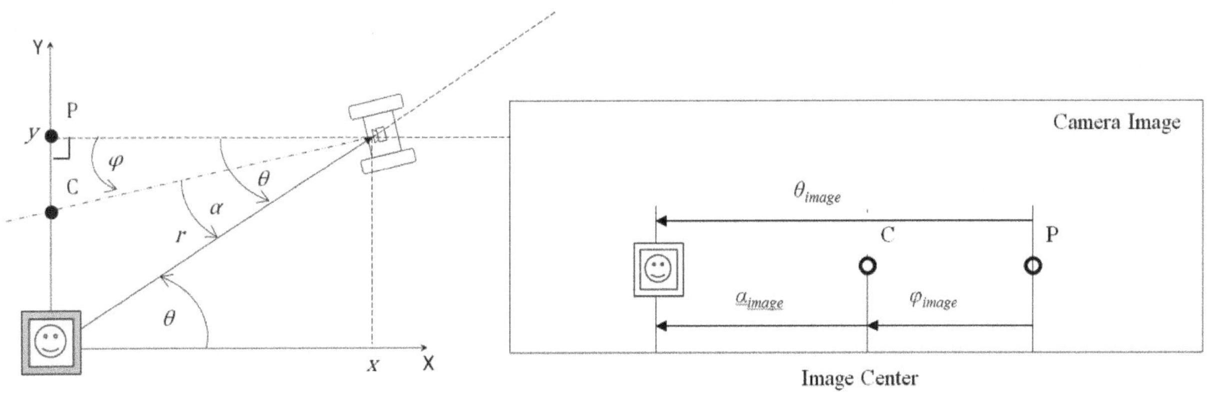

Figure 11. Orientation angle of the robot in the marker coordinate system and in the camera image.

Point C is the image center, which is on a straight heading in front of the robot. Since point P (the projective point of the robot onto the Y axis) is not recognized in the image, the orientation angle φ is not detected directly in the camera image. Hence the angles θ and φ are unknown, even in terms of pixels (θ_{image} or φ_{image}) .

However, the relative angle α between the marker location and robot direction is available in terms of pixels ($\theta_{image} - \varphi_{image} = \alpha_{image}$). In the following discussion, α_{image} and α are not distinguished.

3.4. Control in Distant Region (Motion 1)

Assuming the track speeds v_r and v_l can be manipulated, the feedback controller is expressed using the angular velocity ($\dot{\varphi}$) as the input signal. Throughout the travel, accurate localization can be expected at points A_1, C_1, A_2, C_2,…, because of the camera-marker distance.

Motion 1 navigation is executed by a simple direction controller using only the relative angle α in the feedback law.

$$\dot{\varphi} := K\alpha = K(\theta - \varphi) \tag{5}$$

where K is positive feedback gain. A linear path can be executed when the marker is set to the image center ($\alpha = 0$). Note this control does not provide asymptotic stability of the target path itself, because the tracking error from the target path cannot be included in control Equation (5) without accurate localization. The path-following error depends on the initial robot position. For this reason, the initial positioning is executed in the vicinity region. *Motion 2* control is described later.

Considering the snow-removal task, this method is a reasonable solution because strict tracking is not required. Moreover, the certainty of returning to the marker position is a good advantage.

The basic property of the direction control used in *Motion 1* is described below. The control purpose is to converge the relative angle ($\theta - \varphi$) to zero. This convergence can be checked by a Lyapunov function [32].

$$V_1 = \frac{1}{2}(\theta - \varphi)^2 \tag{6}$$

By differentiating Equation (6) along the solution of Equations (4) and (5),

$$\begin{aligned} \dot{V}_1 &= (\theta - \varphi)(\dot{\theta} - \dot{\varphi}) \\ &= \frac{v}{r}(\theta - \varphi)\, sin(\theta - \varphi) - K(\theta - \varphi)^2 \\ &< \left(\frac{v}{r} - K\right)(\theta - \varphi)^2 \\ &= 2\left(\frac{v}{r} - K\right) V_1 \end{aligned} \tag{7}$$

A sufficient condition for the negative definiteness of $\dot{V}_1$ is:

$$\frac{v}{r} < K \tag{8}$$

When the distance r becomes small, the condition is not satisfied. One solution for this issue is to change the velocity as the distance gets smaller to $v = r$. Under this condition, Equation (7) becomes:

$$\dot{V}_1 \leq (1 - K)(\theta - \varphi)^2 \tag{9}$$

$$\leq 2(1-K)V_1$$

If there exist $\varepsilon > 0$ such that $2(1-K) \leq -\varepsilon$, then:

$$V_1 < V_1(0)e^{-\varepsilon t} \tag{10}$$

Therefore, for any $\delta > 0$, there exists a time T such that for all $t > T$, $\theta - \varphi < \delta$. The reachability of the vehicle to the marker can be understood directly from Equation (4) because the robot is oriented to the marker. The distance r decreases when $|\theta - \varphi| \leq \frac{\pi}{2}$. In appendix C, the behavior of this controller and the conventional PI controller using the available signals is shown.

3.5. *Control for Switching the Target Line (Motion 2)*

In *Motion 2*, the robot changes the target path from the current target line (path i) to the next line (path i + 1). During this phase, the robot trajectory is a curve convergent to the next straight target line (path i + 1). To achieve this, a modification is made to feedback Equation (5).

$$\dot{\varphi} := K(\alpha + \alpha_{ref}) = K(\theta - \varphi + \alpha_{ref}) \tag{11}$$

where α_{ref} is added to change the robot direction. The value of α_{ref} must be designed to move the robot onto the next target line. We define the value of α_{ref} as a function of the camera-marker distance ($\alpha_{ref} = f(r)$) as described below.

Considering a polar coordinate system oriented to the next target line (path i + 1), the curved path is defined to satisfy the relation $r = A\theta$, where A is a constant. This path smoothly converges to the origin of the O-YX coordinate system. Relying on the good accuracy of the camera localization within the vicinity region, the initial position is assumed to be known. Figure 12 shows this path.

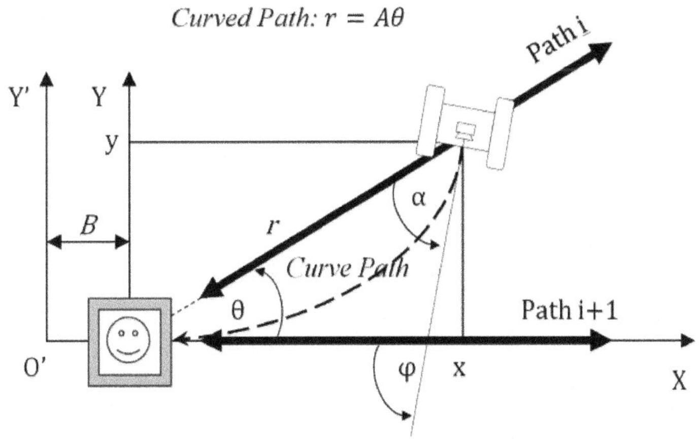

Figure 12. Curved path to reach the marker.

The A value is selected to create the curved path. Because the initial position is known, A can be determined. Next, α_{ref} is related to θ by:

$$\alpha_{ref} + \theta - \varphi = 0 \tag{12}$$

Since the robot motion follows Equation (3), the angle φ can be expressed as:

$$\tan\varphi = \frac{\dot{y}}{\dot{x}} \tag{13}$$

if the motion satisfies the relation $r = A\theta$

$$\dot{y} = \dot{r}\, sin\,\theta + r\, cos\,\theta\,\dot{\theta} = A\dot{\theta}\frac{y}{r} + r\frac{x}{r}\dot{\theta} \tag{14}$$

$$\dot{x} = \dot{r}\cos\theta - r\sin\theta\,\dot{\theta} = A\dot{\theta}\frac{x}{r} - r\frac{y}{r}\dot{\theta} \tag{15}$$

Then, α_{ref} is defined by substituting $\varphi = \tan^{-1}\frac{\dot{y}}{\dot{x}}$ and $\theta = \frac{r}{A}$ into Equation (12)

$$\alpha_{ref} = -\frac{r}{A} + \tan^{-1}\frac{Ay + rx}{Ax - ry} \tag{16}$$

The parameter A has to be selected to satisfy $|\alpha_{ref}| < \frac{vision\ range}{2}$, because the marker must remain within the camera vision range. In this particular case the vision range is limited and $r \cong x$ (because the y value is small compared to x). This approximation makes the implementation much easier. Then Equation (16) can be approximated as:

$$\alpha_{ref} = -\frac{x}{A} + \tan^{-1}\frac{x}{A} \tag{17}$$

Additionally, the minimum marker recognition distance (B) must be considered. The coordinate system is shifted from O-YX to O'-Y'X'; hence Equation (17) becomes:

$$\alpha_{ref} = -\frac{x-B}{A} + \tan^{-1}\frac{x-B}{A} \tag{18}$$

In this method, the asymptotic stability of the path itself is not provided. The stability of the direction control used in *Motion 2* is checked by using a Lyapunov-like function:

$$V_2 = \frac{1}{2}e^2,\ \mathrm{e} = (\theta - \varphi + \alpha_{ref}) \tag{19}$$

By differentiating along the solution of Equation (4) and the feedback equation:

$$\begin{aligned} \dot{V}_2 &= e(\dot{\theta} - \dot{\varphi} + \dot{\alpha}_{ref}) \\ &= e(\dot{\theta} + \dot{\alpha}_{ref} - Ke) \\ &= -Ke^2 + e\xi \\ \xi = \dot{\theta} + \dot{\alpha}_{ref} &= \frac{v}{r}\sin(\theta - \varphi) + \dot{\alpha}_{ref} \end{aligned} \tag{20}$$

Since $\dot{V}_2 < 0$ for $|e| > \frac{sup|\xi|}{K}$, the error e will settle in the region $\left\{ e \,\middle|\, \frac{sup|\xi|}{K} < |e| \right\}$. Hence the expectable accuracy is less than $\frac{sup|\xi|}{K}$. Note that the error can be suppressed by selecting the correct parameter K. The velocity v must be small when the robot is closer to the marker to avoid the growth of ξ.

4. Experimental Results

Experimental verification was carried out using the small version of the *SnowEater* robot. The track motors of the small version are Maxon RE25 20 W motors. Each track speed is PI feedback controlled by using 500 ppr rotary encoders (Copal, RE30E) with 10 ms of sampling period. The control signals are sent through a USB-serial connection between an external PC and the robot microcontrollers (dsPIC30F4012). The PC has an Intel Celeron CPU B830 (1.8 GHz) processor with 4 GB of RAM. The interface is made via Microsoft Visual Studio 2010. Figure 13 shows the system diagram. The track response to different robot angular velocity commands ($\dot{\varphi}$) can be seen in Appendix B.

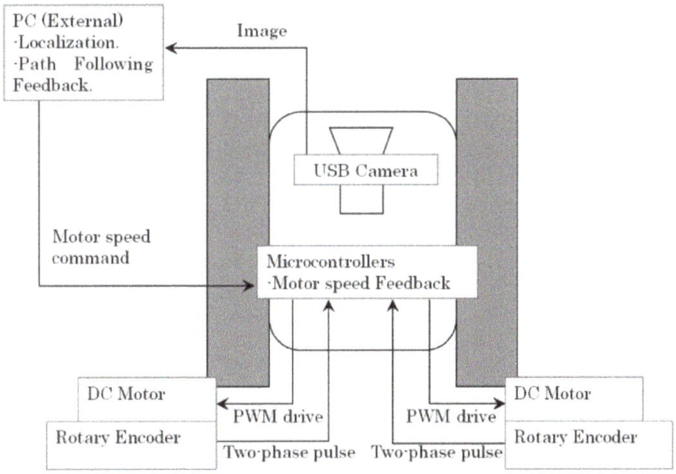

Figure 13. Control system diagram for the *SnowEater* robot.

4.1. Motion 2

In this experiment, *Motion 2* navigation is tested. The control objective is to move the robot to the target line. The experiment is carried out using two different conditions: (1) a hard floor, and (2) a slippery floor of 6 mm polystyrene beads. Figure 14 shows the experimental setups.

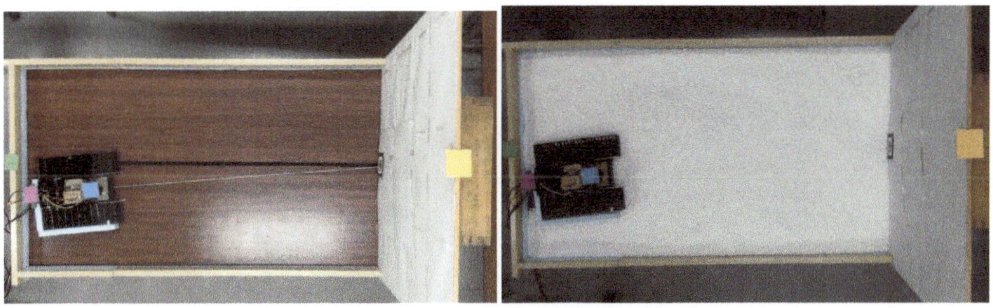

Figure 14. *Motion 2* experimental setup on a hard floor, and on a slippery floor of polystyrene beads.

The robot position results are obtained using a high-resolution camera mounted on the top of the robot. The robot and marker locations are generated by processing the captured image. The initial position was $x_o = 1200$ mm, $y_o = 100$ mm. Figures 15 and 16 show the experimental results.

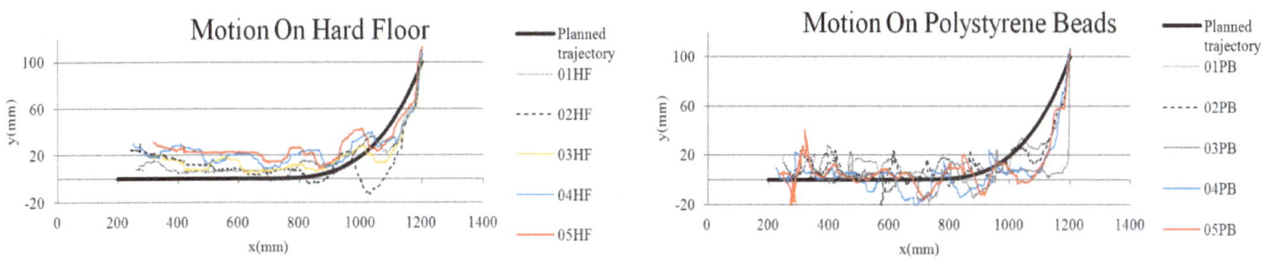

Figure 15. Robot position throughout the experiment.

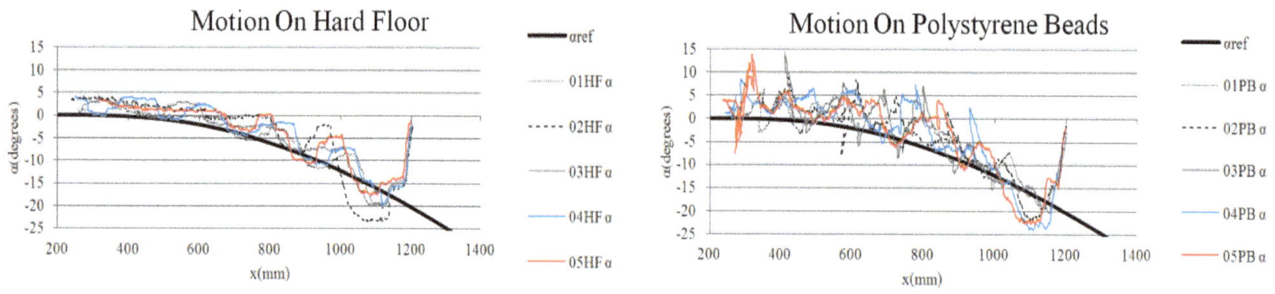

Figure 16. α_{ref} experimental results when *Motion 2* is executed.

Figure 15 shows a comparison of the *Motion 2* experiment on a hard floor and on polystyrene beads repeated five times.

As Figure 16 shows, direction control for the angle α is accomplished.

Because the feedback does not provide asymptotic stability of the path to follow, deviations occur in each experiment. If the final position is not accurate, tracking of the path generated by *Motion 1* cannot be achieved. If the final error is outside the allowable range, a new curve using the *Motion 2* control needs to be generated again. Due to the marker proximity, tracking of the new path is more accurate.

4.2. Motion 1 in Outdoor Conditions

To confirm the applicability in snow environments of this path-following strategy, outdoor experiments were carried out. These experiments were conducted at different times of the day under different lighting conditions. Figure 17 shows the setup of the test area.

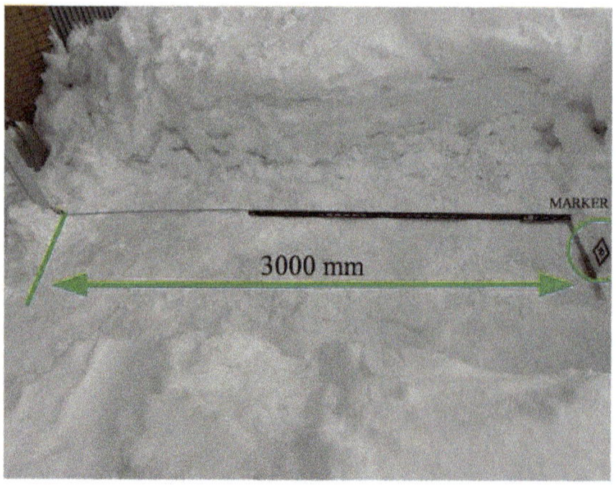

Figure 17. Test area setup.

The test area measured 3000 mm × 1500 mm. The snow on the ground was lightly compacted, the terrain was irregular, and conditions were slippery. For the first experiment, the snow temperature was −1.0 °C and the air temperature is −2.4 °C with good lighting conditions. Figure 18 shows the experimental results for *Motion 1* in outdoor conditions. In both experiments, the robot was oriented toward the marker and the path to follow was 3000 mm long. The color lines in the figures highlight the path followed.

Figure 18. *Motion 1* experimental results on snow in outdoor conditions.

As can be seen in Figure 18, the robot follows a straight-line path while using *Motion 1*.

4.3. Motion 2 in Outdoor Conditions

Figure 19 shows the results for the *Motion 2* experiment conducted under the same experimental conditions as for the *Motion 1* experiment. The color lines highlight the previous path (brown), the path followed (red), and the next path (white).

In the left-hand photograph in Figure 19, the next path is in front of the marker. In the right-hand photograph, the next path has an angle with respect to the marker.

Figure 19. Motion 2 on snow.

4.4. Recursive Back-Forward Motion

Figure 20 shows the results for the recursive back-forward motion in indoor conditions on a hard floor and on snow. The *Motion 2* region is 1200 mm. The test travelled distance is 2000 mm.

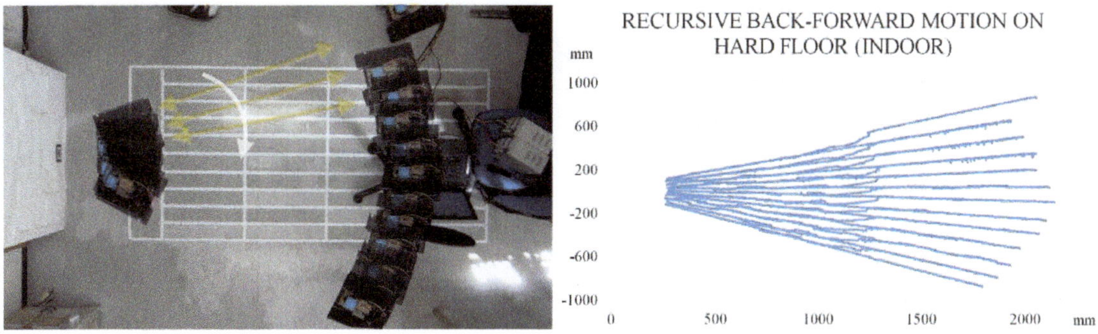

Figure 20. *Cont.*

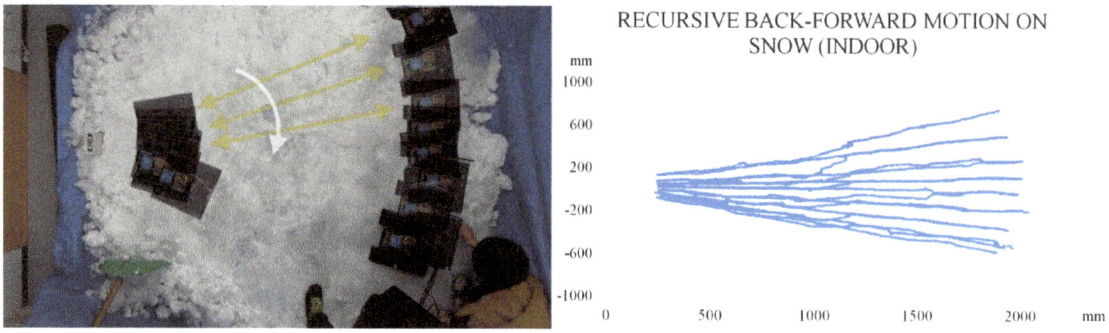

Figure 20. Recursive back-forward motion experimental results in indoor conditions.

Table 1. Recursive back-forward motion experimental results in indoor conditions.

	On Hard Floor	On Snow
Reference angle between paths	4.5°	4.5°
Mean angle between paths	4.0°	4.8°
Max. angle between paths	4.6 °	6.9 °
Min. angle between paths	3 °	1.6 °
Angle between paths deviation	0.49°	1.54°

The robot covered the testing area using the recursive back-forward motion. For a reference angle of 4.5° between line paths, the average results are: 4° for motion on hard floor and 4.8° for motion on compacted snow. The number of paths in the experiment on snow is different because the test area on snow is smaller than the test area on hard floor. Although the motion performance on snow is reduced (compared to the motion on a hard floor), the robot returned to the marker and covered the area. Table 1 shows the results of Figure 20.

In snow removal and cleaning applications, full area coverage rather than strict path-following is required. Therefore, this method can be applied for such tasks.

Figure 21 shows the recursive back-forward motion experiment on snow in outdoor conditions. The snow temperature was −1.0 °C and the air temperature was −1.3 °C.

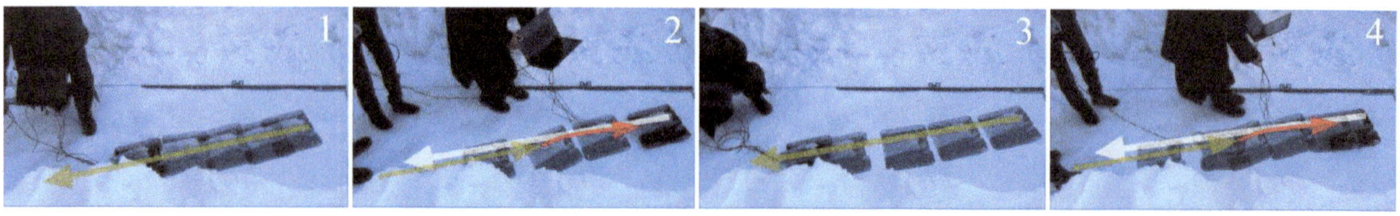

Figure 21. Recursive back-forward motion experimental results in outdoor conditions.

Figure 21 results confirm the applicability of the method under outdoor snow conditions. Due to poor lighting, the longest distance traveled was 1500 mm. For longer routes, a bigger marker is required and to improve the library performance in poor lighting conditions, the luminous markers shown in [31] can be used.

5. Conclusions

In this paper, we presented a new approach for a snow-removal robot utilizing a path-following strategy based on a low-cost camera. Using only a simple direction controller, an area with radially arranged line segments was swept. The advantages of the proposed controller are its simplicity and reliability.

The required localization values, as measured by the camera, are the position and orientation of the robot. These values are measured in the marker vicinity in a stationary condition. During motion or in the marker distant region, only the position of the marker blob in the captured image is used.

With 100 mm × 100 mm square monochromatic marker, a low-cost USB camera (640 × 480 pixels), and the ARToolKit library, the robot followed the radially arranged line paths. For a reference angle of 4.5° between line paths, the average results are: 4° for motion on hard floor and 4.8° for motion on compacted snow. With good lighting conditions, 3000 mm long paths were traveled. We believe with this method our intended goal area can be covered.

Although the asymptotic stability of the path is not provided, our method presents a simple and convenient solution for *SnowEater* motion in small areas. The results showed that the robot can cover all the area using just one landmark. Finally, because the algorithm grants area coverage, it can be applied not only to snow-removal but also to other tasks such as cleaning.

In the future, the method will be evaluated by using the *SnowEater* prototype in outdoor snow environments. Also, the use of natural passive markers (e.g., houses and trees) will be considered.

Appendix A

The camera, marker size and the pattern (inside the marker) selection was done after the following indoor experiments. In these experiments, three different parameters were considered: the camera, the marker size and the pattern inside the marker.

In the first experiment, three different cameras (Buffalo USB camera, SONY PlayStation Eye and Logicool HD Pro webcam) were tested. The cameras were set to 30 fps and 640 × 480 pixels. The marker size was 100 × 100 mm and the pattern inside the marker was the same. Figure A1 shows the results.

In the second experiment, the pattern inside the marker was changed. The camera was Sony PlayStation Eye set to 30 fps and 640 × 480 pixels. The size for each marker was 100 × 100 mm. Figure A2 shows these results.

In the third experiment, the marker size was changed. The camera was Sony PlayStation Eye set to 30 fps and 640 × 480 pixels. The pattern inside each marker was the same. Figure A3 shows these results.

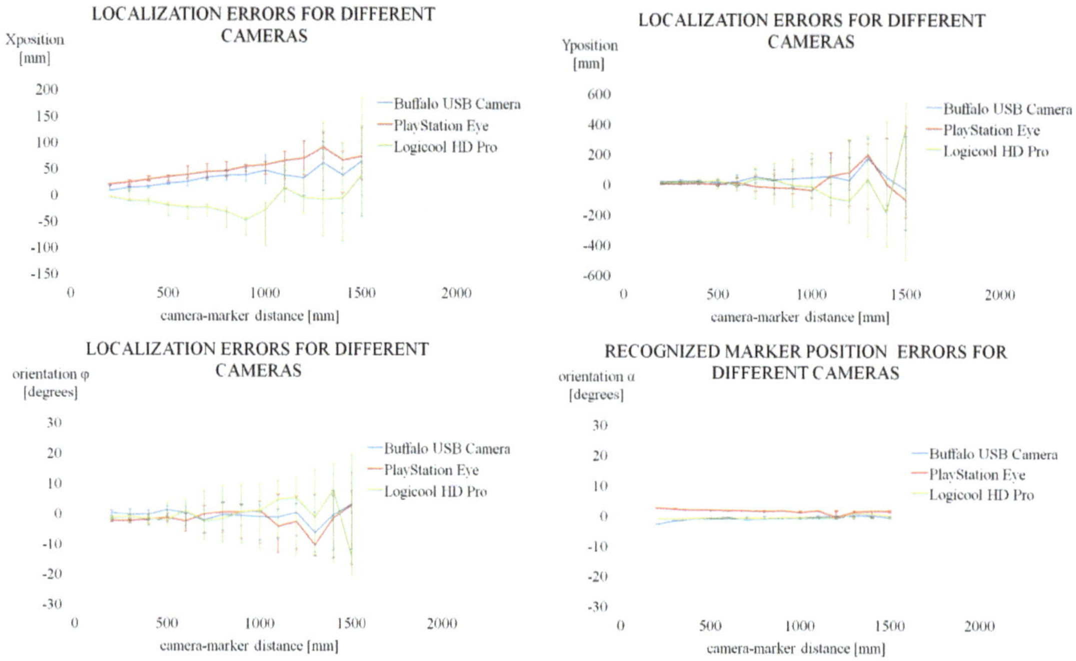

Figure A1. Localization and recognized marker position errors for different cameras.

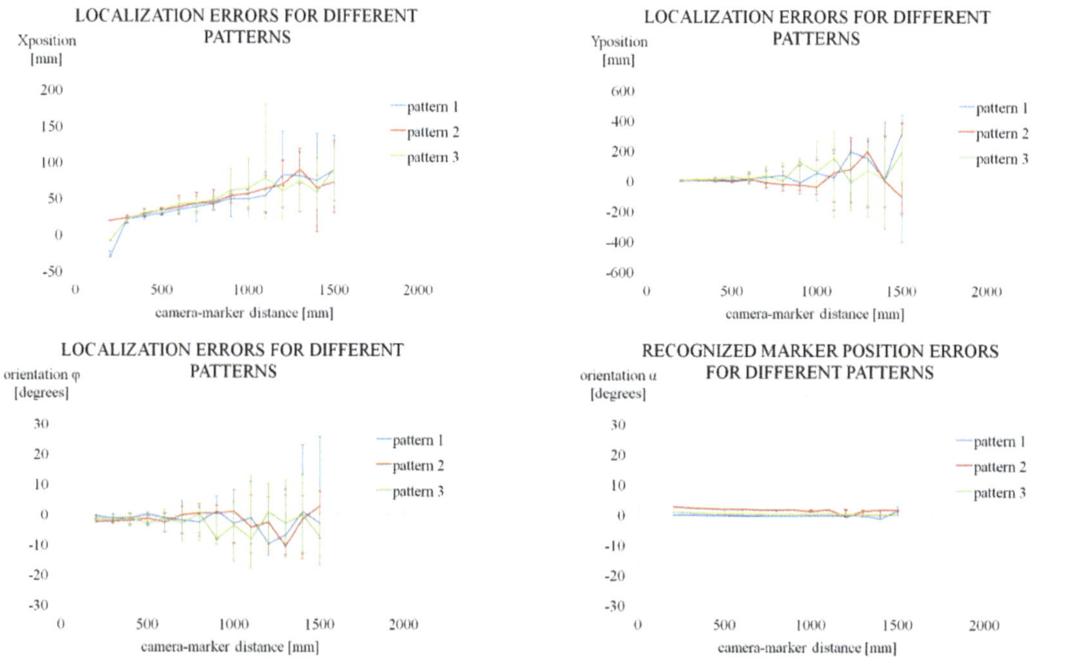

Figure A2. Localization and recognized marker position errors for different patterns.

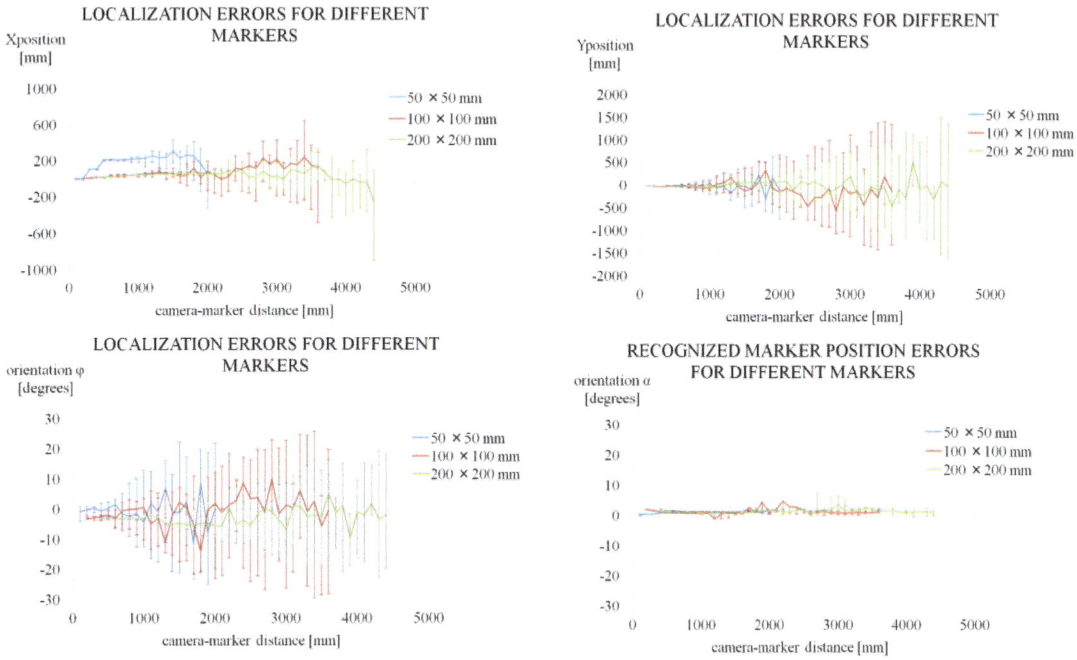

Figure A3. Localization and recognized marker position errors for different marker size.

These are the conclusions from these experiments: The camera hardware has not direct influence in the localization results when set to the same resolution and frame rate. The pattern inside the marker has not relevant influence in the localization results. The marker size is the most relevant factor in the localization accuracy. A larger marker will have a large accurate region but this region is in the vicinity of the marker and proportional to the marker size.

Appendix B

An experiment to check the track response to different robot angular velocity commands ($\dot{\varphi}$) was done. The floor was hard and dry. In the test, the robot rotated 2 revolutions (2π rad) for a given angular velocity, thus the test duration was shorter for higher rotation ratio. Figure B1 shows these results.

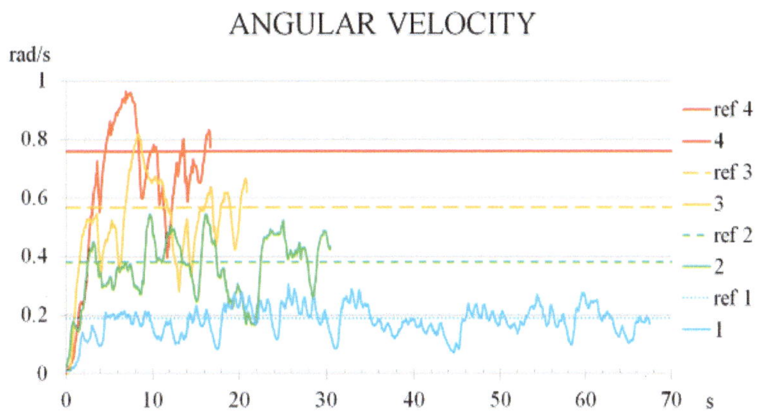

Figure B1. Track response to different angular velocity commands.

Appendix C

An experiment to compare our controller (using the recognized marker position signal) and the conventional path following linear feedback [12] with PI controller (using position/orientation signals from ARToolKit) was done. The control equation of the latter is:

$$\dot{\varphi} = u + \int u dt, u = -K_1\gamma - K_2 l \quad \text{(C1)}$$

where γ and l represents the relative angle and distance from the target trajectory, respectively, and K_1, K_2 are constant feedback gains.

The marker size was 100 × 100 mm and the camera in use is Sony PlayStation Eye. The floor was hard and dry. Figure C1 show the results.

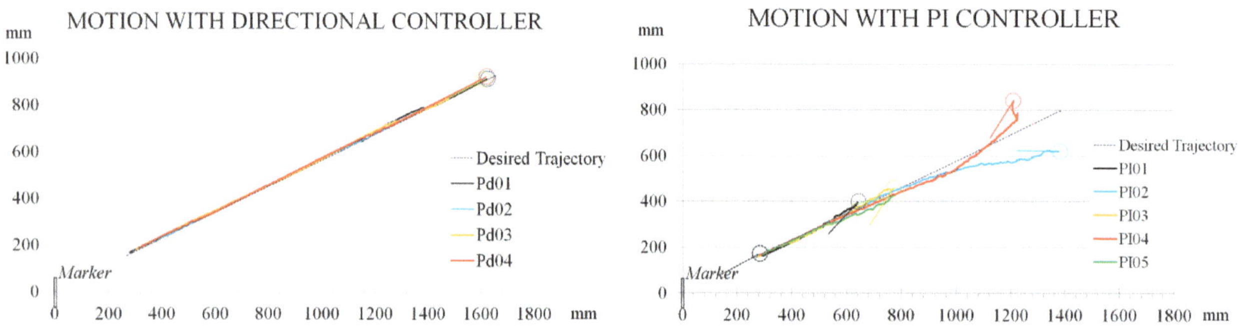

Figure C1. Direction controller (recognized marker position) *vs*. PI controller (inaccurate localization signals).

These results show the behavior of both controllers. In the PI01, PI03 and PI05 experiments the robot lost the marker from the camera vision range; hence the robot stopped.

The PI controller works only in the marker vicinity where the localization signals are accurate. Once the camera-marker distance increases the controller does not work properly due to the poor accuracy of the data. In contrast the direction controller remains stable even if the camera-marker distance increases.

Author Contributions

These authors contributed equally to this work.

Conflicts of Interest

The authors declare no conflict of interest.

References

1. Irobot home page. Available online: http://www.irobot.com/For-the-Home/Vacuum-Cleaning/Roomba (accessed on 14 November 2014).
2. Sato, T.; Makino, T.; Naruse, K.; Yokoi, H.; Kakazu, Y. Development of the Autonomous Snowplow. In Proceedings of ANNIE 2003, St. Louis, MI, USA, 2–5 November 2003; pp. 877–882.

3. Suzuki, S.; Kumagami, H.; Haniu, H.; Miyakoshi, K.; Tsunemoto, H. Experimental study on an Autonomous Snowplow with a Vision Sensor. *Bull. Univ. Electro-Commun.* **2003**, *15*, 2–30.
4. Industrial Research Institute of Niigata Prefecture, *"Yuki-Taro" the autonomous snowplow*; Report on the Prototype Robot Exhibition EXPO 2005 AICHI, Industrial Research Institute of Niigata Prefecture, Niigata, Japan, March 2006 (In Japanese).
5. Mitobe, K.; Rivas, E.; Hasegawa, K.; Kasamatsu, R.; Nakajima, S.; Ono, H. Development of the Snow Dragging Mechanism for an Autonomous Snow Eater Robot. In Proceedings of the 2010 IEEE/SICE International Symposium on System Integration, Sendai, Japan, 22–23 December 2010; pp. 73–77.
6. Rivas, E.; Mitobe, K. Development of a Navigation System for the SnowEater Robot. In Proceedings of the 2012 IEEE/SICE International Symposium on System Integration, Fukuoka, Japan, 16–18 December 2012; pp. 378–383.
7. Moorehead, S.; Simmons, R.; Apostolopoulos, D.; Whittaker, W. Autonomous Navigation Field Results of a Planetary Analog Robot in America. In Proceedings of the International Symposium on Artificial Intelligence, Robotics and Automation in Space, Noordwijk, The Netherlands, 1–3 June 1999.
8. Lever, J.; Denton, D.; Phetteplace, G.; Wood, S.; Shoop, S. Mobility of a lightweight tracked robot over deep snow. *J. Terramech.* **2006**, *43*, 527–551.
9. Lever, J.; shoop, S.; Bernhard, R. Design of lightweight robots for over-snow mobility. *J. Terramech.* **2009**, *46*, 67–74.
10. Wong, J.Y. *Theory of Ground Vehicles*, 4th ed.; John Wiley & Sons, Inc.: Hoboken, NJ, USA, 2008; Chapter 6, Steering of tracked vehicles; pp. 419–449.
11. Dudek, G.; Jenkin, M. Fundamental Problems. In *Computational Principles of Mobile Robotics*, 2nd ed.; Cambridge University Press: New York, NY, USA, 2010; pp. 19–25.
12. Dudek, G.; Jenkin, M. Mobile Robot Hardware. In *Computational Principles of Mobile Robotics*, 2nd ed.; Cambridge University Press: New York, NY, USA, 2010; pp. 36–43.
13. Dudek, G.; Jenkin, M. Visual Sensors and Algorithms. In *Computational Principles of Mobile Robotics*, 2nd ed.; Cambridge University Press: New York, NY, USA, 2010; pp. 123–149.
14. Dudek, G.; Jenkin, M. Pose Maintenance and Localization. In *Computational Principles of Mobile Robotics*, 2nd ed.; Cambridge University Press: New York, NY, USA, 2010; pp. 240–243.
15. Canudas, C.; Siciliano, B.; Bastin, G. Modelling and structural properties. In *Theory of Robot Control*, 1st ed.; Springer: London, 1997; p. 278.
16. Canudas, C.; Siciliano, B.; Bastin, G. Nonlinear feedback control. In *Theory of Robot Control*, 1st ed.; Springer: London, 1997; pp. 331–341.
17. Morin, P.; Samson, C. Motion control of wheeled mobile robots. In *Springer Handbook of Robotics*; Siciliano, K., Ed.; Springer-Verlag: Berlin Heidelberg, Germany, 2008; pp. 799–826.
18. Campion, G.; Chung, W. Wheeled Robots. In *Springer Handbook of Robotics*; Siciliano, K., Ed.; Springer-Verlag, Berlin Heidelberg, Germany, 2008; pp. 391–410.
19. Gonzalez, R.; Rodriguez, F.; Guzman, J. Modelling Tracked Robots in Planar Off-Road Conditions. In *Autonomous Tracked Robots in Planar Off-Road Conditions*; Springer International Publishing: Cham, Switzerland, 2014; pp. 11–33.

20. Normey-Rico, J.; Alcala, I.; Gomez-Ortega, J.; Camacho, E. Mobile robot path tracking using robust PID controller. *Control Eng. Pract.* **2001**, *9*, 1209–1214.
21. Coelho, P.; Nunes, U. Path-following Control of Mobile Robots in Presence of Uncertainties. *IEEE Trans. Robot.* **2005**, *21*, 252–261.
22. Low, C.B.; Wang, D.W. GPS-Based Path Following Control for a Car-Like Wheeled Mobile Robot With Skidding and Slipping. *IEEE Trans. Control Syst. Technol.* **2008**, *16*, 340–347.
23. ARToolKit home page. Available online: http://www.hitl.washington.edu/artoolkit/ (accessed on 14 November 2014).
24. Kato, H.; Bullinghurst, M. Marker tracking and HMD calibration for a video-based augmented reality conferencing system. In Proceedings of the 2nd IEEE and ACM International Workshop, San Francisco, CA, USA, 20–21 October1999; pp. 85–94.
25. Gonzalez, R.; Rodriguez, F.; Guzman, J.; Pradalier, C.; Siegward, S. Control of off-road mobile robots using visual odometry and slip compensation. *Adv. Robot.* **2013**, *27*, 893–906.
26. Mirkhani, M.; Forsati, R.; Shahri, A.M.; Moayedikia, A. A novel efficient algorithm for mobile robot localization. *Robot. Auton. Syst.* **2013**, *61*, 920–931.
27. Savino, S. An algorithm for robot motion detection by means of a stereoscopic vision system. *Adv. Robot.* **2013**, *27*, 981–991.
28. Yoshida, T.; Shiozawa, H.; Fukao, T.; Yokokohji, Y. Dense 3D Reconstruction Using a Rotational Stereo Camera and a Correspondence with Epipolar Transfer. *J. Robot. Soc. Jpn.* **2013**, *31*, 1019–1027 (In Japanese).
29. Wang, K.; Liu, Y.; Li, L. Visual Servoing Trajectory Tracking of Nonholonomic Mobile Robots Without Direct Position Measurement. *IEEE Trans. Robot.* **2014**, *30*, 1026–1035.
30. Shiozawa, H.; Yoshida, T.; Fukao, T.; Yokokohji, Y. 3D Reconstruction using Airbone Velodyne Laser Scanner. *J. Robot. Soc. Jpn.* **2013**, *31*, 992–1000 (In Japanese).
31. Ishida, M.; Shimonomura, K. Marker Based Camera Pose Estimation for Underwater Robots. In Proceedings of the 2012 IEEE/SICE International Symposium on System Integration, Fukuoka, Japan, 16–18 December 2012; pp. 629–634.
32. Haddad, W.M.; Chellaboina, V. Stability Theory for Nonlinear Dynamical Systems. In *Nonlinear Dynamical Systems and Control*; Princenton University Press: New Jersey, USA, 2008; pp. 135–182.

Aerodynamic Bio-Mimetics of Gliding Dragonflies for Ultra-Light Flying Robot

Akira Obata *, Shotarou Shinohara, Kyohei Akimoto, Kakeru Suzuki and Miyuki Seki

Micro Flying Robot Laboratory, Nippon Bunri University, 1727 Itigi Oita City 870-0397, Japan; E-Mails: shinohara19841026@gmail.com (S.S.); e-fd3s.type.rb@ezweb.ne.jp (K.A.); k_s_hokkaido_infinite@yahoo.jp (K.S.); seki@nbu.ac.jp (M.S.)

* Author to whom correspondence should be addressed; E-Mail: obata@nbu.ac.jp

Abstract: A detailed investigation including a low-speed flow study is presented on the development of ultra-light dragonfly mimetic flying robots with a focus on the dragonfly's remarkable gliding capability. It is revealed that the dragonfly's corrugated wing structure and cruciform configuration provide superior flying characteristics for fixed wing robots in low Reynolds number flight. It was also found that the dragonfly configuration has additional merit in its compatibility with propellers or high lift devices. This combination with such classic aero-engineering makes possible robots with broader flight envelope than conventional fixed-wing flying robots.

Keywords: dragonfly; gliding; corrugated wing; bio-mimetics; bio-mimicry; low reynolds No. flying robot

1. Introduction

The remarkable flying capabilities of the dragonfly, particularly its flapping flight, have caught the eye of many scientists, and through the great efforts of leading researchers [1,2], many pioneering inventors are now able to develop ultra-small flapping-wing flying robots. Up to date information regarding the development of flying robots is easily obtainable from the internet. Some teams have now succeeded in creating hovering or floating robots; however, perfect imitation of the flapping small dragonfly has not yet been achieved though only relatively large size flying dragonfly shaped robot was successfully developed [3], in spite of researchers efforts [4,5]. There might be difficulty in four wings motion control system to decrease their weight. On the other hand, the trial of referring to two wings flapping system of humming bird get the crown of ultra-light flapping flying robot development race recently. This is achieved by Aero-Vironment, and the name of this tiny flying robot is called Nano Hummingbird [6]. It has only 19 g of weight, which is less than a real hummingbird and shows astonishing controlled performances in flight, including hover, even in outdoor conditions.

The technical progress of flapping wing type flying robots, made by Nano Hummingbird, is so great and seems to reach to far higher stages than fixed wing type flying robots, as we are not able to get a fixed wing type flying robot which is controllable in windy condition and has a weight around 20 g.

One of the reasons why an ultra-light fixed wing flying robot, weighing less than 20 g, has not been successfully constructed is considered due to increasing viscous force effect in low Reynolds number flow [7]. Second barrier to fixed wing type robot for outdoor flight is considered to be gust influence. For small size fixed wing type aircraft, the outdoor flight condition becomes far too tough compared to normal airplanes.

The ratio of gust speed to insects' flying speed in outdoor flight is relatively high (these speeds are of comparative magnitudes) compared to that of normal airplanes, which fly much faster. It is well known that dragonflies' superior gliding ability allows them to fly over the ocean, as demonstrated by their migration from China to Japan. This suggests that we do not yet fully understand their aerodynamics as they have to glide during this time to reduce energy consumption and are continually met with aerodynamic disturbances. We can also deduce that dragonflies do not rely solely on flapping to alleviate the influence of gusts; however, they may utilize intermittent flapping during gliding.

In previous work we have shown that dragonflies' corrugated wings exhibit not only outstanding lift generation, but also superior anti-turbulence characteristics compared to conventional wing profiles developed for low speed flight [8]. It is also well known that fixed-wing type flying robots that are less than a few grams in mass are inferior in outdoor flight at low speeds because of their sensitivity to gusts. It seems to be reasonable to consider the possibility of utilizing insect mimicry to achieve a gliding mode for stable flight, supposing that the reason for the poor flight stability of conventional wings is due to aerodynamic defects, and it also seems to be reasonable to consider that mimicry of the dragonfly's gliding shape might lead us to a new kind of turbulence-resistant ultra-lightweight flying robot. The mimicry of gliding has an advantage in that it does not require complex control or actuation systems, and it might make possible a kind of flying robot that is tolerant of gusts in outdoor flight. We attempted to develop a new kind of flying robot, which is no more than a few tens of grams in mass and has corrugated fixed-wings.

In this paper, a new mimetic technique and some new flight characteristics for a dragonfly mimetic robot are introduced. Although mimetic procedures for fixed-wing/gliding are far simpler than those for flapping flight mimicry, there remain barriers associated with the materialization of numerical measurement methods for robot flying characteristics. Considering this essential problem, we set the purpose of this research to make clear the degree to which flight envelopes are improved as compared to conventional flying robots by adopting dragonfly mimicry.

2. Materials and Method

In this research, experimental techniques were used to clarify issues and visualization methods, namely, a large circulating water tank with flow visualizing devices, developed by MFRL, were used to reduce development time. In this research and development process we produced various types of free flight paper aircraft around 2 g in mass; an RC flying robot 4 g in mass with mimicry of the dragonfly's corrugated wing and two-channel control; and a dragonfly shaped flying RC robot with two-channel control and a mass of around 20 g.

The main experimental apparatuses used in the research are given below, all of which belong to NBU.

- Large scale circulating water tank with flow visualization devices (Figure 1)
- Two-dimensional low speed wind tunnel for streamline visualization using smoke
- 3D digital microscope
- Catapult for paper aircraft
- 3D-CAD CATIA and a machine tool coupled to CATIA;
- Heating and decompression device (1 atm. on gauge) for CFRP forming
- Speed-gun (The STAKER SOLO 2 Rader)
- Hi-vision movie camera (HVR-Z1J)

Figure 1. Flow visualization circulating water tunnel at NBU.

3. Experiment

3.1. Review of Aerodynamic Characteristics of Corrugated Wing through Visualization

The fact that dragonfly wings are corrugated is well known, and by using the profile data for Lesser Emperors given in [9], we attempted to clarify the two-dimensional flow around the wings of the Lesser Emperor by using the original visualization method by using aluminum powder. As shown in [8], we found that the dragonfly's corrugated wing plays an important role in producing lift and stability during flight for Reynolds (Re.) numbers between 2000 and 7000. An actual dragonfly seems to glide in around Re.2000 in calm condition, but the flow phenomena in a higher Re. number condition is interesting from the view point of application, so the experiment was achieved to maximum Re. flow in which our visualization technique is effective.

The two-dimensional flow experiments showed that in gliding flight, the dragonfly produces vortex trains on the upper surfaces of their corrugated wings, which blow off congested air that forms due to the viscosity at the upper surfaces. The result is streamline airflow around the corrugated wing. The typical airflow pattern of a dragonfly is shown in Figure 2 where the yellow area denotes vortex flow.

Figure 2. Flow around a Lesser Emperor's wing at Re.7000 (hind wing, 75% half-span section, angle of attack is 5°).

The flow around NACA 4418 is shown in Figure 3, in which the Re. number is 7000 and the angle of attack is 5°.

Figure 3. Flow around NACA4418 wing (angle of attack is 5°).

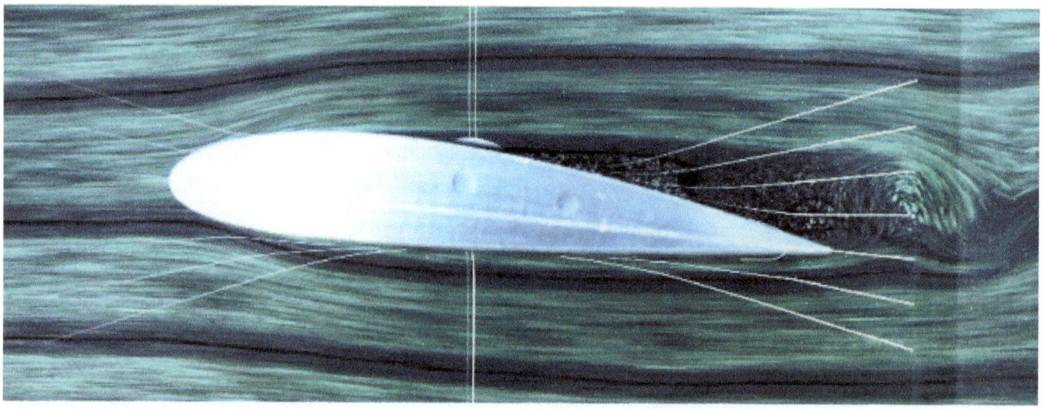

The difference between this wing and the dragonfly wing is notable. In low Re. number flow, the growth of the congested area is especially apparent on the upper surface of NACA 4418, generating undesirable flow pattern characteristics and ultimately reducing lift. On the other hand, the congested flow on the dragonfly wing is blow away by the vortex train and reforms the streamline airfoil shape so as to generate lift.

The modeled 2-D wing is not identical to an actual dragonfly's 3-D wing, but flow around the 2-D wing gives enough information when application to fixed wing flying object is considered.

Now, it is apparent that NACA 4418 is not suitable to low Re. number flight, but it is possible to improve by using thinner streamlined wing [9]. Aerodynamic characteristics of thin curved plate seem not inferior to corrugated wing as far as the polar curve is considered [9].

To understand the behaviors of wings in turbulent condition, we set the curved plate wing and corrugated wing in a steady flow, and once the condition of the congested region has stabilized, we rapidly accelerated the wing and observed the resulting change in airflow. In the case of the curved plate wing, half of the grown congested area was blown off by the acceleration and the shape of the streamline flow changed. In the case of the corrugated wing, however, the shape of the streamline did not change at all, but the strength of vortices increased to compensate for the accelerated flow [8]. The flow mechanism for a curved plate wing is shown schematically in Figure 4.

Figure 4. Schematic diagram of wing acceleration effect.

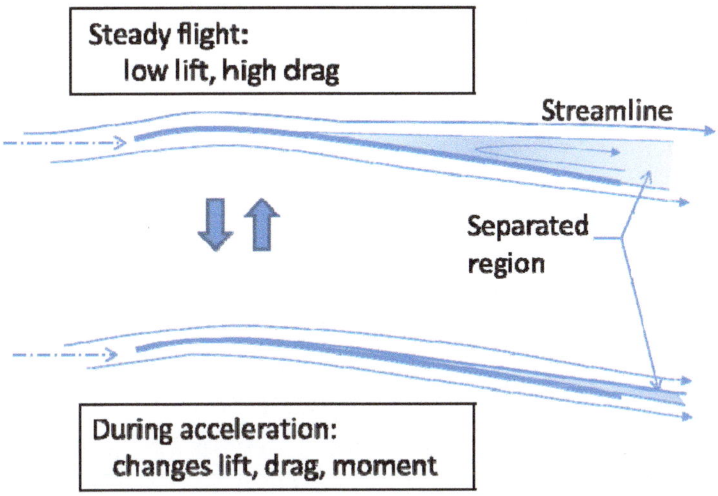

The actual flight of such a wing is complicated, because after accelerating the congested area grows again to a steady state, resulting in pilots being unable to understand the aerodynamic situation as they cannot see the change in air flow.

Figure 4 suggests that for the curved plate wing or a thin streamlined wing with a positive camber, it is very difficult to manage this unstable aerodynamic behavior. The corrugated wing, however, is unaffected by these changes of circumstance and is able to maintain steady flight. This suggests that fixed wing robots with corrugated wings are able to fly in turbulent air at low Re. numbers in the same way that conventional aircraft fly at high Re. numbers. The Re. number in the above experiment was 7000 but the Re. number expected in the actual flights of developed robots is in the order of tens of thousands and it remains some degree of discrepancy in Re. number.

As a means of precaution against unexpected behavior, we tried to visualize the flow using a wind tunnel generating smoke streamlines with a Re. number of 70,000 (Figure 5). The flow did not deteriorate and we did not find any critical phenomena. We concluded that in the region of 7000 to 70,000, the performance of the corrugated wing may degrade gradually but does not show any abrupt change.

For reference, the flow around NACA 4418 was found to drastically improve at the Re. number of 70,000, and produces a better flow pattern than the corrugated wing (Figure 6).

Figure 5. Flow around a Lesser Emperor's wing at Re.70,000 (hind wing, 75% half span section, angle of attack is 5°, photo does not show pure 2D section of wing).

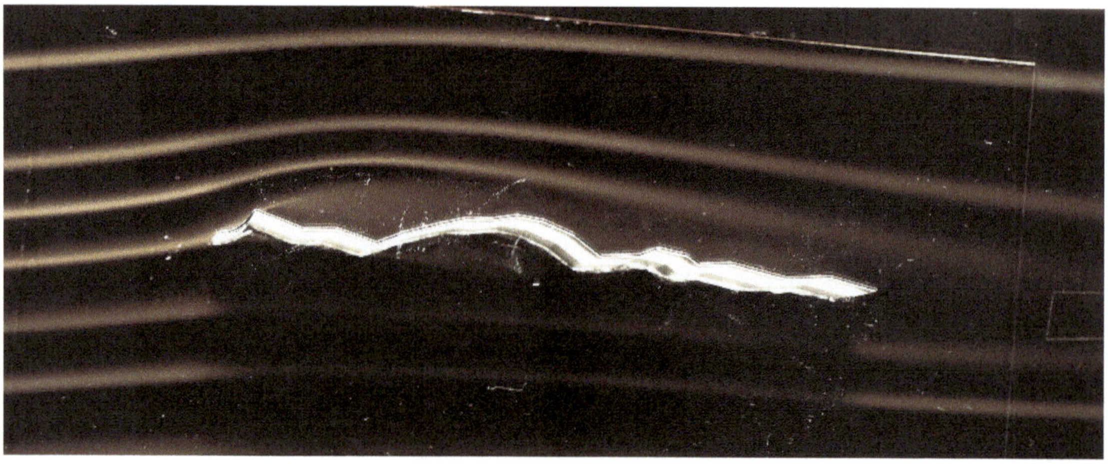

Figure 6. Flow around NACA4418 and Lesser Emperor's wing at Re.70,000 (angle of attack is 13°, photo does not show pure 2D section of wing).

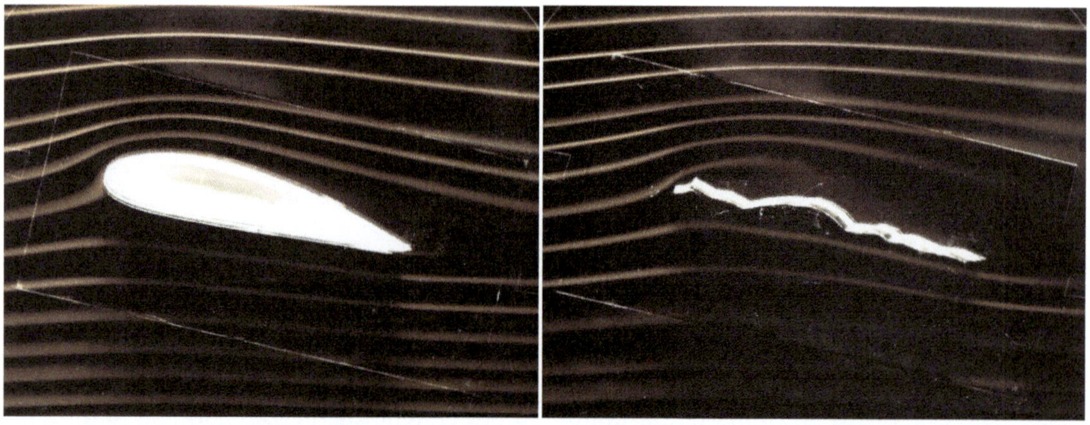

The outcome of these experiments encouraged us to try to make a flying robot with corrugated wings for flight at Re. numbers in the order of tens of thousands. At this stage, special fluid dynamics for measuring lift, drag, and moments was not required because it is sufficient to know the best position for the center of gravity and to use statically stable characteristics for each aircraft configuration to achieve flight.

We obtained the profile of the Wandering Glider's wing through 3-D digital microscopy and tried the flow visualization for the wandering glider. The shape of the Wandering Glider's wing is very different to that of the Lesser Emperor's. However, there were few differences in the flow pattern and vortex train generation processes (Figure 7), thus, the corrugated wings of the Lesser Emperor and Wandering Glider were chosen for the first dragonfly mimetic corrugated wing flying robots to find comparative merits and demerits.

Figure 7. Flow around Wandering Glider's wing at Re.7000 (forewing, 75% half span section, angle of attack is 5°).

3.2. Preliminary Mimetic Study of Dragonfly Gliding at Low Re. Numbers

Photographs show that when gliding dragonflies use around 15° of forewing dihedral angle and either the same order for their hind wing or none at all (Figures 8 and 9).

Figure 8. A Wandering Glider gliding.

Figure 9. A Rhyothemis fuliginosa gliding.

We made aircrafts, 12 cm in wing-span, using copying paper that consist of only wing and no fuselage applying general outline we have got about dragonfly (Figure 10). At this stage, 3-D mimetics was intended. It weighed 0.6 g and showed remarkable straight flight tendency and the robustness to deformation with passage of time. Even after half a year of leaving, the plane in Figure 10 flew straight without any adjustment work. A normal thin paper airplane is very subtle to deformation of the wing and delicate adjustment is required in each flight. This led us to further investigation of dragonfly's gliding ability.

Figure 10. Dragonfly paper plane without fuselage.

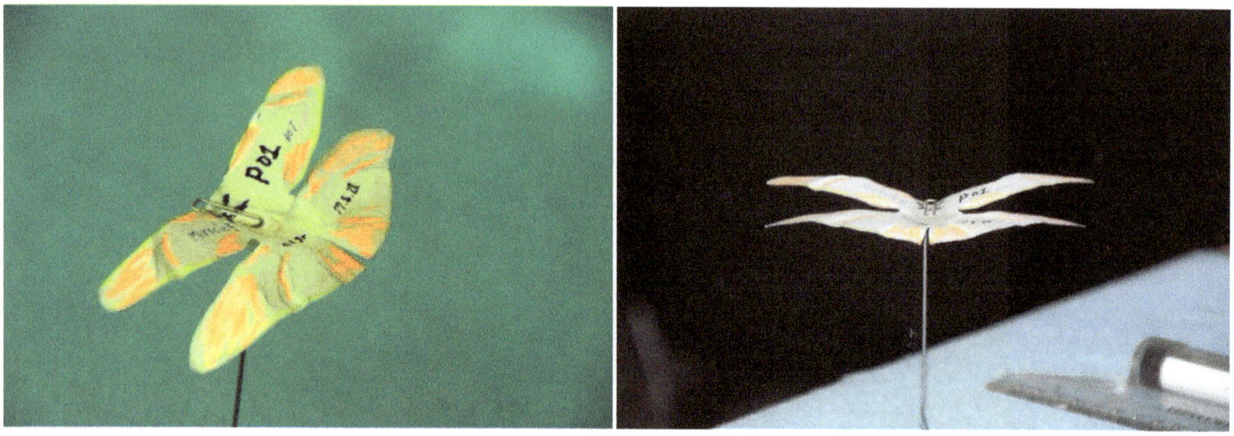

Considering the technical difficulty of 3-D mimetics of real dragonfly's wings, we decided to investigate the effect of paper aircraft having corrugated but uniform wing section. To realize the tail boom of dragonfly, we had to make up sized airplanes of 15 cm in wing span and around 2.5 g in weight (Figure 11).

Wing section profiles of the aircraft in Figure 11 are different to dragonflies. This is one of artificial models obtained from flow visualization test at Re.2000 (Figure 12), but later it is replaced with a dragonfly-like section because of its toughness to high speed and turbulences.

Figure 11. Dragonfly paper plane having head and fuselage and catapult.

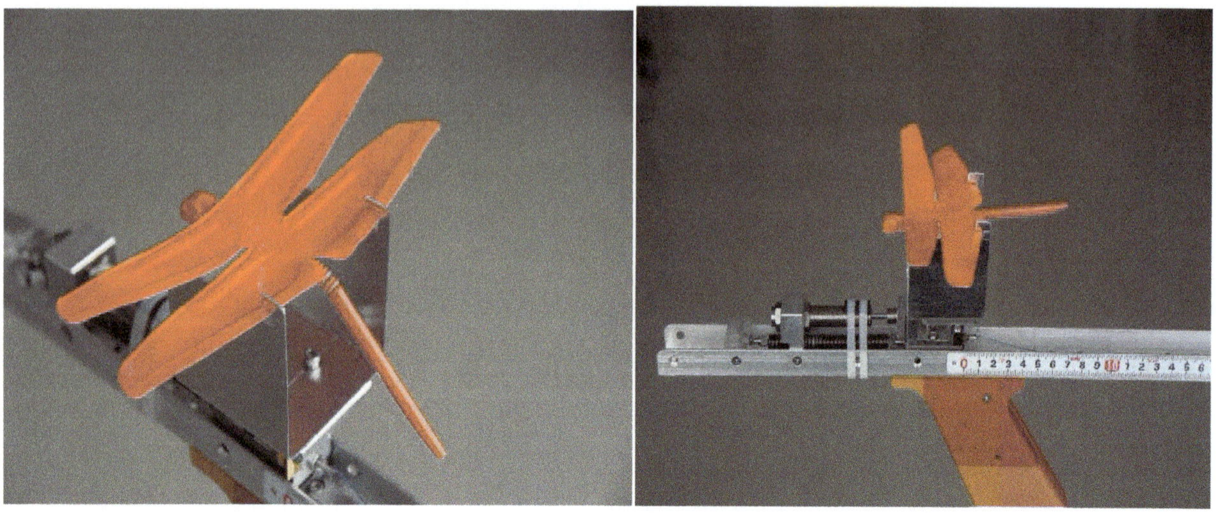

Figure 12. Flow around simplest corrugated wing at Re.2000.

The aspect ratio was made to be the same as that of a gliding dragonfly, and the position of the center of gravity and the mounting angle of the hind wing were determined through flight experiments. The mounting angle used for the wings of normal tailless airplanes was equated for the case of twin wings, like those of a dragonfly, to the hind wing leading edge is lower than trailing edge. The best performance was achieved with 30% whole wings chord position of center of gravity, wherein the chord is the distance from the leading edge of the forewing to the trailing edge of the hind wing, and achieved at the 15° of the leading edge lower installation angle of the hind wing.

Flight testing of best 15 cm span model revealed that the dihedral forewing and horizontal hind wing configuration achieved a lift-drag ratio of 6, which is better than the configuration in which both wings have a non-zero dihedral angle. Thus, we adopted this configuration as standard for gliding models thereafter. Experiments with wing-only models in outdoor conditions suggested that the effect of wind is too strong for stable flight.

A paper aircraft with a head and fuselage achieved more stable flight compared to the wing-only model. The addition of a head and fuselage changes the balance of the moment of inertia of the flying object.

The mass distribution of the paper plane with no fuselage was wide, whereas those with a head and fuselage had an almost square distribution. The later models have the potential to resist the large rolling moments induced by gusts. When the moment of inertia around the principal axis is divided into moments large, medium, and small in magnitude, and when they start to rotate, it is known that the rotating motion around the axis having the medium size of moment of inertia diverges for almost every configuration [10]. This information leads us to conclude that, for small-sized fixed-wing flying robots for outdoor use, we have to adopt the cruciform configuration of dragonflies, to avoid unstable dynamic movement about the rolling axis, because the rolling movements caused by gusts are far more significant than those of conventional large-sized aircraft. Foregoing discussion leads us to the design principle that the basic configuration of small fixed-wing aircraft with corrugated wings should have cruciform mass distribution.

The outdoor flying tests of paper aircraft with tandem corrugated wings and a head and tail showed superior robustness to side-on gusts compared to the wing-only type, as expected. New differences between the corrugated wing and the curved wing were found for the cruciform configuration through catapult flight tests against wind produced by a fan.

The latter aircraft was blown upwards or turned suddenly when it approached the fan, but the corrugated wing aircraft was able to maintain a straight flight until getting very close. This makes apparent an additional advantage of the corrugated wing.

3.3. Remotely Controlled Flying Robot with a Single Corrugated Wing

Ultra-lightweight flying toys with single wings that are controllable using infrared rays were easily realizable at the time we have started research. The next step in our experiment was in successfully exchanging the toy's fuselage and wing with a single corrugated wing.

All the structural parts were manufactured by university students through this research. These models were mainly used to confirm the flight performance of corrugated wing in rather large Re. numbers than that of paper aircraft or of flow visualization experiment.

The span of the prototype robot was 18 cm and the mass was 4 g. The single pusher propeller and rudder configuration of the original model remained intact. The periphery of the wing was a CFRP frame and the inner surface of the wing was made from thin but rigid Japanese paper which allowed us to form the corrugated profile. Position of the center of gravity was set at nearly 30% of longest wing chord. We prepared two types of wing section, Lesser Emperor's and Wandering Glider's profiles at the 75% half unit span. The Lesser Emperor's model was shown in Figure 13. The black part of structure was made of CFRP and the white part of wing is made of Japanese paper. Base data of the wing section is digitalized on CATIA, and after the mold is machined, a heating and decompression device (1 atm. on gauge) was used for final CFRP forming.

Figure 13. Ultra-light RC flying robot with a single corrugated wing.

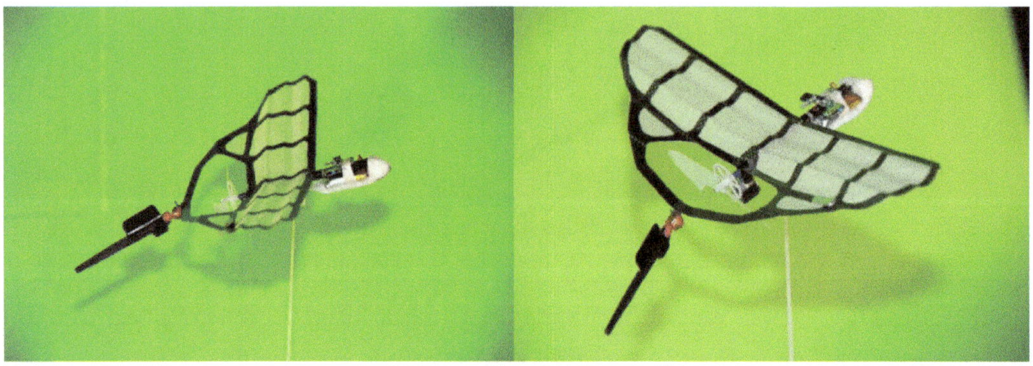

Longitudinal stability was ensured by adopting the tail up reflection of the wing. The main experiment was conducted in a wide lecture room at NBU that was 10 m^2 in size and 3 m in height.

Both wing types showed almost same flying characteristics for pilots. Lesser Emperor's wing (Figure 2) has larger maximum height (5% of chord) than that (3% of chord) of Wandering Glider's wing (Figure 7) and the position of maximum height is near the center chord, and is considered to have healthy structure for rough operation by pilots and selected as our standard wing section for robot.

The prototype robot showed surprising stability at the speed of 4 m/s. The chord length referenced Re. number was around 14,000 and, although it differed from that investigated using the water tunnel, the influence of the Re. number difference was not apparent. Two-channel remote control flying robots do not have control around the pitching axis so it is normal for such robots to exhibit oscillatory attitude immediately after changing altitude. The corrugated wing model, however, did not show any oscillatory movement and was easily able to fly close to the ceiling or desktops.

Figure 14. RC flying robot hitting a balloon.

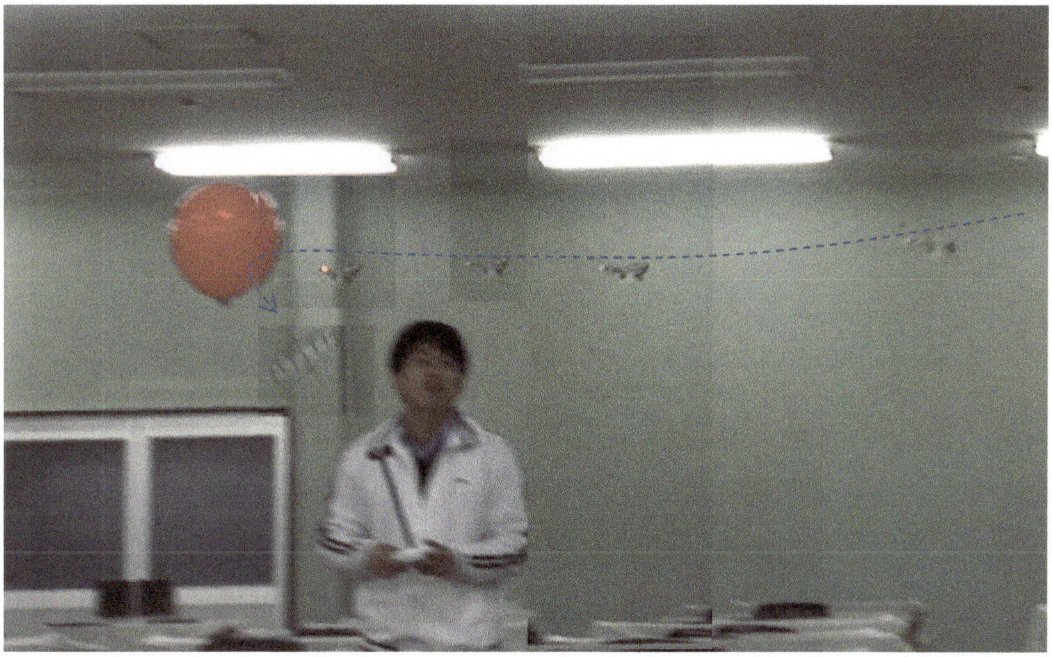

A flight speed of 4 m/s is relatively high for flight in the above mentioned lecture room, which would causes pilots difficulty if the airplane did not have excellent dynamic stability. However, it was possible to pilot the prototype robot into a balloon moored in the room. Figure 14 shows the balloon-hitting trajectory of the robot digitally pasted over frames of the recorded video. This experiment confirmed the stability of the corrugated wing and suggested the flying robot was ready for outside testing. Since the infrared control system is too sensitive to sunlight for outdoor use, we developed a radio controlled flying robot.

3.4. Remotely Controlled Dragonfly Shaped Flying Robot with Corrugated Wings

The experimental model used for the flight tests had tandem wings like a dragonfly of which the wingspan was 27 cm and the mean chord was 3.5 cm. The total wing area was 189 cm^2 and the mass was around 20 g.

The position of the center of gravity was around 30% of whole wing chord, which means the distance from the leading edge of the forewing to the trailing edge of the hind wing.

The installation angles of the wing to fuselage were determined using the results of the paper airplane experiments. Pusher type counter rotating propellers were installed under both forewings. Fortunately, propeller position between forewing and hind wing gives propellers a high lift generating function. Prop wash activates the dispirited upper surface flow of the hind wing in a high angle of attack and produce high lift and delays stall of the hind wing.

Velocity control was achieved by the collective operation of the rotational speed of both propellers and yawing was achieved by the differential operation of the propellers.

The command and control system and the power driving system were cannibalized from a flying toy (X-TWIN PRO STORM by Silverlit). Static thrust of both propellers was 18 g. The wing profile of the Lesser Emperor at 75% half unit span was adopted with a forewing dihedral angle of 15° and a horizontal hind wing. The early model of the wing was made from CFRP, which weighed 25 g, and was later improved to a periphery CFRP structure with a thin plastic film inner surface to reduce the weight to 17 g (Figure 15). Fuselage structure, including the nose, was made of Japanese cypress considering experimental convenience and the nose was covered by EPP (expanded polypropylene).

Figure 15. Dragonfly shaped RC flying robot with corrugated wings.

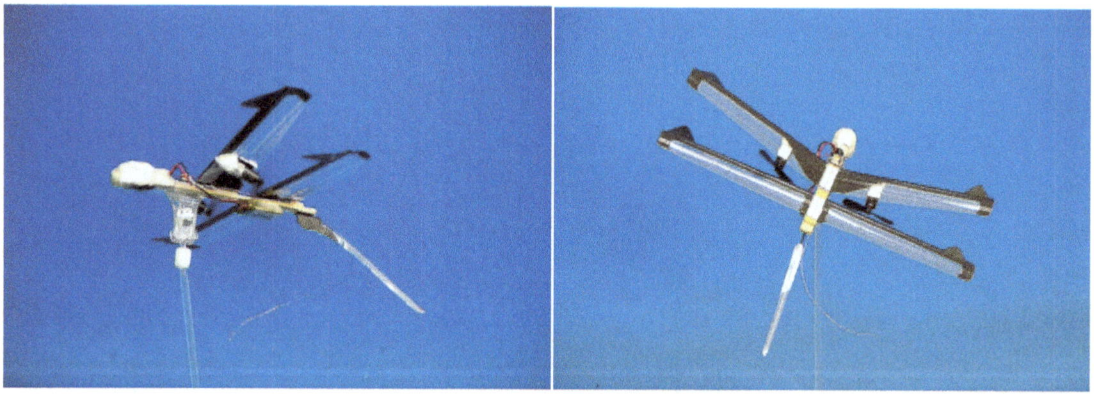

Structure is simple and sturdy and born many times of unintentional collision shocks.

The battery (200 mAh, 5.47 g) was installed on the cypress bar just under the hind wing and control circuit board for radio control was attached to nose part of the cypress and covered by EPP.

Forward facing delta winglets installed on the leading edges of the wing tips generated strong tip vortices and delayed stall.

The relatively large ventral fin under hind wing is attached when continuous high angle of attack flight is necessary and it was effective to attenuate the rolling oscillation occurring at a high angle of attack (Figure 16).

The angle of the tail to the wing's lower surface was made adjustable. It was mainly the university gymnasium that was used for indoor flight and the university athletic track and field on hilly terrain that was used for outdoor flight. This field receives a northerly wind during almost every day from Setonaikai Bay, located two miles north of the track. The dragonfly configured flying robot showed excellent stability, as expected. The speed at level flight was 6 m/s with the configuration of easiest control, obtained by adjusting the position of the center of gravity. The chord based Re. number was 14,000, the same as that for the former single winged ultra-lightweight flying robot. The fact that this robot encountered no problems in steep descents may indicate that corrugated wings were able to maintain the vortex chain and its effects up to an Re. number of 20,000.

Figure16. Dragonfly shaped RC flying robot for high angle of attack.

In indoor flight, the pilot was able to fly as desired using a two-channel control from low speed to relatively high speed. Both loops and pulling up near the ground were easy for skilled pilots. In outdoor flight, the robots kept their high flight stability, but it was necessary to set the center of gravity somewhat forward to decrease the angle of attack, since the change of angle of attack in outdoor flight due to the gusts and turbulence was far greater than that in indoor conditions. A flight over the track and field area is shown in Figure 17. We found that if the tail was fitted at a greater angle to the wings (tail down), the robot seems to have increased stability. It is possible that the nose down moment induced by the bending of the tail reduces the angle of attack and lowers the likelihood of stalling. Furthermore, the tail-down configuration restricts the rolling movements of the robot caused by side-on gusts and alleviates the descent tendency by lift loss accompanied with rolling.

Figure 17. Flying dragonfly robot over the NBU field.

Forward facing delta winglets installed on the leading edges of the wing tips were expected to delay stall by their strong tip vortex.

The dragonfly-shaped robot with this delta winglet was able to achieve controlled flight at an angle of attack of 40°. Figure 18 shows the robot turning at high angles of attack.

Figure 18. Controlled flight of dragonfly shaped robot in the gymnasium (angle of attack is 40°).

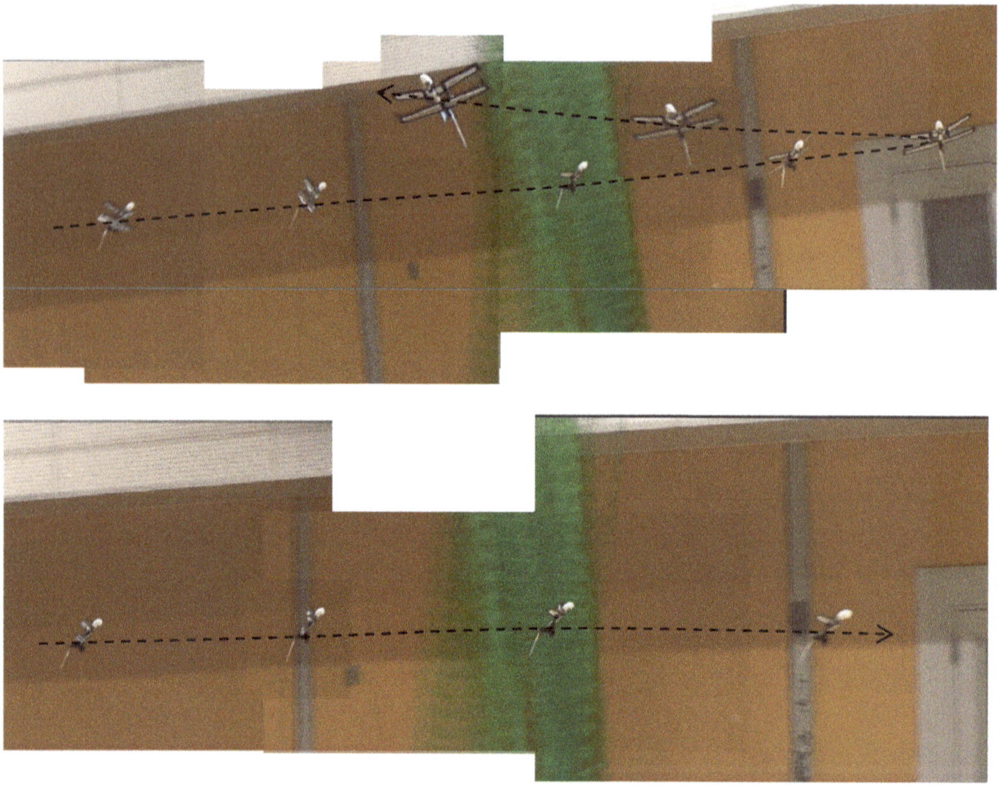

It is worthwhile to note that the rectangular plan formed dragonfly wings accept these simple high lift devices with minimum loss of weight or performance.

This image was made in the same way as that shown in Figure 14. As would be expected, partial stalling was unavoidable and rocking behavior was exhibited, but total loss of control did not occur. The flight speed at this attack configuration as obtained by a speed gun was 2.2 m/s, meaning the dragonfly shaped flying robot was able to operate with normal controllability up to a lift coefficient (C_L) of 3.6 for the whole airplane.

The fact that the aircraft, which has a relatively large aspect ratio, attains a high C_L such as 3.6, suggests that dragonfly gliding mimetics has great potential. Not only is the dragonfly robot capable of flight at high angles of attack, but it also allows us to select a wide range of C_L which can be used to extend the flight time. The maximum flight time of fixed wing aircraft is greatest when $C_L^{1.5}/C_D$ is maximized (where C_D is drag coefficient), which clearly requires the highest possible value of C_L. The main drive system was cannibalized from a toy aircraft of the same size, with which one of the developed dragonfly robots, in which the weight was 23 g, attained a flight time of 13 min and 30 s, although the original toy aircraft's maximum flight time was around 10 min as written in the manual.

In this flight, the flight speed was 5 m/s, and C_L was calculated to be 1.5, a result not easily achieved by conventionally configured aircraft in which it is difficult to install a stall prevention system.

It is noteworthy that the dragonfly-shaped robot has good compatibility with artificial devices, such as propellers and delta winglets. For example, the dragonfly robot did not lose altitude even when entering a forewing stall, since the propwash always flows over some area of the hind wings, meaning the hind wings never enter a full stall. Thus, if a forewing-stall does occur, the continuing hind wing lift produces a nose down pitching movement, allowing quick recovery and little loss of altitude. This is extremely valuable in outdoor flight because weak turbulence causes high angles of attack for low speed flying objects. These merits are considered to be achieved by the configuration of rectangular wings in a tandem arrangement.

A summary of the superiorities of the dragonfly flying robot is given below. Controlled flight over the track and field when the wind speed was less than 4 m/s was possible. Even with a wind speed of 6 m/s, it was possible to hover like a kite for at least 3 s if the wind flow was uniform. As will be explained later, we found that a cruciform configuration in which the forewings have a dihedral angle and the hind wings have an anhedral angle is the most resistant to rolling into sideslip. This configuration also shows considerably good flight characteristics and does not easily lose altitude in turns or with side-on wind, making it suitable for low-level flying. This configuration does not require a large fin installed in the middle of the body to stabilize the robot, a feature that was previously considered necessary to maintain controllability at high angles of attack.

This cruciform dragonfly robot has now become the standard for gliding mimetics in MFRL.

4. Discussion

4.1. Corrugated Wing

It was demonstrated that the corrugated wing, which does not change the form of the outer flow during acceleration, shows good stability in unsteady wind conditions and may be the only wing profile for fixed wing aircraft that allows outdoor ultra-low-speed flight.

The reason for the strong straight flight tendency shown by the corrugated wing aircraft is still not well understood and must be investigated further. As this stage it is hypothesized that the on-wing vortices resist acceleration and yawing moments and the drag induced by tail boom works positive to the direction stability.

4.2. The Dynamic Characteristics of the Dragonfly Configuration

It is suggested that one of the reasons for the robustness of the dragonfly configuration to outdoor flight conditions stems from the cross mass distribution that causes a gentle response to large rolling movements. Dragonfly consists of wing, head, fuselage, and tail has same order of moment of inertia regarding body and wing axis.

The reason for the positive effect of increasing the tail boom angle on the longitudinal stability is considered as a result of the nose down moment lowering the angle of attack and decreasing the likelihood of a stall occurring and as the drag induced by tail boom works positive to the direction stability. The reason that the steep angle of the tail boom provides the dragonfly robot with a robust ability for straight flight is still unclear and is considered to be an area for further investigation.

4.3. Flying Response of Ultra-Small Flying Objects

High response movement to control inputs is serious problem for maneuvering Ultra-small flying object.

In level flight, the following equation holds.

$$W = \frac{1}{2}\rho V^2 S C_L \tag{1}$$

Here, W is the weight of the object, ρ is the air density, V is the velocity, S is the wing area, and C_L is the lift coefficient.

The angular acceleration around the lateral axis is as follows for the conventional configuration which has tail wing.

$$I\ddot{\theta} = \frac{1}{2}\rho V^2 S_t C_{Lt} l = W \frac{S_t}{S}\frac{C_{Lt}}{C_L} l \propto \rho_{AC} \frac{S_t}{S}\frac{C_{Lt}}{C_L} l^4 \tag{2}$$

$$\ddot{\theta} \propto \frac{1}{\rho_{AC} l^5} \cdot \rho_{AC} \frac{S_t}{S}\frac{C_{Lt}}{C_L} l^4 = \frac{S_t}{S}\frac{C_{Lt}}{C_L}\frac{1}{l} \tag{3}$$

Here, C_{Lt} is the lift coefficient of the tail wing, S_t is the area of the tail wing, θ is the pitching angle, and ρ_{AC} is the density of aircraft.

Equation (3) shows that the angular acceleration around the lateral axis is inversely proportional to the size l of the flying object so far as the form similarity consists. As miniaturization progresses to an extreme level, pitching moments occur at a rate that a human controller cannot follow.

This phenomenon requires a new control system to overcome these problems for ultra-lightweight and small fixed wing flying robots.

The implementation of tiny but complex control systems that can manage the motion of ultra-small flying objects is considered to be difficult even with advanced technology; thus, a robot that moves sedately at low speed is, not only necessary, but advantageous. In this sense, the dynamic response characteristics of the dragonfly robot produced by the vortices are attractive. For our developed dragonfly robot, a three-channel model was successfully produced and flown, but the pilot found it difficult to use the third stick to adjust the pitching attitude of the robot. The pilot's comments suggested it was far easier to use the two-channel model than the three-channel variant.

4.4. Matching of Propellers and High Lift Devices to the Dragonfly Configuration

Since the dragonfly configuration uses a rectangular wing shape, it is easy to mount power or aerodynamic devices on the wings, and such devices can successfully achieve their function without degrading the original performance of the dragonfly robot.

4.5. Evolutionary Dragonfly Shaped Flying Robot

Thus far, it has been suggested that the dragonfly-shape robot is the optimal model for objects flying at low speed, but it is worth investigating more advanced shapes, including configurations impossible for dragonflies due to biological reasons. For example, dragonflies' hind wings are not able to take an anhedral angle in a leading edge down position. A dragonfly model with such a configuration showed good flight characteristics and was robust to side-on gusts. This is beneficial to actual flying, since rolling results in a loss of altitude and is not desirable when flying close to the ground. Since the forewings have a dihedral angle, the hind wing anhedral model is considered not to have overall dihedral angle effect and considered to be a questionable design in the sense of flight dynamics, but the prototype showed any inconvenience natures.

4.6. Applications of the Dragonfly Configuration

Refinement of the dragonfly shaped robot will provide simple solutions for the observation of living things from air and a candidate for flights in the atmosphere of Mars, where the Re. numbers are one order of magnitude lower than those in Earth's atmosphere.

5. Conclusions

Through the developmental research of dragonfly mimetic flying robots, the following conclusions were drawn.

(1) Corrugated wing profiles provide us with new wings for flying robots that can realize continued stable flight at low Re. numbers. The maximum Re. number tolerable is around 20,000.
(2) It is possible to achieve robustness to outdoor turbulence by using corrugated wing robots with a cruciform dragonfly configuration.
(3) The dragonfly configuration with propellers makes possible a continued high angle of attack flight, improves resistance to gusts, and reduces induced by drag.
(4) By adding delta winglets to the leading edges of the wing tips, the robot achieved controlled flight with a lift coefficient of 3.6.
(5) It is necessary to investigate further the stable characteristics of flying objects with corrugated wings which produce lateral vortex chains on their upper surfaces while flying.
(6) It seems to be more practicable to employ bio-mimetic technology to utilize insect forms such as the dragonfly for specific functional parts of conventional robots only, or to comprehensively refine existing engineering, instead of directly imitating entire complex functions of living creatures.

Acknowledgments

This work was supported by the Ministry of Education, Science, and Culture of Japan, and Nippon Bunri University.

Author Contributions

Akira Obata invented the water flow visualization method and conducted whole research program, Shotarou Sinohara achieved water tunnel and flying robot experiments, Kyohei Akimoto contributed in manufacturing CFRP ultra-light flying robot structure, and Kakeru Suzuki assisted the Sinohara's water tunnel experiments, Miyuki Seki worked to develop experimental tool which is necessary to fulfill this experimental research. Akira Obata wrote the paper.

Conflicts of Interest

The authors declare no conflict of interest.

References

1. Ellington, C.P. The aerodynamics of hovering insect flight. IV. Aerodynamic mechanisms. *Phtl. Trans. R. Soc. Lond.* **1984**, *B305*, 79–113.
2. Azuma, A.; Azuma, S.; Watanabe, I. Flight mechanics of a dragonfly. *J. Exp. Biol.* **1985**, *116*, 79–107.
3. BionicOpter—Inspired by dragonfly flight. Available online: http://www.festo.com/cms/en_corp/13165.htm (accessed on 10 March 2014).
4. Isogai, K.; Kamisawa, Y.; Sato, H. Optimum aeroelastic design of a resonance type flapping wing for micro air vehicles. *Trans. Jpn. Soc. Aeronaut. Space Sci.* **2010**, *52*, 238–245.
5. Nagai, H.; Nishi, A.; Fujimoto, T.; Isogai, K. Study on flow interaction mechanism of fore- and hind-wings of dragonfly. In Proceedings of the 4th International Symposium of Aero Aqua Bio-Mechanisms, Shanghai, China, 31 August 2009.
6. Nano Air Vehicle (NAV). Available online: http://www.avinc.com/nano (accessed on10 March 2014).
7. Petricca, L.; Ohlckerd, P.; Grinde, C. Micro-and nano-air vehivles: State of the art. *Int. J. Aerosp. Eng.* **2011**, doi:10.1155/2011/214549.
8. Obata, A.; Shinihara, S. Flow visualization study of the aerodynamics of modeled dragonfly wings. *AIAA J.* **2009**, *47*, 3043–3046.
9. Okamoto, M.; Yasuda, K.; Azuma, A. Aerodynamic characteristics of the wings and body of dragonfly. *J. Exp. Biol.* **1996**, *199*, 281–294.
10. Barger, V.D.; Olsson, M.G. *Classical Mechanics A Modern Perspective*; Mcgraw-Hill, Inc.: New York, NY, USA, 1973.

11

Robust Bio-Signal Based Control of an Intelligent Wheelchair

Xiaodong Xu [1,*], Yi Zhang [2], Yuan Luo [2] and Dongyi Chen [1]

[1] School of Automation Engineering, University of Electronic Science and Technology of China, Chengdu 611731, China; E-Mail: chenxigt@hotmail.com

[2] Engineering Research & Development Center for Information Accessibility, Chongqing University of Posts and Telecommunications, Chongqing 400065, China; E-Mails: zhangyi@cqupt.edu.cn (Y.Z.); luoyuan@cqupt.edu.cn (Y.L.)

* Author to whom correspondence should be addressed; E-Mail: cupfire@gmail.com

---

Abstract: In this paper, an adaptive human-machine interaction (HMI) method that is based on surface electromyography (sEMG) signals is proposed for the hands-free control of an intelligent wheelchair. sEMG signals generated by the facial movements are obtained by a convenient dry electrodes sensing device. After the signals features are extracted from the autoregressive model, control data samples are updated and trained by an incremental online learning algorithm in real-time. Experimental results show that the proposed method can significantly improve the classification accuracy and training speed. Moreover, this method can effectively reduce the influence of muscle fatigue during a long time operation of sEMG-based HMI.

Keywords: intelligent wheelchair; sEMG; incremental support vector machine; human-machine interaction

---

1. Introduction

The deployment of electric powered wheelchairs (EPWs) has been increased rapidly for better quality of life for handicapped and elderly people over the last 20 years [1]. However, most of these EPWs are traditionally controlled by joysticks, which are not suitable for people with severe physical disabilities like spinal cord injury or hemiplegia. Therefore, alternative HMI methods for wheelchair operation have been studied, such as voice recognition [2], eye moment [3], head gesture [4] and biopotential signals interface [5–8]. Bio-potential signals such as Electrooculargraphic (EOG), electromyographic (EMG) and electroencephalographic (EEG) are also deployed to control such kind of smart service robot [9–11]. Compared to other signals, EMG signal is more easily obtained and has less signal interference, especially for the weak groups, such as patients with upper limb amputation and elderly people for which other methods are inconvenient for their daily life.

The EMG signal, which is sampled from the skin's surface, can directly reflect the movement intention of human body and could be very suitable to serve as the controller. Englehart and Hudgins investigated a continuous segmentation or continuous classification by using different analysis window lengths [12]. Moon used a preset threshold value method to distinct the scapulae muscle signal at fairly high recognition rates [13]. Hiroki Tamura *et al.* obtained s-EMG signals from facial movements and six patterns were recognized to control the wheelchair movements [14]. Lai Wei combined forehead sEMG signals and color face image information to control the smart robot [5]. However, it suffers from poor reliability and robustness in long-term operations, which were mainly caused by muscle fatigue. These may arise not only because of peripheral changes at the level of the muscle, but also because the central nervous system fails to drive the motoneurons adequately [15]. Usually, the amplitude of the sEMG signals from muscles we used without fatigue is lower than 400 μV. However, after fatigue occurs in muscles, there are significant rises to 600 μV, which can be seen Figure 1. Qin and Mitsuhiro investigate a torque estimation method for muscle fatigue tracking, using stimulus evoked electromyography in the context of a functional electrical stimulation rehabilitation system [16]. However, how to choose the best signal evaluation parameter according to specific muscle condition and method of reducing the influence of muscle fatigue is still not very clear, and has been discussed during recent decades [17,18].

Figure 1. The amplitude of surface electromyography (sEMG) signals.

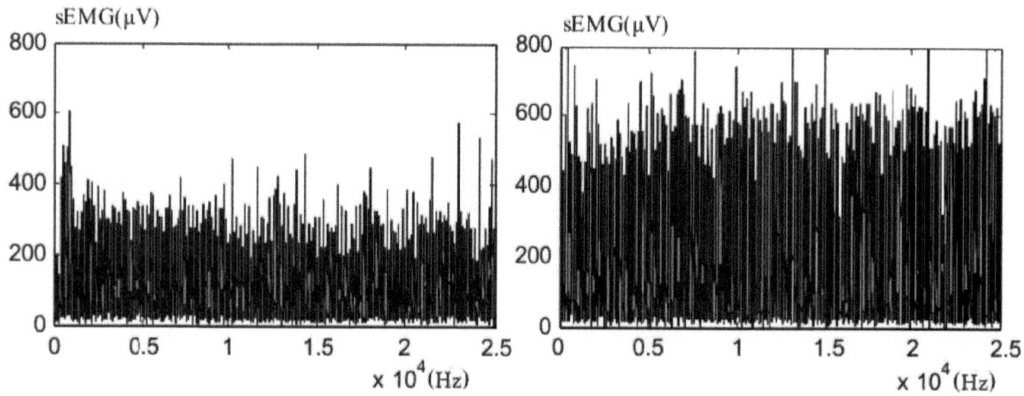

(a) Before fatigue occurring (b) After fatigue occurring

To investigate the real time and long time performance of the proposed system in this paper, forehead sEMG signals are first obtained and analyzed by using three dry electrodes sensor. Then, a four-order autoregressive model parameter algorithm is deployed to acquire data compression and extract four coefficients of the signals. An incremental online support vector machine (SVM) is employed to classify sEMG patterns that have the systematic parameters corresponding to the changes of the signals during the running time. Furthermore, the improved incremental training algorithm is presented and applied on the HMI system.

2. The Research Objects and Methods

The sEMG signal controlled intelligent wheelchair should provide a high degree of intuitive and dexterous control, and offer a convenient and high level of performance for disabled people. The human face has plenty of facial muscles to make expressions. The frontalis muscle, located on the front of the head, is one of the largest muscles which can provide continual and distinct, recognizable signals. The main components of the system include sEMG signals collected by electrodes placed on the skin over a user's forehead muscle, electrodes accompanied by miniature pre-amplifiers to differentiate small signals of interest, signal features detected after amplified, filtered, digitized via standard EMG instruments, and finally transferred to a wheelchair controller. Five subjects voluntarily participated in the experiment and have independent consciousness without facial disease, deformity or trauma. In addition, there was no muscle fatigue before the experiments were performed.

The experimental system consists of an intelligent wheelchair with two independently driven motors, a laptop computer for human-machine interface, and a data acquisition device. A Cyberlink or NIA system is used to acquire the sEMG signals of the subjects. The interface unit has an amplifier with a gain of 50,000, and a filter system provides a band pass filter with a bandwidth of 0.5 to 45 Hz. The weak bioelectricity signals from forehead surface skin are adopted from the user by wearing a sensory headband. The interface box processes the signals and connects to a computer through a serial or USB port. Figure 2 shows the headband devices we used in the experimental system.

Figure 2. The headband EMG device in the human-machine interaction (HMI) system.

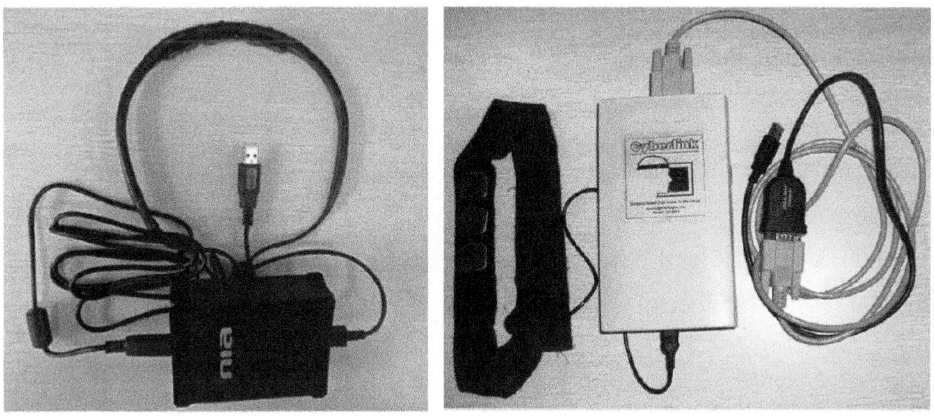

(a) NIA (b) Cyberlink

The muscle movements are defined to be control signals for a HMI system, namely single jaw click (SJC) and double jaw click (DJC). They are generated by the subjects contracting the masseter muscle and buccinator muscle with a jaw clenching and chewing-like movement. Every subject uses a designed sEMG pattern based HMI system to control an intelligent wheelchair, and the control signals are recorded at the same time. There are five control states designed for the wheelchair, namely: "Forward", "Backward", "Left", "Right", and "Stop". Note that the first four states are scanned in turn. Users can use a single jaw click movement to switch the direction to whereas they want. If the wheelchair is in the control state of "Forward", "Backward", "Left", or "Right", a double jaw click movement will be used to switch the wheelchair to the "Stop" state. The complete structure of the HMI system is depicted in Figure 3.

Figure 3. The diagram of the proposed HMI system.

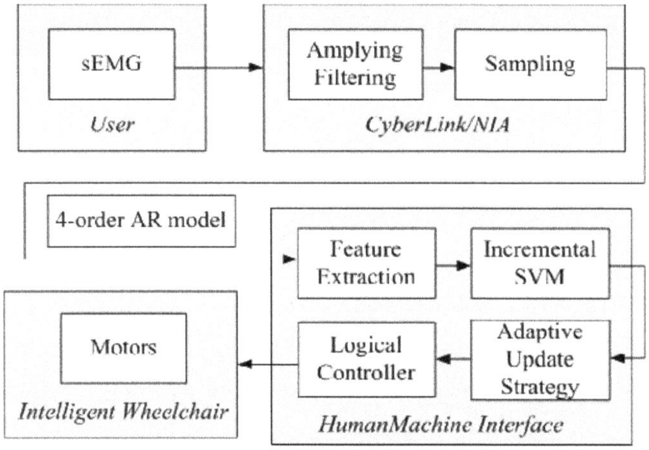

3. Feature Extraction

Myoelectric control systems can be divided into two groups. In the first group, the desired classes of functions are discriminated from signal patterns by classifiers, and the variety of functions depends directly on classification performance. In the second group, recognition-based controllers, which are mainly constructed on threshold control and/or finite state machines, are merely output limited and pre-defined control commands based on a sequence of input signal patterns [19]. The latter gives a direct sense sometimes but has a non-adaptive shortage. After the forehead sEMG signals are obtained by using the dry electrodes sensor, an autoregressive model parameter algorithm is deployed to extract the signal feature.

sEMG signals can be considered as the output signals of a linear sequence with zero mean white noise. The white noise reflects the randomness of signals, while the deterministic part (the order and parameters of the model) reflects the predictability of the process. The four-order AR model is chosen to analyze the sEMG signals; the mathematical expression is as follows:

$$x(n) = -\sum_{k=1}^{p} a_k x(n-k) + w(n) \tag{1}$$

where $x(n)$ is the current sampling value of sEMG signals. $w(n)$ is the current incentive value, *i.e.*, the residual of the white noise. p is the order of the model. a_k is the kth coefficient of the AR model.

The meaning of Equation (1) is that $x(n)$ is generated by the linear combination of several past values $x(n-k)$ and the current incentive value $w(n)$. The model function is defined as follows:

$$H(z) = \frac{1}{1 + \sum_{k=1}^{p} a_k z^{-k}} \tag{2}$$

For the AR model, the selection of p is critical. If the order is too low, like a high-order curve fitted with a low-order curve, it will produce a smooth result; if the order is too high, like a low-order curve fitted with a high-order curve, it will produce a dramatic change and vibration. Many experiments show that the analysis and recognition results of sEMG signals are the best when the order of AR model is 4, so the four-order AR coefficient vector $A = [a1, a2, a3, a4]$ is the input to the SVM algorithm. Table 1 shows four groups of all four-order AR coefficients of Single jaw click and Double jaw click. The AR coefficients corresponding to two kinds of facial movements have obvious differences, so the variations of coefficients can be used as the basis for judging the changes of movement patterns. Meanwhile, it verifies whether the AR coefficients are reasonable as sEMG features.

Table 1. AR coefficients of Single jaw click and Double jaw click.

		1	**2**	**3**	**4**
Single jaw click	a1	−1.6459	−1.4644	−0.84630	−1.6434
	a2	0.77630	1.0654	−0.44390	0.72990
	a3	−0.15580	−1.1692	−0.10020	−0.11680
	a4	0.039300	0.57750	0.40900	0.048200
Double jaw click	a1	−0.96300	−0.99530	−1.2525	−0.98050
	a2	−0.22440	−0.38880	0.12190	−0.32100
	a3	0.087300	0.088600	0.31680	0.22390
	a4	0.12780	0.31810	−0.16790	0.097000

Since EMG is a random signal, its power spectral density (PSD) should be estimated prior to calculating the time-course of the spectral variables. The mean (MNF) and median (MDF) spectral frequencies, which provide some basic information about signal spectral, have been the most popular such variables both in academic studies and in clinical practice, owing to their relevance to underlying physiological processes that control fatigue [20,21]. To measure muscle fatigue, the widely used Fast Fourier transform technique (FFT) is utilized to analyze mean frequency of sEMG.

$$MNF = \frac{\sum_{i=1}^{M} f_i P_i}{\sum_{i=1}^{M} P_i} \tag{3}$$

$$MDF = \frac{1}{2}\sum_{i=1}^{M} f_i P_i \tag{4}$$

where P_i is the ith line of the power spectrum; f_i is the frequency variable; and M is the highest harmonic considered.

MNF and MDF were calculated as previously described [22]. Statistical analysis is applied to find meaningful differences of sEMG from different subjects during various periods. The five subjects are required to perform repetitive movements and the average MNF and MDF calculated by before muscle fatigue, run 20 min, 40 min, and 60 min is showed in Figure 4. Each period lasted 20 s. Both MNF and MDF will decrease with time when muscle fatigue takes place. The MNF and MDF of forehead muscle decreased 10.8%, 17.2% separately while the average err fluctuation was almost unchanged after 60 min muscle contractions. It is experimentally proven that the manifestation of fatigue in myoelectric signals is significant during long-term muscular activities.

Figure 4. Mean (MNF) and median (MDF) shifts during different periods. Each period lasted 20 s.

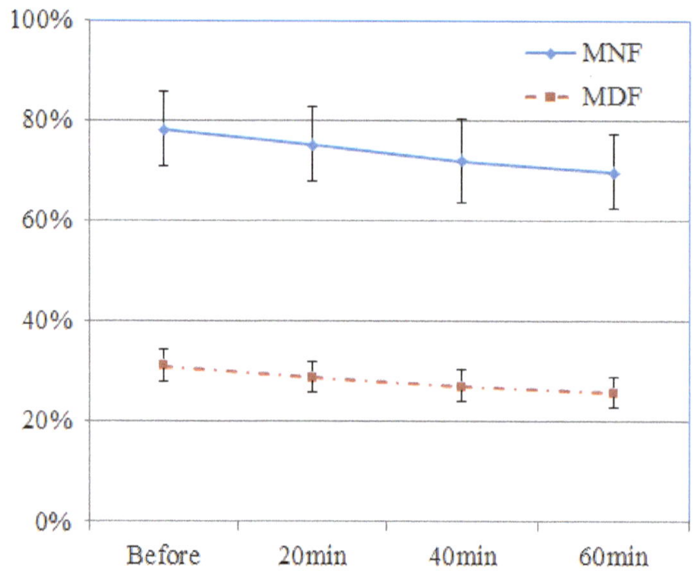

4. Adaptive Incremental Online SVM

sEMG patterns gradually change during long durations of muscle activities due to muscle fatigue, so the threshold method will lose its effectiveness. As we know, a SVM classifier constructs an optimal separating hyperplane in a high dimension feature space of training data that is mapped using a nonlinear kernel function [23,24]. The key ideas of SVM are: getting a maximized classification margin by solving dual optimization problem, and then obtaining an optimal separating hyperplane.

If the data is linearly separable, the decision function is:

$$g(x) = (w \cdot x) + b = \sum_{i=1}^{n} y_i (x_i \cdot x) + b \tag{5}$$

where $\{x_i, y_i\}$, $i = 1,\ldots, l$, are the training samples, y_i is the class label of the training sample x_i, l is the number of the training samples, b is the offset.

When the data is nonlinearly separable, the original data is mapped into a high dimensional feature space by kernel function $K(x_i, x)$, in which the mapped data is linearly separable. Then, the decision function is as follows:

$$K(x, x_i) = \exp\left(\frac{-\|x - x_i\|^2}{2\sigma^2} \right) \tag{6}$$

$$g(x) = \sum_{i=1}^{n} \alpha_i K(x_i, x) + b \tag{7}$$

According to the Karush-Kuhn-Tucke (KKT) condition [25], there is only one solution for the optimization problem based on the quadratic optimization theory, and $\alpha = [\alpha_1, \alpha_2,\ldots, \alpha_l]$ is the optimal solution for the dual problem. When it satisfies the KKT condition, the problem is divided into the following three situations:

$$\begin{cases} \alpha_i = 0 \Rightarrow y_i g(x_i) > 1 \\ 0 < \alpha_i < c \Rightarrow y_i g(x_i) = 1 \\ \alpha_i = c \Rightarrow y_i g(x_i) \le 1 \end{cases} \tag{8}$$

where α is Lagrangian dual, c is an upper bound for samples that lie on the wrong side of the hyperplane. Nonzero constant α_i is the sample corresponding to support vector. Samples are located on the classification interval $0 < \alpha_i < c$. When $\alpha_i = c$, samples lie in the classification interval.

Incremental support vector machines are instrumental in practical applications of online learning. The purpose of the incremental algorithm is to add the new sample data to the three sets and to delete the redundant data, while maintaining all other samples that still belong to the three sets. In the classical SVM, the hyperplane in m dimensions can be shown as $w^T \cdot x + b = 0$. In our updating samples method, only the samples that are closest to the current boundary are chosen to train the classifier. There is a maximum distance of all update samples to boundary, which is defined as 2δ.

$$i^* = \arg|g(x_i)| \le \delta \tag{9}$$

If the new added training sample violates the KKT conditions, the Lagrange multiplier α that is obtained from the previous sample will not be the optimum solution. The proposed improved online incremental training algorithm with the updating method is as follows. The standard SVM is used to initialize the classification and a training set is obtained. The next step is choosing the most informative patterns that have the best possibility of becoming the support vectors in the in-put sEMG sequence, and checking whether these samples meet the KKT conditions. If not, try the next data until KKT conditions are satisfied. Then construct samples including the update data, and train samples online to form a new classifier. The process repeats until there is no in-put sEMG data.

5. Experiment Results

Traditional SVM, such as LIBSVM, employed a radial basis function kernel to classify data, with the same parameter ($C = 1$ and $\gamma = 0.5$) [26]. The data is normalized and limited to the range of −1 to 1. 800 samples are randomly selected from 2,000 pre-collected signals as a training set. After that, choose 800 samples in the rest samples randomly as a testing set. When the result is compared with the method given above, the classification accuracy is shown in Table 2. The training time was generated at 300 samples.

Table 2. Comparison real time performance of the LIBSVM and the Incremental online support vector machine (SVM).

	LIBSVM	**Incremental online SVM**	
Single jaw click	73%	90%	recognition rate
	92 ms	58 ms	training time
Double jaw click	81%	93%	recognition rate
	103 ms	62 ms	training time

Note that five subjects in the experiment are required to control the intelligent wheelchair running throughout a specified path in the lab. A map of the experiment environment with the designed path is given in Figure 5. The size of the map is 8,000 × 6,000 mm and the actual size of the wheelchair is 1,000 × 700 mm. "Forward", "Backward", "Left", "Right", and "Stop": we tried to generate an average for each control state over two experiment days. For each experiment, the five subjects are required to control the intelligent wheelchair for 90 min without a rest.

Figure 5. The path designed for experiment.

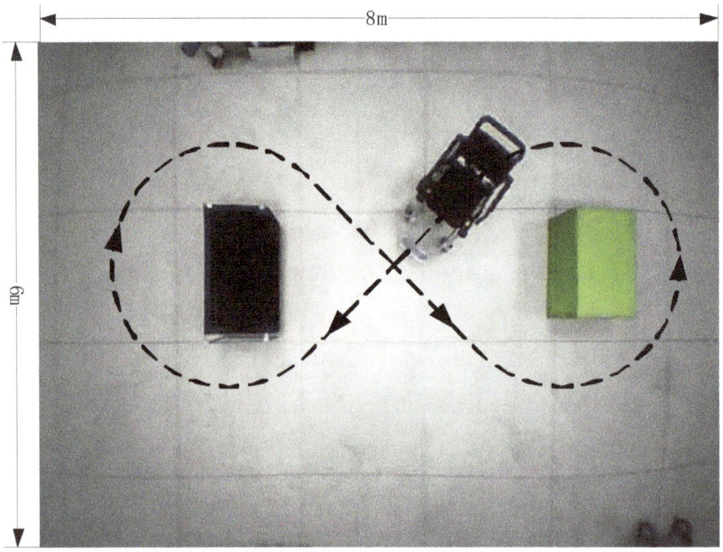

Figure 6 shows the average time consumptions of five subjects in repetitive experiments during deferent periods. The average consuming time is 103 s for one round at the beginning, but the average time increased to 154 s at the last period when muscle fatigue occurred. After using the new method, the average time at the last period could decrease to 110 s. Table 3 shows the average recognition rate of five subjects controlling the wheelchair using the proposed HMI system with the two online SVM in different periods. The recognition of jaw click, or what is called "single jaw click" and "double jaw click", is the most important factor of the experiment.

Figure 6. The five subjects' average time consumptions of repetitive experiments during different periods.

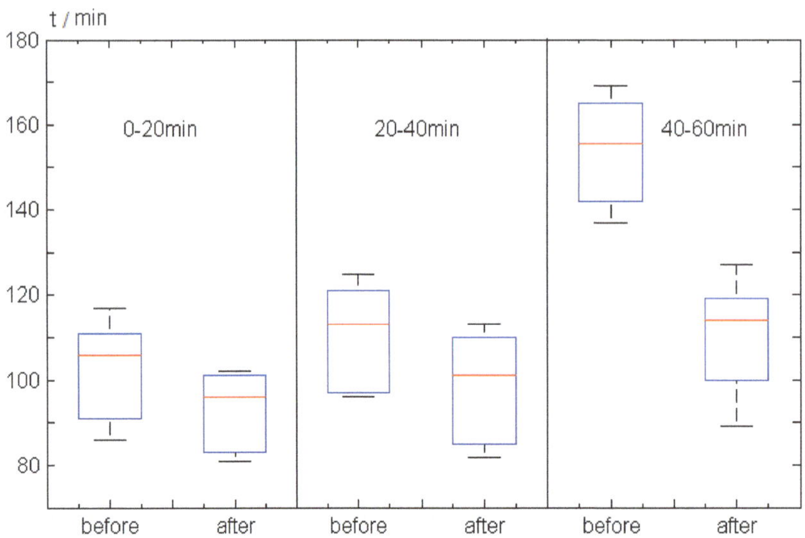

Table 3. Comparison long time performance of the LIBSVM and the Incremental online SVM.

Subjects	Recognition rate for five state using LIBSVM			Recognition rate for five state using incremental online SVM in seconds		
	0–20 min	20–40 min	40–60 min	0–20 min	20–40 min	40–60 min
A	86.5	80.1	73.8	92.6	88.4	86.7
B	88.7	81.6	73.3	92.8	89.5	87.4
C	83.1	79.2	70.9	86.7	82.8	79.3
D	90.4	85.5	79.2	93.6	90.1	88.3
E	80.8	75.3	70.1	84.9	81.6	78.4
Average	85.9	80.34	73.46	90.12	86.48	84.02

Experimental results indicate that the performance of the sEMG based HMI system with LIBSVM gradually declines as time increases with the occurrence of muscle fatigue. The subjects control the intelligent wheelchair by this system using more time in different periods, as shown in Table 3. Also, one should note that, although the threshold EMG method has a pretty high recognition rate before muscle fatigue, the performance after that declines quickly. The incremental online SVM proposed showed good properties in both convergence and optimality.

6. Conclusions

Muscle fatigue can negatively influence the performance of human-machine interaction using sEMG signals. This paper presents a novel sEMG based HMI system for the hands-free control of an intelligent wheelchair both in real time and in long time operations. The system is used to control the simple movements of the intelligent wheelchair, *i.e.*, forward, turn-left, turn-right, backward and stop. The experimental results show that the proposed HMI is easy to operate in real-time and has a high recognition rate. It could be deployed in many aspects of real world applications, such as helping the elderly and disabled, rehabilitation, and so on.

The speed and accuracy of pattern recognition of forehead sEMG signals are to be further improved in our next stage of research. By improving the incremental online SVM algorithm, we hope to reduce the false recognition rate, and obtain a fast response time. In addition, we will investigate how to effectively reduce the noisy signals from facial movements when a user is talking and looking around in the practical application. Moreover, Electrooculography (EOG) signals will be combined with EMG signals to generate the control signals for wheelchair motion control so that the control method is more flexible and robust.

Finally, it is worth remarking that Brain Computer Interface (BCI) based on the Electrooculography (EEG) signals have been used to control the electric-powered wheelchair during the last decade [17,18]. However, these EEG based HMIs are not performing well in the real world since they are non-stationary and low in signal to noise ratio. This was mainly caused by the subject's brain conditions or dynamically changing environments. Therefore, it is necessary to carefully design signal processing algorithms for feature extraction, classification and adaptation. Alternatively, EEG signals could be fused with EMG and EOG signals for accurate human intension detection in real time.

Acknowledgments

This research is financially supported by International Science and Technology Cooperation Program of China, 2010DFA12160 and Science & Technology Research Project of Chongqing CSTC, KJ100516.

Conflicts of Interest

The authors declare no conflict of interest.

References

1. Nihei, M.; Fujie, M.G. Proposal for a New Development Methodology for Assistive Technology Based on a Psychological Model of Elderly People. In *Assistive Technologies*; InTech: New York, NY, USA, 2012.
2. Pires, G.; Nunes, U. A wheelchair steered through voice commands and assisted by a reactive fuzzy-logic controller. *J. Intell. Rob. Syst.* **2002**, *34*, 301–314.
3. De Santis, A.; Iacoviello, D. Robust real time eye tracking for computer interface for disabled people. *Comput. Methods Progr. Biol.* **2009**, *96*, 1–11.
4. Jia, P.; Hu, H.; Lu, T.; Yuan, K. Head gesture recognition for hands-free control of an intelligent wheelchair. *Int. J. Ind. Rob.* **2007**, *34*, 60–68.
5. Wei, L.; Hu, H. A hybrid human-machine interface for hands-free control of an intelligent wheelchair. *Int. J. Mechatron. Autom.* **2011**, *1*, 97–111.
6. Galan, F.; Nuttin, M.; Lew, E.; Ferrez, P.W.; Vanacker, G.; Philips, J.; Millan, J.R. A brain-actuated wheelchair: Asynchronous and non-invasive Brain-computer interfaces for continuous control of robots. *Clin. Neurophysiol.* **2008**, *119*, 2159–2169.
7. Rebsamen, B.; Guan, C.; Zhang, H.; Wang, C.; Teo, C.; Ang, M.J.; Burdet, E. A brain controlled wheelchair to navigate in familiar environments. *IEEE Trans. Neural Syst. Rehabil. Eng.* **2010**, *18*, 590–598.
8. Maskeliunas, R.; Simutis, R. Multimodal wheelchair control for the paralyzed people. *Electron. Electr. Eng.* **2011**, *111*, 81–84.
9. Hashimoto, M.; Takahashi, K.; Shimada, M. Wheelchair Control Using an EOG-and EMG-Based Gesture Interface. In Proceedings of IEEE/ASME International Conference on Advanced Intelligent Mechatronics, Singapore, 14–17 July 2009; pp. 1212–1217.
10. Santana, A.; Yang, C. Robotic control using physiological EMG and EEG signals. *Adv. Auton. Rob.* **2012**, *7429*, 449–450.
11. Brando, A.S.; Felix, L.B.; Bastos-Filho, T.F.; Sarcinelli-Filho, M. Controlling devices using biological signals. *J. Adv. Rob. Syst.* **2011**, *8*, 22–33.
12. Englehart, K.; Hudgins, B. A robust, real-time control scheme for multifunction myoelectric control. *IEEE Trans. Biomed. Eng.* **2003**, *50*, 848–854.
13. Moon, I.; Lee, M.; Chu, J.; Mun, M. Wearable EMG-Based HCI for Electric-Powered Wheelchair Users with Motor Disabilities. In Proceedings of the 2005 IEEE International Conference on Robotics and Automation (ICRA 2005), Barcelona, Spain, 18–22 April 2005; pp. 2649–2654.

14. Gandevia, S.C. Spinal and supraspinal factors in human muscle fatigue. *Physiol. Rev.* **2001**, *81*, 1725–1789.
15. Tamura, H.; Murata, T.; Yamashita, Y.; Tanno, K.; Fuse, Y. Development of the electric wheelchair hands-free semi-automatic control system using the surface-electromyogram of facial muscles. *Artif. Life Rob.* **2012**, *17*, 300–305.
16. Zhang, Q.; Hayashibe, M.; Fraisse, P.; Guiraud, D. FES-induced torque prediction with evoked EMG sensing for muscle fatigue tracking. *ASME Trans. Mechatron.* **2011**, *16*, 816–826.
17. VØllestad, N.K. Measurement of human muscle fatigue. *J. Neurosci. Methods* **1997**, *74*, 219–227.
18. Williams, C. *Human Muscle Fatigue*; Routledge: New York, NY, USA, 2009.
19. Oskoei, M.; Hu, H. Myoelectric control systems: A survey. *J. Biomed. Signal Process. Control* **2007**, *2*, 275–294.
20. Nikooyan, A.A.; Zadpoor, A.A. Effects of muscle fatigue on the ground reaction force and soft-tissue vibrations during running: A model study. *IEEE Trans. Biomed. Eng.* **2012**, *59*, 797–804.
21. Alty, S.R.; Georgakis, A. Mean Frequency Estimation of Surface EMG Signals Using Filterbank Methods. In Proceedings of the 19th European Signal Processing Conference (EUSIPCO 2011), Barcelona, Spain, 29 August–2 September 2011.
22. Gerdle, B.; Elert, J. The temporal occurrence of the mean power frequency shift of the electromyogram during maximum prolonged dynamic and static working cycles. *Int. J. Sports Med.* **1994**, *15*, 32–37.
23. Lucas, M.F.; Gaufriau, A.; Pascual, S.; Doncarli, C.; Farina, D. Multi-channel surface EMG classification using support vector machines and signal-based wavelet optimization. *J. Biomed. Signal Process. Control* **2008**, *3*, 169–174.
24. Kawano, S.; Okumura, D.; Tamura, H.; Tanaka, H.; Tanno, K. Online learning method using support vector machine for surface-electromyogram recognition. *J. Artif. Life Rob.* **2009**, *13*, 483–487.
25. Vapnik, V. *The Nature of Statistical Learning Theory*; Springer: Heidelberg, Germany, 1999.
26. Chang, C.C.; Lin, C.J. LIBSVM: A library for support vector machines. *ACM Trans. Intell. Syst. Technol.* **2011**, *2*, 27:1–27:27.

12

Moving Object Localization Using Sound-Based Positioning System with Doppler Shift Compensation

Slamet Widodo [1,*], Tomoo Shiigi [1], Naoki Hayashi [1], Hideo Kikuchi [2], Keigo Yanagida [2], Yoshiaki Nakatsuchi [3], Yuichi Ogawa [1] and Naoshi Kondo [1]

[1] Graduate School of Agriculture, Kyoto University, Kitashirakawa, Oiwake-cho, Sakyo-ku, Kyoto 606-8502, Japan; E-Mails: tshiigi@kais.kyoto-u.ac.jp (T.S.); naoki.h@kw4.ecs.kyoto-u.ac.jp (N.H.); ogawayu@kais.kyoto-u.ac.jp (Y.O.); kondonao@kais.kyoto-u.ac.jp (N.K.)

[2] System Watt Co., Ltd. 4-1-38, Isobedori, Chuo-ku, Kobe 651-0084, Japan; E-Mail: kikuchi@watt.co.jp (H.K.); yanagida@yanago.co.jp (K.Y.)

[3] The New Industry Research Organization (NIRO), 1-5-2, Minatojima-minamimachi, Chuo-ku, Kobe 650-0047, Japan; E-Mail: nakatsuchi@niro.or.jp

* Author to whom correspondence should be addressed; E-Mail: widodo@kais.kyoto-u.ac.jp

---

Abstract: Sound-based positioning systems are a potential alternative low-cost navigation system. Recently, we developed such an audible sound-based positioning system, based on a spread spectrum approach. It was shown to accurately localize a stationary object. Here, we extend this localization to a moving object by compensating for the Doppler shift associated with the object movement. Numerical simulations and experiments indicate that by compensating for the Doppler shift, the system can accurately determine the position of an object moving along a non-linear path. When the object moved in a circular path with an angular velocity of 0 to 1.3 rad/s, it could be localized to within 25 mm of the actual position. Experiments also showed the proposed system has a high noise tolerance of up to −25 dB signal-to-noise ratio (SNR) without compromising accuracy.

Keywords: Doppler shift; moving object localization; spread spectrum; agricultural robot

---

1. Introduction

Automated robotic systems in agriculture have been studied extensively in the last few decades [1,2]. It has been applied to various agricultural operations, such as weeding [3,4], plant monitoring [5], application of fertilizer and chemicals [6] and harvesting [7–9]. However, to date, there have been limited commercial applications of these agricultural robots; mainly due to their high cost. Among the components that make up an automated agricultural robot, the navigation system is a major contributor to the total development cost. A reduction in the price of the navigation system, combined with optimum operating time, will enable autonomous agricultural vehicles to become economically feasible [10]. Thus, the development of a low cost navigation system is critical to realizing the wide spread use of commercial agricultural robots.

In existing agricultural vehicle navigation systems (see [11] for a brief review), a Real Time Kinematics Global Positioning System (RTK-GPS) is the most widely used for determining vehicle position. It can provide very high positioning accuracy. However, due to the high price of RTK-GPS, it is also considered to be the main component contributing to the increased cost of these robotic systems [12]. Therefore, finding an alternative low cost positioning system is highly sought after. The use of a sound-based local positioning system has the potential to achieve this. Compared to other alternative positioning systems, it offers a number of merits. It can provide high positioning accuracy at a relatively low cost. There have been many reports of the development of sound-based positioning systems in an indoor environment, such as Active Bat [13], Cricket [14,15], Dolphin [16–18] and 3D-Locus [18,19]. However, only a limited number have documented the use of such a system in an outdoor environment.

In the current research, we develop an alternative outdoor navigation system for agricultural autonomous vehicles using a sound-based positioning system. Although ultrasound has been widely used in previous research, we use a spread spectrum sound. There are a number of characteristics that recommend such an approach for this application. It has high noise tolerance, as well as high signal identification properties [20]. Also, as mentioned in [21], high measurement accuracy can be achieved using this kind of system, due to the large bandwidth and continuous nature of the spread spectrum approach.

To date, this approach has been applied to non-agricultural settings [18,19,22–25]. In the current research, we optimize the system to an agricultural setting, where conditions are substantially different from previous applications. For instance, in order to obtain high positioning accuracy, there are many factors that need to be taken into account, such as temperature gradient, wind, background noise and the presence of obstacles. We used spread spectrum sound in which part of its spectrum is in the range of audible sound. Hence, it gives an advantage of the low frequency sound waves for handling the obstacles issue [22,26]. Although it will generate additional noise, it is considered to be more applicable to an agricultural setting, where many obstacles are present.

In previous work [27], we found that high positioning accuracy can be achieved using this positioning system for a stationary object. Within a 30 m × 30 m outdoor area, on average, the position of a stationary object was estimated with accuracy around 20 mm. This accuracy level is sufficient for many applications, including navigation of autonomous agricultural vehicles. The next step is to develop a system for localization of a dynamic/moving object. Such a sequential development process has been used effectively to develop other systems [21,28].

To accurately locate an object, it is first necessary to accurately measure distances to that object. In the proposed sound-based positioning system, distance is calculated by measuring the time of flight (TOF) of the sound wave to or from the object and then multiplying by the sound velocity. When an object is in motion, there is a frequency shift of the transmitted sound wave, due to the Doppler Effect. Thus, the correlation calculation between received signal and replica signal will be affected, resulting in an inaccurate estimation of the TOF and, therefore, distance estimation. This will in turn mean a decreased accuracy in localization. Even more, for faster moving objects, a system without Doppler compensation cannot be expected to accurately localize the object.

There are alternative solutions for moving object localization by using sound wave proposed in previous works. Among them, filtering methods, such as the Kalman filter [21] and particle filter [28,29], are the most popular. In [25], a tracking method with a limited range of correlation calculation was proposed for moving object localization using a spread spectrum ultrasonic wave. In their work, they used a limited correlation range instead of full range to prevent decreasing of the correlation value. As stated, using this approach, a moving object with a speed of 0.2 m/s could be realized within an average error of 50 mm. Another approach was proposed in [30]. They develop a Doppler-tolerant receiver for an ultrasonic positioning system by using the Kasami sequence. Simulation results showed the capability of the system to detect the signal emitted by moving a device with a velocity up to 3 m/s. The real experiment results unfortunately did not show the actual best performance that can be achieved, due to manufacturing defect and problems for performance evaluation when the robot is turning.

In this paper, we propose an alternative approach for accurate localization of moving objects by compensating for the Doppler shift associated with the movement. This approach is similar to that used in [31,32]. Beside the methodology for compensating for the Doppler shift associated with moving objects, we also present performance evaluation results using this proposed localization system. Furthermore, we also investigated the potential of noise to interfere with the performance of the proposed system.

2. Spread Spectrum Sound-Based Positioning System

2.1. System Configuration

Figure 1(a) shows the configuration of the spread spectrum sound-based positioning system used. The proposed system applies an inverse-GPS configuration, where four microphones are installed at known positions as fixed nodes and one omni-directional speaker is installed on a mobile platform (mobile station). Photos of the microphones and omni-directional speaker used in this work are shown in Figure 1(b,c).

This equipment consisted of silicon microphones (MP0404UD, Knowles Electronics, Tokyo, Japan), a speaker (FT28D, Fostex Company, Tokyo, Japan), a sound interface (Octa-Capture, Roland Corporation, Hamamatsu, Japan), an amplifier (Kama Bay Amp Rev. B, Scythe Inc., Ichikawa, Japan) and a personal computer (Windows XP, Core 2 duo processor 2.66 GHz and 3 GB RAM). For simplicity, a wired system was used here. In a real life application, this would be replaced by a wireless system. Each microphone and mobile station is equipped with a digital thermometer to monitor surrounding temperature, so that sound velocity could be adjusted for variation with temperature.

Figure 1. (**a**) Configuration of spread spectrum sound-based local positioning system. (**b**) Microphone. (**c**) Omni-directional speaker.

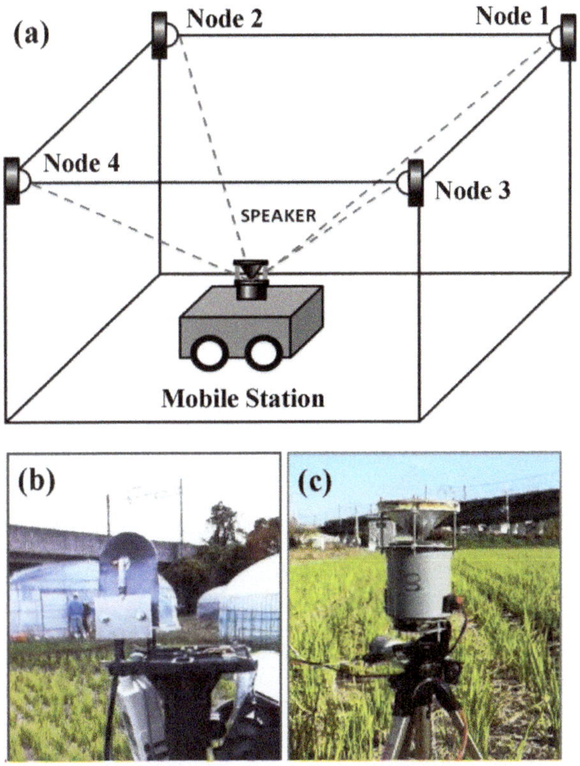

The spread spectrum sound was created from a maximum length sequence (M-sequence), pseudo random code with Binary Phase Shift Keying (BPSK) modulation [33]. The main parameters of the emitted signal were the carrier frequency, the length of the sequence and the number of carrier cycles per code chip (chip rate). For signal detection, the sampling frequency and sampling bit, as well as other parameters were set as shown in Table 1. To synchronize signal reception for all microphones, especially for the arrival time calculation, a trigger signal is also sent at the time of the spread spectrum sound emitted from the speaker, as depicted in Figure 2.

Table 1. Property of spread spectrum sound. M-sequence, maximum length sequence; BPSK, Binary Phase Shift Keying.

Property	Value/Remark
Number of sound wave	1
Sampling frequency	96 kHz
Sampling bit	16 bits
M-sequence length	1,023
Modulation	BPSK
Carrier wave frequency	24 kHz

Figure 2. Spread spectrum sound emission and trigger signal for synchronization.

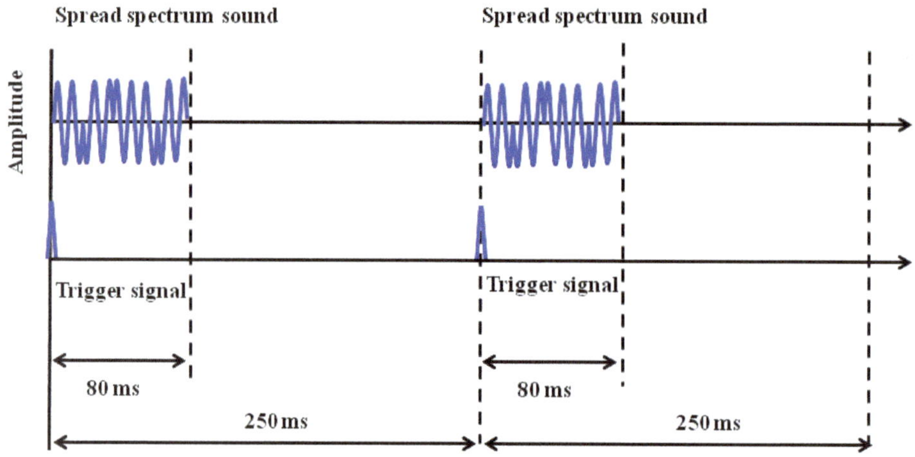

2.2. Position Estimation Method

To estimate the position of an object in three-dimensional space, it is necessary to have at least three distances from that object to known fixed nodes. Because our system used an inverse GPS configuration, position estimation was as follows: when the speaker on the mobile station is located at an unknown position (x_m, y_m, z_m) and the four nodes are located at known positions, $R_i = (x_i, y_i, z_i)$, with i denoting each of the four nodes. Using the estimated distance from the speaker to each node (d_{mi}):

$$d_{mi} = \sqrt{(x_m - x_i)^2 + (y_m - y_i)^2 + (z_m - z_i)^2} \tag{1}$$

$$d_{mi} = v_s \times \Delta t_{mi} \tag{2}$$

where v_s is sound velocity and Δt_{mi} is the TOF of the sound wave, the position of the speaker can be obtained by solving these equations. Here, a least squares method was used to solve the equation. In this current system, it takes 150 ms to estimate the position. However, to accommodate longer distance measurement, the period of updating position is set as 250 ms.

To calculate TOF, the trigger and arrival time of the signal were first estimated using cross-correlation. Then, TOF was calculated as the difference between these two times. Thus, determining detection time, obtained by detecting the peak correlation value, was critical, as this determines the accuracy of the distance measurement, as well as positioning accuracy. This will be elaborated on below.

3. Doppler Shift Compensation

Signal arrival time was obtained by calculating a cross-correlation value, C_i, from the received signal, $r_i(n)$, and a replica of the transmitted signal, $s(n)$, using Fourier transform:

$$C_i = \sum_{n=0}^{N-1} s(n)\, r_i(n+t) \tag{3}$$

where $n = 0, 1, 2, \ldots, N-1$ with N is the number of samples used in the Fourier transform calculation. For efficient calculation, a Fast Fourier Transform (FFT) was used to calculate the correlation value as follows:

$$C_i = IFFT\left(conj.\left(S(n)\right) R_i(n)\right) \tag{4}$$

where $conj.(S(n))$ and $R(n)$ are the complex conjugate of the FFT of $s(n)$ and the complex spectrum of FFT for $r_i(n)$, respectively. IFFT denotes the Inverse Fast Fourier Transform.

After calculating cross-correlation, a threshold operation was applied to get peak values. The signal arrival time is then estimated from the first detected peak. The threshold value C_{th} is set as:

$$C_{th} = \frac{C_{\max} + C_{ave} + 3\,\sigma_{corr}}{2} \tag{5}$$

where C_{max}, C_{ave} and σ_{corr} are the maximum value, average and standard deviation of cross-correlation. Based on the three-sigma rule, for normally distributed data, 99.7% data can be represented by $C_{ave} + 3\ \sigma_{corr}$. However, there is no information about the data distribution of the correlation signal. Therefore, C_{max} value was added. We used half of this value as the threshold, and it works well for our system.

To compensate for the Doppler shift, another sound wave is added to the spread spectrum sound. Thus, the transmitted sound wave $s(n)$ is changed from Equation (6) into Equation (7):

$$s(n) = \sin\left(\frac{2\pi f_c n}{fs}\right) \times M(m) \tag{6}$$

$$s(n) = \sin\left(\frac{2\pi f_c n}{fs}\right) \times M(m) + \sin\left(\frac{2\pi f_{ds} n}{f_s}\right) \tag{7}$$

with:

$$m = round\left(\frac{f_c}{2f_s} n\right) \tag{8}$$

where f_c = frequency of carrier wave, f_s = sampling frequency and f_{ds} = frequency of sound wave used for detecting Doppler shift. Here, f_{ds} is set to 36 kHz.

The Doppler shift is then estimated by detecting the maximum value of the power spectrum (f_{ds_max}) in the range 35.84–36.16 kHz. This range can be used to detect a frequency shift caused by an object moving at a speed of up to 1.5 m/s. This speed is thought to be commensurate with general agricultural operations. FFT was calculated from 16,384 samples for the FFT calculation, giving a 5.86 Hz frequency resolution for Doppler shift estimation. The Doppler shift can be calculated as follows:

$$\Delta f_{ds} = f_{ds\_\max} - f_{ds} \tag{9}$$

Using this information, a new replica of the transmitted signal is re-generated using Equation (7), replacing f_c with the new value (f_{c_new}), as follows:

$$f_{c\_new} = f_c + \left(\frac{f_c}{f_{ds}}\right) \Delta f_{ds} \tag{10}$$

A cross-correlation is then calculated using this new replica signal and the detected signal using Equation (4). A summary of the proposed Doppler shift compensation is depicted in Figure 3.

Figure 3. Schematic of the proposed Doppler-shift compensation method.

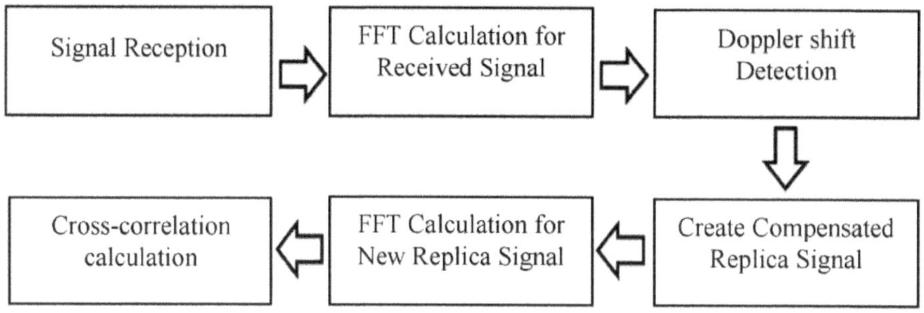

4. Experimental Setup

To evaluate the performance of the proposed method, we conducted several experiments to localize a moving object in one dimension (1D) and two dimensions (2D).

4.1. The Influence of Moving Speed

In the first experiment (Figure 4), the influence of moving speed on the performance of the proposed method was examined. An omni-directional speaker was placed on a conveyor as the object to be localized. The conveyor could move at variable speeds in both directions: toward and away from the microphone. In this experiment, the distance to the moving speaker (with respect to the microphone) was measured periodically using the spread spectrum sound system, with and without Doppler shift compensation.

We compared the localization performance in terms of position error, as well as signal identification, as indicated by the signal-to-noise ratio of cross-correlation (correlation SNR or SNR_{Corr}). First, the incoming signal (C^*_{sig}) was detected for the case of Doppler shift compensation, and correlation SNR was calculated as:

$$Noise^* = \frac{\left(\sum_{i=0}^{N-1} C_i^*\right) - C^*_{sig}}{N-1}\text{; for } C_i > 0 \tag{11}$$

$$SNR^*_{Corr} = \frac{C^*_{sig}}{Noise^*} \tag{12}$$

The asterisk symbol (*) denotes a system with Doppler shift compensation. Using the same sample number as C^*_{sig} (the i-th sample, where $C^*_i = C^*_{sig}$), then the same procedure was used to calculate the correlation SNR of the non-compensated system. The experiments were conducted indoors to prevent other undesirable influences, such as wind and noise, from interfering. We also compared the result with theoretical values obtained from numerical simulations.

Figure 4. Experimental setup for the moving object localization in one dimension (1D).

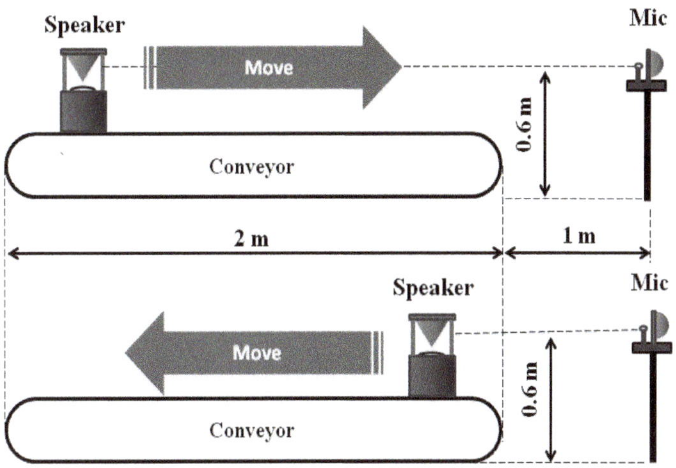

4.2. Moving Object Localization Test

To evaluate the performance of the proposed method to localize an object moving along a non-linear path, we developed a testbed as illustrated in Figure 5. An omni-directional speaker was mounted on a mobile platform that could move in a circular path with an angular velocity (ω) between 0–1.3 rad/s. A stepping motor (Plexmotion SSA-PR-42D4, Shinano Kenshi Co., Ltd., Nagano, Japan) was used to generate the motion, and the desired angular velocity was achieved by controlling the input signal sent to the motor. As the floor was relatively flat, only positioning in 2D space was considered in this evaluation. Trials at different angular velocities were performed to evaluate the influence of object speed on localization performance.

Figure 5. A testbed for moving object localization using spread spectrum sound.

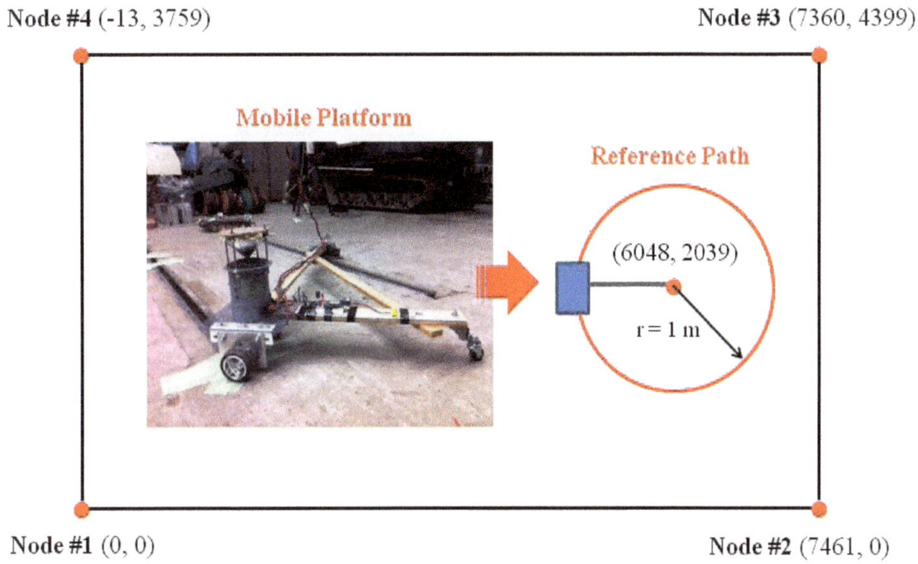

4.3. The Influence of Noise

Background noise is common in an agricultural setting, especially in open fields. Therefore, we also investigated the influence of background noise. Using the same setup as in Figure 4, white noise sound with a different sound signal-to-noise ratio (sound SNR) was introduced to simulate background noise. Sound SNR is calculated as:

$$sound\ SNR = 20\log_{10}\left(\frac{sound}{noise}\right) \tag{13}$$

Seven noise levels were trialed: 0, −5, −10, −15, −20, −25 and −30 (all values in dB). The performance was then evaluated by using correlation SNR, as described in Equations (11) and (12).

5. Results and Discussion

In this section, both simulation and experimental results are presented. The simulation results were conducted to give an extended perspective on the experiment results obtained.

5.1. The Influence of Moving Speed

Figure 6 shows a profile of the distance to the moving speaker with respect to the microphone in a 1D case. From a start position, the speaker moves at a set speed for a few seconds and then stops. While stationary, the distance to the object was accurately determined. However, when the object is moving, the localization estimates obtained without compensating for the Doppler shift were quite inaccurate. When the Doppler shift was compensated for, there was accurate localization of the object. Due to limitations with the conveyor's speed controller, only two speeds were used in this experiment: 0.2 m/s and 0.8 m/s.

To further analyze the influence of object speed, numerical simulations in Matlab were conducted. As shown in Figure 7, the correlation value of the detected signal obtained from the numerical simulation was in good agreement with the experimental results. This indicates simulations can be used to extend the analysis of the influence of object speed on the proposed localization method.

Figure 6. Profile of distance travelled by the moving speaker: (**a**) speaker moves toward the microphone; (**b**) speaker moves away from the microphone.

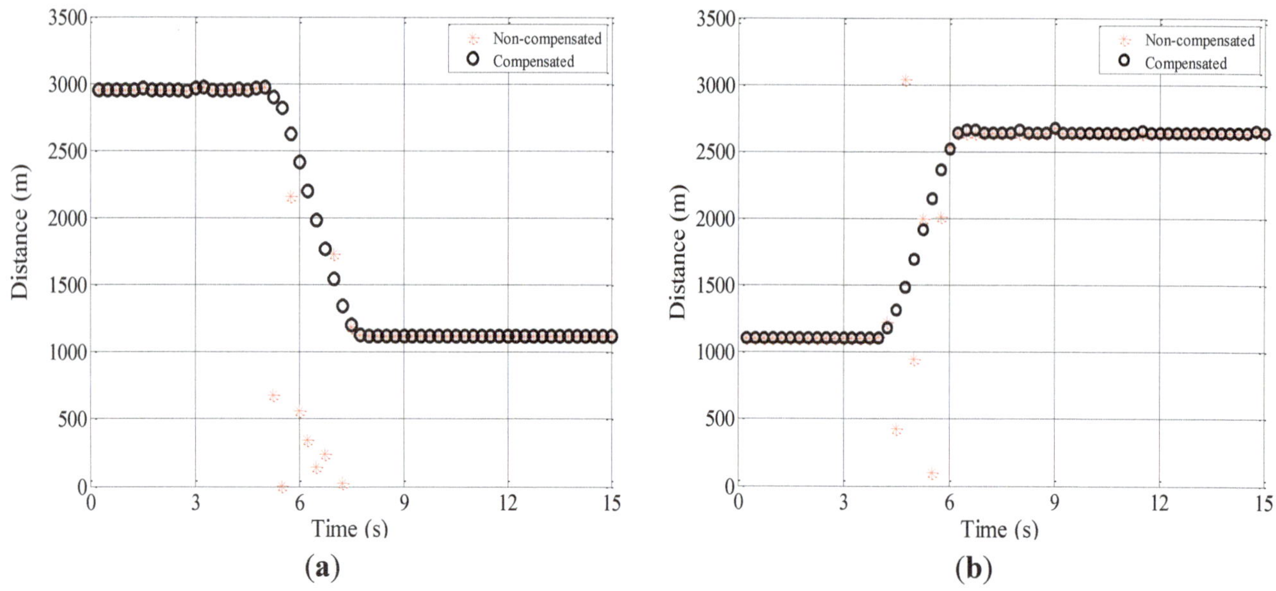

Figure 7. Correlation value of received signal for moving speed 0.2 m/s. (**a**) without Doppler shift compensation; (**b**) without Doppler-shift compensation; (**c**) with Doppler shift compensation; and (**d**) with Doppler shift compensation. Real experiment: (a,c); simulation (b,d).

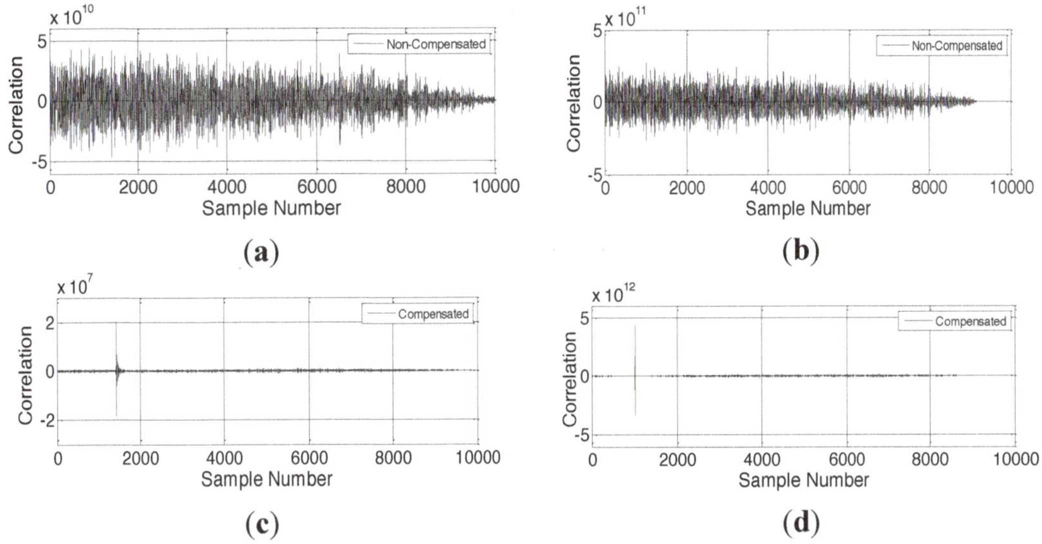

The correlation SNR, as described in Equations (11) and (12), is used to compare localization performance with and without Doppler shift compensation. This value indicates the ease with which peak correlation could be detected, which is critical for distance measurements using sound waves. The higher the correlation SNR value, the easier the incoming signal could be detected. As shown in Figure 8, when attempting to localize without compensation, the correlation SNR values decreased with increasing object speed. For localization with Doppler shift compensation, there was no significant influence of object speed. This result concurs with the experimental results shown in Figure 6. When the object (speaker) was moving at 0.8 m/s, the correlation SNR was almost zero. Thus, it was difficult to detect the incoming signal correctly, and as a result, it failed to provide a correct position for the object when there was no Doppler shift compensation.

Figure 8. Comparison of correlation signal-to-noise ratio (SNR) for different moving speed with and without Doppler shift compensation.

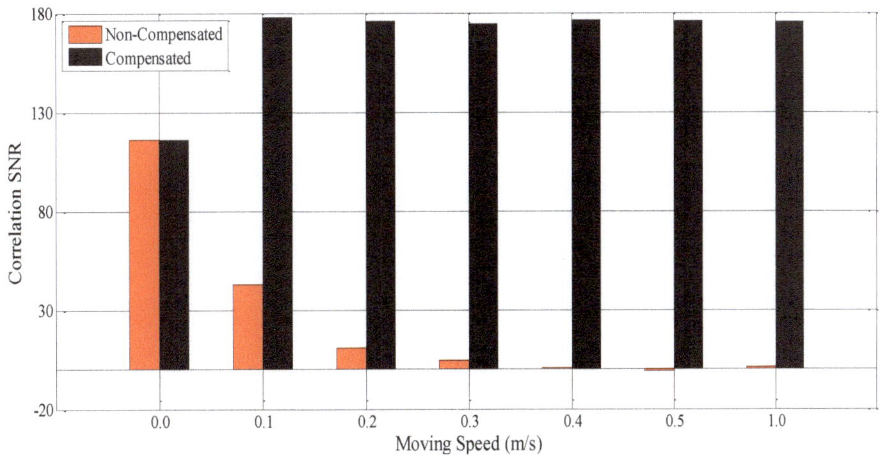

5.2. Moving Object Localization Test

The localization results for an object moving in a circular path at various angular velocities are shown in Figure 9. Generally, the position was accurately estimated. No performance degradation was observed, even when the object was moving at a relatively high speed (1.3 rad/s).

However, there was an issue with regards to outliers. An occasional incorrect measurement was obtained, presumably due to a disturbance, such as reflection, interference and so on (see Figure 9). Further analysis showed that this problem was generated by erroneous distance measurements. Figure 10 shows a profile of the measured distances to the node 3, where erroneous distance measurements were observed.

Figure 9. Examples of outlier (marked with red circle) as the result of inaccurate distance measurement: (**a**) ω = 0.3 rad/s and (**b**) ω = 1.3 rad/s.

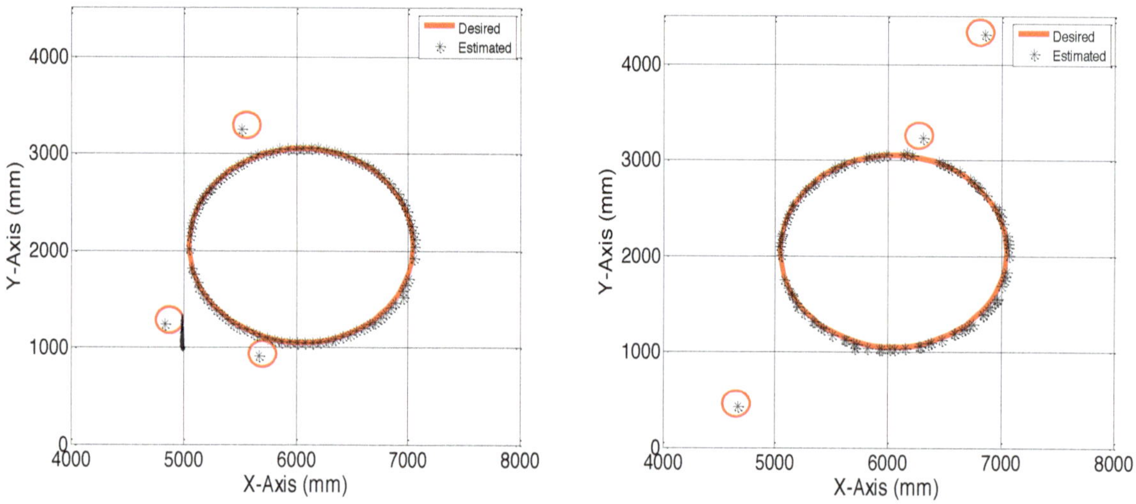

(**a**) (**b**)

Figure 10. Examples of wrong distance measurements (marked with red circle) at different angular velocity: (**a**) ω = 0.3 rad/s and (**b**) ω = 1.3 rad/s.

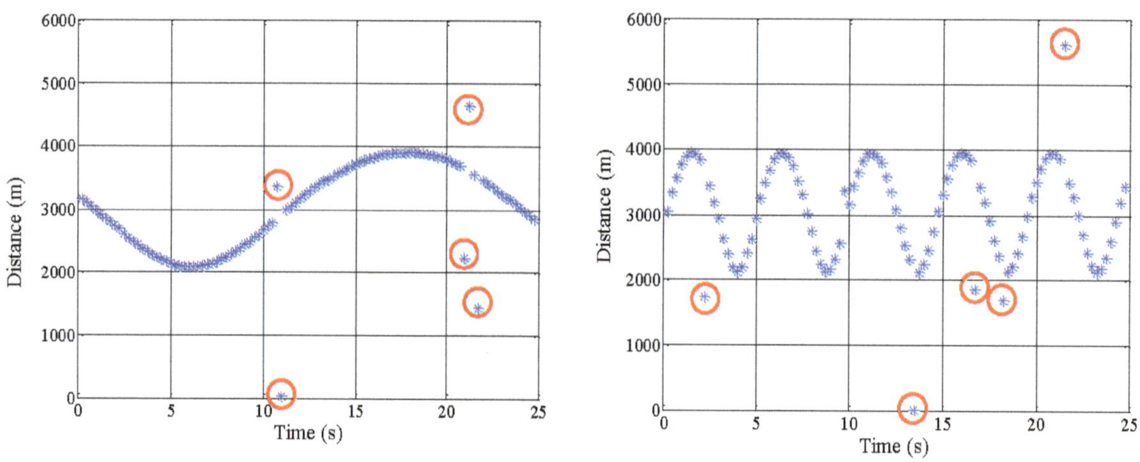

(**a**) (**b**)

This problem was also seen in other development systems [21,25]. To overcome this problem, outlier rejection is necessary, regardless of the cause. We implemented an outlier rejection procedure as follows:

(1) Get the current distance from the speaker to each fixed microphones (nodes): $d_{m/k} = [d_{m1/k}, d_{m2/k}, d_{m3/k}, d_{m4/k}]$ and Doppler-shift $\Delta f_k = [\Delta f_{1/k}, \Delta f_{2/k}, \Delta f_{3/k}, \Delta f_{4/k}]$.

(2) Estimate the velocity the object is moving towards or away from each microphone $v_{m/k} = [v_{m1/k}, v_{m2/k}, v_{m3/k}, v_{m3/k}]$ using the following equation:

$$\Delta f_{i/k} = \left(\frac{v_s}{v_s - v_{mi/k}} - 1 \right) f_0 \tag{14}$$

where $f_0 = 36$ kHz and c is the velocity of sound, which depends on temperature (T) as follows:

$$v_s = 331.5 + 0.61\,T \tag{15}$$

(3) From the previous valid distance, $d_{m/k-1}$ and $v_{s/k}$, the predicted distance at the next time step, $d^*_{m/k} = [d^*_{m1/k}, d^*_{m2/k}, d^*_{m3/k}, d^*_{m4/k}]$ can be estimated using a kinematics model:

$$d^*_{mi/k} = d_{m/k-1} + v_{si/k}\,\Delta t_s \tag{16}$$

where Δt_s is sampling time.

(4) In order to validate a new data point, a standard deviation of measurement error associated with each fixed microphone (σ_i) is set, and the measured distance ($d_{m/k}$) is compared with the predicted distance ($d^*_{m/k}$). If $d^*_{mi/k} - 5\sigma_i \leq d_{mi/k} \leq d^*_{m/k} + 5\sigma_i$ then $d_{mi/k}$ is a valid value, otherwise it is considered an outlier and rejected.

Note: The value of σ_i is obtained from measurement results in previous research. Here, the tolerance value is set to $\pm 5\sigma_i$ to moderate outlier rejection (statistically, for normally distributed data, 99.7% of data will be in the range of $d^*_{mi/k} \pm 3\sigma_i$).

(5) To estimate the position of an object in 3D space, three or more valid distances are required. Therefore, after validating the measured distances, if the number of valid distances is ≥3, then the position of the speaker is updated. Otherwise, the update is postponed until the required number of valid distances can be obtained.

Figure 11 shows localization performance after implementing the outlier rejection procedure. In these figures, only the data of estimated position for one revolution are shown. Despite this, the performance was evaluated from several revolutions data. For real life navigation of autonomous vehicles, this system can be combined with an odometer to provide additional information to validate and determine the position in between data updates. Thus, data gaps, due to outlier rejection, can be accommodated. This approach is also widely used for GPS-based navigation system, e.g., the system used in [34].

Position errors relative to object moving speed are shown in Figure 12. Where ω was between 0.3–1.3 rad/s, the position error was around 20 mm. This performance is better than that reported in another experiment [28]. Even for faster moving objects, the measured errors were smaller than that previously reported. Beside the advantage of using a spread spectrum approach, the main reason for this difference in performance was related to the system configuration. In that previous report, they used a GPS configuration (passive mobile system), while in our system, we used an inverse-GPS configuration (active mobile system). The main limitation of passive systems for localization of a moving object is that simultaneous measurement of distance cannot be guaranteed [21]. Usually, the distances are measured one at a time; in between measurements, the object has moved on from one measuring point to another. As a result, there is a high dependency between positioning accuracy and object moving speed. The faster the object moves, the more positioning accuracy deteriorates.

Figure 11. Localization of an object moving in a circular path with variable velocity. (**a**) ω = 0.3 rad/s; (**b**) ω = 0.7 rad/s; (**c**) ω = 1.0 rad/s; (**d**) ω = 1.3 rad/s.

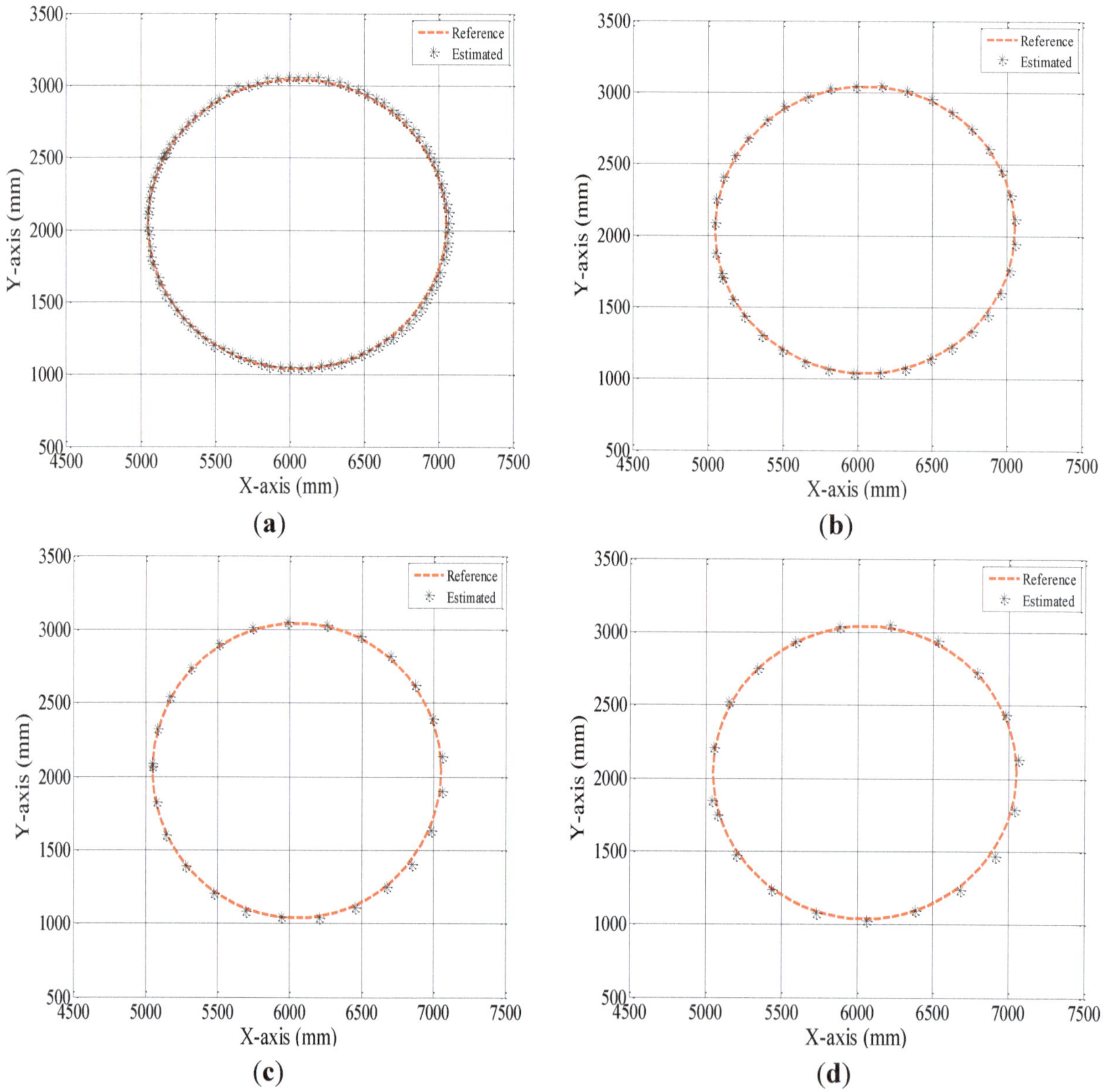

In [28], two different frequencies were used for the ultrasonic positioning system. This system has both the advantage of a faster measurement time and minimized problems with simultaneous measurement. However, because four beacons were used, it means that two successive measurements were still needed to get all four distances, thus the problem of simultaneous measurements is retained. This could contribute to increased position inaccuracy in such a system, as the angular velocity of the object increases.

Figure 12. Relation of angular velocity of moving object and the localization error.

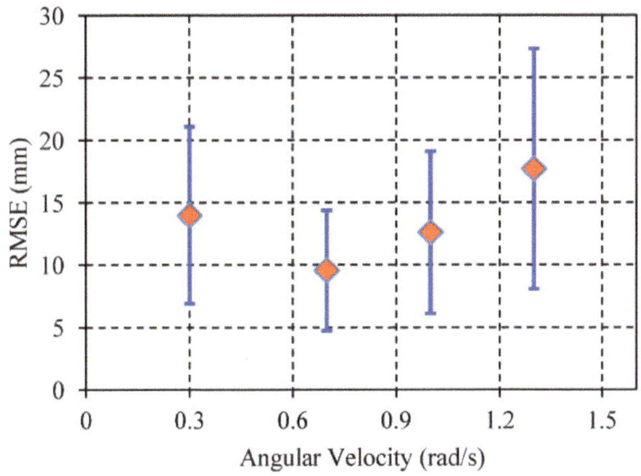

An active system does, however, have its limitations. There is a problem related to network infrastructure as the number of mobile station increases and also a privacy issue [21]. This, however, is not an issue here, because the system is used for navigation or other possible applications with limited numbers of mobile stations. In any event, privacy may not be a critical issue for those applications.

Another important issue is related with the possibility of performance degradation for application at a wider area and at a higher moving speed. Even though performance evaluation of the proposed method was conducted in a relatively small area, the obtained results show that the accuracy of the moving object localization is close to the accuracy of the distance measurement. Based on this result and also the results from our previous work [27], the proposed method is expected to work well at a wider area without significantly decreasing the performance. With respect to moving speed, there is no clear trend from the plot shown in Figure 12. However, considering high accuracy at the current highest speed (~1.3 m/s), it is plausible that the proposed method also has a possibility to be applied at a higher speed. Indeed, the speeds used in this experiment are sufficient for general agricultural application and, therefore, may not be a critical issue here.

5.3. The Influence of Noise

For real life applications in agriculture, background noise is likely to be present. Intense noises may come from a variety of sources, especially agricultural machinery [35]. Thus, it is necessary to evaluate the performance of the proposed method in the presence of noise. We tested localization performance at various noise levels (*i.e.*, at variable sound SNR values) ranging from 0 to −30 dB. We also compared the performance for both stationary and moving objects.

When the object was stationary (Figures 13(a) and 14(a)), the object could be accurately localized, with or without compensation, up to a −25 dB noise level. Clear identification of the signal was achieved as the high correlation SNR indicates. However, for a moving object (Figures 13(b) and 14(b)), for the non-compensated system, there was a very low correlation SNR. On the other hand, although there was some decrease in the correlation SNR for the compensated system compared to when the object was not moving, it still accurately localized the moving object. This indicates that the presence of background noise did not significantly interfere with the performance of the system with the Doppler shift compensation. The background noise limit that the system could tolerate was −25 dB.

Beyond this value, interference to the system was too large, regardless of whether the object was stationary or moving.

Figure 13. Correlation SNR for compensated and non-compensated system: (**a**) stationary; (**b**) moving object (0.2 m/s).

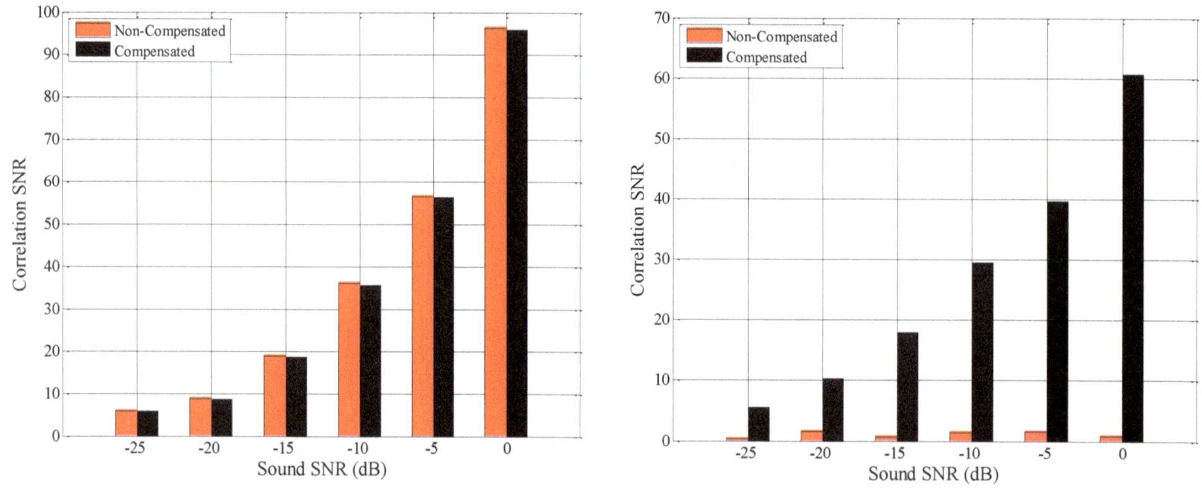

(**a**) (**b**)

Figure 14. Correlation SNR for compensated and non-compensated system: (**a**) stationary; (**b**) moving condition (0.8 m/s).

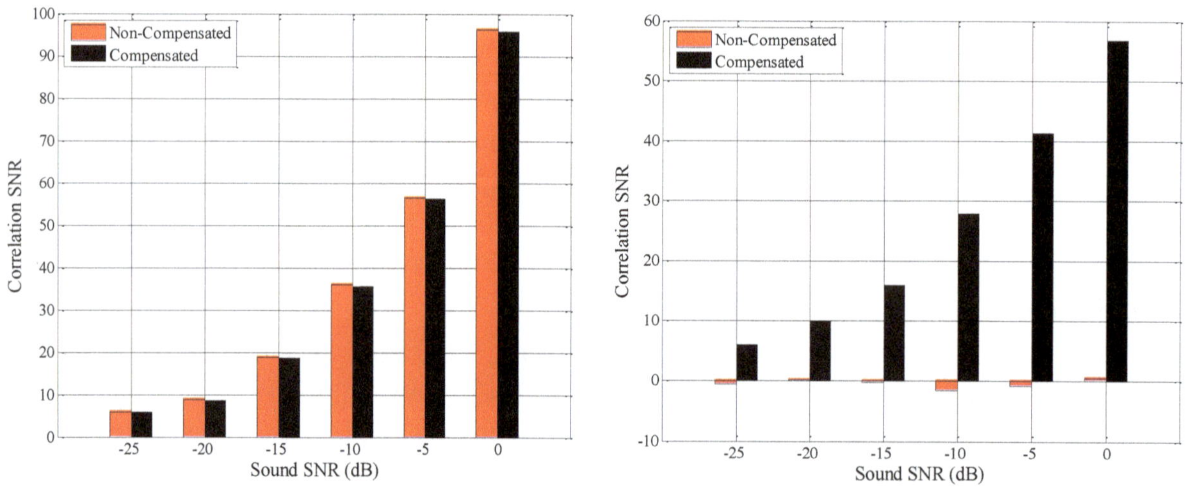

(**a**) (**b**)

6. Conclusions

This paper evaluates an alternative solution for localizing a moving object by using a spread spectrum sound-based positioning system with Doppler shift compensation. By taking advantage of a spread-spectrum approach, which is capable of precise distance measurements; and an active mobile architecture, experimental results demonstrate that the proposed system can achieve high positioning accuracy. Investigations also indicate that the proposed system can tolerate background noise up to −25 dB without compromising accuracy. To further improve the current system, estimation of the position using filtering methods, such as a Kalman filter or a particle filter, could be investigated. For practical applications, the limitation of the proposed system in terms of the position updating rate could be overcome by integration with an odometer and/or inertial measurement unit. Another important issue is about the scalability of the system for wider area (*i.e.*, using more devices). In this case, the development of an auto-calibrated system is highly required and will be a point of interest for further improvement.

Acknowledgments

The authors would like to express their gratitude to Tateshi Fujiura and Josse De Baerdemaeker for their valuable discussion and suggestion. We are also grateful to Garry Piller for checking this manuscript. This work is supported by MAFF (the Ministry of Agriculture, Forestry and Fisheries of Japan).

References

1. Kondo, N.; Ting, K.C. *Robotics for Bioproduction Systems*; American Society of Agricultural Engineers: Ann Arbor, MI, USA, 1998.
2. Edan, Y.; Han, S.; Kondo, N. Automation in agriculture. In *Springer Handbook of Automation*; Springer: Berlin, Germany, 2009; pp. 1095–1128.
3. Bakker, T.; van Asselt, K.; Bontsema, J.; Müller, J.; van Straten, G. An autonomous weeding robot for organic farming. In *Field and Service Robotics*; Corke, D.P., Sukkariah, D.S., Eds.; Springer: Berlin, Germany, 2006; pp. 579–590.
4. Slaughter, D.C.; Giles, D.K.; Downey, D. Autonomous robotic weed control systems: A review. *Comput. Electron. Agr.* **2008**, *61*, 63–78.
5. Acaccia, G.M.; Michelini, R.C.; Molfino, R.M.; Razzoli, R.P. Mobile Robots in Greenhouse Cultivation: Inspection and Treatment of Plants. In Proceesings of the the 1st International Workshop on Advances in Service Robotics, Bardolino, Italy, 13–15 March 2003.
6. Søgaard, H.T.; Lund, I. Application accuracy of a machine vision-controlled robotic micro-dosing system. *Biosyst. Eng.* **2007**, *96*, 315–322.
7. Van Henten, E.J.; van Tuijl, B.A.J.; Hemming, J.; Kornet, J.G.; Bontsema, J.; van Os, E.A. Field test of an autonomous cucumber picking robot. *Biosyst. Eng.* **2003**, *86*, 305–313.
8. Tanigaki, K.; Fujiura, T.; Akase, A.; Imagawa, J. Cherry-harvesting robot. *Comput. Electron. Agr.* **2008**, *63*, 65–72.
9. Hayashi, S.; Shigematsu, K.; Yamamoto, S.; Kobayashi, K.; Kohno, Y.; Kamata, J.; Kurita, M. Evaluation of a strawberry-harvesting robot in a field test. *Biosyst. Eng.* **2010**, *105*, 160–171.
10. Goense, D. The Economics of Autonomous Vehicles. In Proceedings of the VDI-MEG Conference on Agricultural Engineering, VDI-Tagung Landtechnic, Hannover, Germany, 7–8 November 2003.
11. Li, M.; Imou, K.; Wakabayashi, K.; Yokoyama, S. Review of research on agricultural vehicle autonomous guidance. *Int. J. Agric. Biol. Eng.* **2009**, *2*, 1–16.
12. Pedersen, S.M.; Fountas, S.; Have, H.; Blackmore, B.S. Agricultural robots—System analysis and economic feasibility. *Precis. Agr.* **2006**, *7*, 295–308.

13. Harter, A.; Hopper, A.; Steggles, P.; Ward, A.; Webster, P. The anatomy of a context-aware application. *Wireless Netw.* **2002**, *8*, 187–197.
14. Priyantha, N.B.; Chakraborty, A.; Balakrishnan, H. The Cricket Location-Support System. In Proceeding of the 6th Annual International Conference on Mobile Computing and Networking, Boston, MA, USA, 6–11 August 2000; pp. 32–43.
15. Balakrishnan, H.; Priyantha, N.B. The Cricket Indoor Location System: Experience and Status. In Proceedings of the 2003 Workshop on Location-Aware Computing, Seattle, WA, USA, 12 October 2003; pp. 7–9.
16. Hazas, M.; Ward, A. A High Performance Privacy-Oriented Location System. In Proceeding of the 1st IEEE International Conference on Pervasive Computing and Communications, Fort Worth, TX, USA, 23–26 March 2003; pp. 216–233.
17. Hazas, M.; Hopper, A. Broadband ultrasonic location systems for improved indoor positioning. *IEEE Trans. Mobile Comput.* **2006**, *5*, 536–547.
18. Prieto, J.C.; Jiménez, A.R.; Guevara, J.I.; Ealo, J.L.; Seco, F.A.; Roa, J.O.; Ramos, J.X. Subcentimeter-Accuracy Localization through Broadband Acoustic Transducers. In Proceeding of IEEE International Symposium on Intelligent Signal Processing 2007, Alcala de Henares, Spain, 3–5 October 2007; pp. 1–6.
19. Prieto, J.C.; Jiménez, A.R.; Guevara, J.I.; Ealo, J.L.; Seco, F.A.; Roa, J.O.; Ramos, F. Performance evaluation of 3D-LOCUS advanced acoustic LPS. *IEEE Trans. Instrum. Meas.* **2009**, *58*, 2385–2395.
20. Girod, L.; Estrin, D. Robust Range Estimation Using Acoustic and Multimodal Sensing. In Proceeding of the 2001 IEEE/RSJ International Conference on Intelligent Robots and Systems, Maui, HI, USA, 29 October–3 November 2011; pp. 1312–1320.
21. Smith, A.; Balakrishnan, H.; Goraczko, M.; Priyantha, N. Tracking Moving Devices with the Cricket Location System. In Proceeding of MobiSYS'04, Boston, MA, USA, 6–9 June 2004.
22. Vallidis, N.M. WHISPER: A Spread Spectrum Approach to Occlusion in Acoustic Tracking. Ph.D. Thesis, The University of North Carolina, Chapel Hill, NC, USA, 2002.
23. Suzuki, A.; Iyota, T.; Choi, Y.; Kubota, Y.; Watanabe, K.; Yamane, A. Measurement Accuracy on Indoor Positioning System Using Spread Spectrum Ultrasonic Waves. In Proceeding of the 4th International Conference on Autonomous Robots and Agents, Wellington, New Zealand, 10–12 February 2009; pp. 294–297.
24. Ito, N.; Suzuki, A.; Iyota, T. Verification of CDMA and Accuracy on Echo Ranging System Using Spread Spectrum Ultrasonic Signals. In Proceeding of the 5th International Conference on Automation, Robotics and Applications, Wellington, New Zealand, 6–8 December 2011; pp. 476–479.
25. Itagaki, Y.; Suzuki, A.; Iyota, T. Indoor Positioning for Moving Objects Using a Hardware Device with Spread Spectrum Ultrasonic Waves. In Proceeding of the 2012 International Conference on Indoor Positioning and Indoor Navigation, Sydney, Australia, 13–15 November 2012; pp. 1–6.
26. Mak, L.C.; Furukawa, T. A time-of-arrival-based positioning technique with non-line-of-sight mitigation using low-frequency sound. *Adv. Robot.* **2008**, *22*, 507–526.

27. Widodo, S.; Shiigi, T.; Than, N.M.; Kikuchi, H.; Yanagida, K.; Nakatsuchi, Y.; Ogawa, Y.; Kondo, N. Wind compensation using base station for spread spectrum sound-based positioning system in open field. *EAEF* **2013**, in press.
28. Ko, S.I.; Choi, J.S.; Kim, B.H. Indoor mobile localization system and stabilization of localization performance using pre-filtering. *Int. J. Contr. Autom. Syst*. **2008**, *6*, 204–213.
29. Muller, H.L.; McCarthy, M.; Randell, C. Particle Filters for Position Sensing with Asynchronous Ultrasonic Beacons. In Proceeding of the Second International Workshop, Dublin, Ireland, 10–11 May 2006; pp. 1–13.
30. Álvarez, F.J.; Hernández, Á.; Moreno, J.A.; Pérez, M.C.; Ureña, J.; de Marziani, C. Doppler-tolerant receiver for an ultrasonic LPS based on Kasami sequences. *Sens. Actuator. A Phys.* **2013**, *189*, 238–253.
31. Kusy, B.; Ledeczi, A.; Koutsoukos, X. Tracking Mobile Nodes Using RF Doppler Shifts. In Proceeding of the 5th ACM Conference on Embedded Networked Sensor Systems, Sydney, Australia, 6–9 November 2007.
32. Amundson, I.; Koutsoukos, X.; Sallai, J. Mobile Sensor Localization and Navigation Using RF Doppler Shifts. In Proceedings of the 1st ACM international Workshop on Mobile Entity Localization and Tracking in GPS-Less Environments, San Francisco, CA, USA, 14–19 September 2008.
33. Shiigi, T.; Kondo, N.; Tsuzuki, S.; Okada, S.; Maekawa, A.; Nobara, T.; Sakakibara, M.; Watanabe, K.; Naniwada, Y.; Okada, K. Position Detecting Method Using Spread Spectrum Sound: Correction Method of Measurement Error by Compensating Wind and Temperature. In Proceedings of the 3rd IFAC International Conference Agricontro, Kyoto, Japan, 6–8 December 2010.
34. Ohno, K.; Tsubouchi, T.; Shigematsu, B.; Yuta, S. Differential GPS and odometry-based outdoor navigation of a mobile robot. *Adv. Robot.* **2004**, *18*, 611–635.
35. McBride, D.I.; Firth, H.M.; Herbison, G.P. Noise exposure and hearing loss in agriculture: A survey of farmers and farm workers in the southland region of New Zealand. *J. Occup. Environ. Med.* **2003**, *45*, 1281–1288.

Permissions

All chapters in this book were first published in Robotics, by MDPI; hereby published with permission under the Creative Commons Attribution License or equivalent. Every chapter published in this book has been scrutinized by our experts. Their significance has been extensively debated. The topics covered herein carry significant findings which will fuel the growth of the discipline. They may even be implemented as practical applications or may be referred to as a beginning point for another development.

The contributors of this book come from diverse backgrounds, making this book a truly international effort. This book will bring forth new frontiers with its revolutionizing research information and detailed analysis of the nascent developments around the world.

We would like to thank all the contributing authors for lending their expertise to make the book truly unique. They have played a crucial role in the development of this book. Without their invaluable contributions this book wouldn't have been possible. They have made vital efforts to compile up to date information on the varied aspects of this subject to make this book a valuable addition to the collection of many professionals and students.

This book was conceptualized with the vision of imparting up-to-date information and advanced data in this field. To ensure the same, a matchless editorial board was set up. Every individual on the board went through rigorous rounds of assessment to prove their worth. After which they invested a large part of their time researching and compiling the most relevant data for our readers.

The editorial board has been involved in producing this book since its inception. They have spent rigorous hours researching and exploring the diverse topics which have resulted in the successful publishing of this book. They have passed on their knowledge of decades through this book. To expedite this challenging task, the publisher supported the team at every step. A small team of assistant editors was also appointed to further simplify the editing procedure and attain best results for the readers.

Apart from the editorial board, the designing team has also invested a significant amount of their time in understanding the subject and creating the most relevant covers. They scrutinized every image to scout for the most suitable representation of the subject and create an appropriate cover for the book.

The publishing team has been an ardent support to the editorial, designing and production team. Their endless efforts to recruit the best for this project, has resulted in the accomplishment of this book. They are a veteran in the field of academics and their pool of knowledge is as vast as their experience in printing. Their expertise and guidance has proved useful at every step. Their uncompromising quality standards have made this book an exceptional effort. Their encouragement from time to time has been an inspiration for everyone.

The publisher and the editorial board hope that this book will prove to be a valuable piece of knowledge for researchers, students, practitioners and scholars across the globe.

List of Contributors

John Oyekan
Centre for Research in Distributed Technologies, University of Bedfordshire, Park Square, Luton, LU1 3JU, UK

Huosheng Hu
School of Computer Science and Electronic Engineering, University of Essex, Colchester CO4 3SQ, UK

Tüze Kuyucu
Engineering Department, Griffith Textile Machines, Washington NE38 8QA, UK

Ivan Tanev
Department of Information Systems Design, Doshisha University, Kyotanabe 610-0321, Japan

Katsunori Shimohara
Department of Information Systems Design, Doshisha University, Kyotanabe 610-0321, Japan

Ralf Möller
Computer Engineering, Faculty of Technology, Bielefeld University, Bielefeld D-33594, Germany Center of Excellence "Cognitive Interaction Technology", Bielefeld University, Bielefeld D-33594, Germany

Michael Horst
Computer Engineering, Faculty of Technology, Bielefeld University, Bielefeld D-33594, Germany

David Fleer
Computer Engineering, Faculty of Technology, Bielefeld University, Bielefeld D-33594, Germany

Lorenza Tiberio
CNR—Italian National Research Council, ISTC, Via San Martino Della Battaglia 44, Rome 00185, Italy
ECONA (Interuniversity Centre for the Research on Cognitive Processing in Natural and Artificial Systems), Rome 00185, Italy

Amedeo Cesta
CNR—Italian National Research Council, ISTC, Via San Martino Della Battaglia 44, Rome 00185, Italy

Marta Olivetti Belardinelli
ECONA (Interuniversity Centre for the Research on Cognitive Processing in Natural and Artificial Systems), Rome 00185, Italy
Department of Psychology, Università degli Studi di Roma —La Sapienza , Rome 00185, Italy

Natalie Ramos Pedroza
Department of Physical Sciences, Embry-Riddle Aeronautical University, 600 S Clyde Morris Blvd, Daytona Beach, FL 32114, USA

William MacKunis
Department of Physical Sciences, Embry-Riddle Aeronautical University, 600 S Clyde Morris Blvd, Daytona Beach, FL 32114, USA

Vladimir V. Golubev
Department of Aerospace Engineering, Embry-Riddle Aeronautical University, 600 S Clyde Morris Blvd, Daytona Beach, FL 32114, USA

Yogo Takada
Graduate School of Engineering, Osaka City University, 3-3-138, Sugimoto, Sumiyoshi-ku, Osaka 558 8585, Japan

Keisuke Koyama
Graduate School of Engineering, Osaka City University, 3-3-138, Sugimoto, Sumiyoshi-ku, Osaka 558 8585, Japan

Takahiro Usami
Graduate School of Engineering, Osaka City University, 3-3-138, Sugimoto, Sumiyoshi-ku, Osaka 558-8585, Japan

Emanuela Elisa Cepolina
Snail Aid—Technology for Development, Via Montallegro5/3, Genoa 16145, Italy

Michał Przybyłko
PMAR Robotics Group, Department of Mechanics, University of Genoa, Via All'Opera Pia 15/A, Genoa 16145, Italy

Giovanni Battista Polentes
Pierre Trattori snc, Via Novi 19, 15060 Silvano D'orba (AL), Italy

Matteo Zoppi
PMAR Robotics Group, Department of Mechanics, University of Genoa, Via All'Opera Pia 15/A, Genoa 16145, Italy

Giuseppe Tortora
The BioRobotics Institute, Scuola Superiore Sant'Anna, viale R. Piaggio 34, 56025, Pontedera (Pisa) 56025, Italy

Tommaso Ranzani
The BioRobotics Institute, Scuola Superiore Sant'Anna, viale R. Piaggio 34, 56025, Pontedera (Pisa) 56025, Italy

Iris De Falco
The BioRobotics Institute, Scuola Superiore Sant'Anna, viale R. Piaggio 34, 56025, Pontedera (Pisa) 56025, Italy

Paolo Dario
The BioRobotics Institute, Scuola Superiore Sant'Anna, viale R. Piaggio 34, 56025, Pontedera (Pisa) 56025, Italy

Arianna Menciassi
The BioRobotics Institute, Scuola Superiore Sant'Anna, viale R. Piaggio 34, 56025, Pontedera (Pisa) 56025, Italy

Ernesto Rivas
Department of Mechanical Systems Engineering, Yamagata University, Yamagata Prefecture, Yonezawa 992-8510, Jo-nan 4-3-16, Japan

Koutarou Komagome
Department of Mechanical Systems Engineering, Yamagata University, Yamagata Prefecture, Yonezawa 992-8510, Jo-nan 4-3-16, Japan

Kazuhisa Mitobe
Department of Mechanical Systems Engineering, Yamagata University, Yamagata Prefecture, Yonezawa 992-8510, Jo-nan 4-3-16, Japan

Genci Capi
Graduate School of Science and Engineering, University of Toyama, Toyama 930-8555, Gofuku 3190, Japan

Akira Obata
Micro Flying Robot Laboratory, Nippon Bunri University, 1727 Itigi Oita City 870-0397, Japan

Shotarou Shinohara
Micro Flying Robot Laboratory, Nippon Bunri University, 1727 Itigi Oita City 870-0397, Japan

Kyohei Akimoto
Micro Flying Robot Laboratory, Nippon Bunri University, 1727 Itigi Oita City 870-0397, Japan

Kakeru Suzuki
Micro Flying Robot Laboratory, Nippon Bunri University, 1727 Itigi Oita City 870-0397, Japan

Miyuki Seki
Micro Flying Robot Laboratory, Nippon Bunri University, 1727 Itigi Oita City 870-0397, Japan

Xiaodong Xu
School of Automation Engineering, University of Electronic Science and Technology of China, Chengdu 611731, China

Yi Zhang
Engineering Research & Development Center for Information Accessibility, Chongqing University of Posts and Telecommunications, Chongqing 400065, China

Yuan Luo
Engineering Research & Development Center for Information Accessibility, Chongqing University of Posts and Telecommunications, Chongqing 400065, China

Dongyi Chen
School of Automation Engineering, University of Electronic Science and Technology of China, Chengdu 611731, China

Slamet Widodo
Graduate School of Agriculture, Kyoto University, Kitashirakawa, Oiwake-cho, Sakyo-ku, Kyoto 606-8502, Japan

Tomoo Shiigi
Graduate School of Agriculture, Kyoto University, Kitashirakawa, Oiwake-cho, Sakyo-ku, Kyoto 606-8502, Japan

Naoki Hayashi
Graduate School of Agriculture, Kyoto University, Kitashirakawa, Oiwake-cho, Sakyo-ku, Kyoto 606-8502, Japan

Hideo Kikuchi
System Watt Co., Ltd. 4-1-38, Isobedori, Chuo-ku, Kobe 651-0084, Japan

Keigo Yanagida
System Watt Co., Ltd. 4-1-38, Isobedori, Chuo-ku, Kobe 651-0084, Japan

Yoshiaki Nakatsuchi
The New Industry Research Organization (NIRO), 1-5-2, Minatojima-minamimachi, Chuo-ku, Kobe 650-0047, Japan

Yuichi Ogawa
Graduate School of Agriculture, Kyoto University, Kitashirakawa, Oiwake-cho, Sakyo-ku, Kyoto 606-8502, Japan

Naoshi Kondo
Graduate School of Agriculture, Kyoto University, Kitashirakawa, Oiwake-cho, Sakyo-ku, Kyoto 606-8502, Japan